MW00449777

Firefighting Strategy and Leadership

Second Edition

CHARLES V. WALSH
Deputy Chief, New York Fire Department, retired
Instructor, Rutgers, The State University, New Brunswick, New Jersey

DR. LEONARD G. MARKS
Deputy Chief, San Jose Fire Department, retired
Instructor, Santa Ana College, Santa Ana, California

Gregg Division/McGraw-Hill Book Company

New York • St. Louis • Dallas • San Francisco • Auckland
Bogotá • Düsseldorf • Johannesburg • London • Madrid
Mexico • Montreal • New Delhi • Panama • Paris
São Paulo • Singapore • Sydney • Tokyo • Toronto

Library of Congress Cataloging in Publication Data

Walsh, Charles V
Firefighting strategy and leadership.

Includes index.
1. Fire extinction. 2. Fire-departments. I. Marks, Leonard G., joint author. II. Title.
TH9310.5.W33 1977 628.9'25 76-23246
ISBN 0-07-068026-4

Firefighting Strategy and Leadership,
Second Edition

2 3 4 5 6 7 8 9 0 DODO 7 8 3 2 1 0 9 8

The editors for this book were Susan H. Munger and Claire Hardiman, the designer was Dennis G. Purdy, the art supervisor was George T. Resch, and the production supervisor was Regina R. Malone. It was set in Times Roman by Monotype Composition. Printed and bound by R. R. Donnelley & Sons Company

CREDITS

Cover: Photograph by John Bassano. **Page i:** Photograph courtesy of the New York Fire Department. **Pages 1, 2, 8, 17, 18, 22, 28:** Photographs courtesy of *WNYF* magazine. **Page 35:** Photograph courtesy of the New York Fire Department. **Page 52:** Photograph courtesy of the New York Board of Fire Underwriters. **Page 61:** Photograph courtesy of *WNYF* magazine. **Pages 68, 69:** Charts from *The Contemporary Curtain Wall* by William Dudley Hunt, McGraw-Hill, 1958. **Page 81:** Photograph by John Bassano. **Pages 98, 102, 114:** Photographs courtesy of *WYNF* magazine. **Page 116:** Illustration from *The Neutron Story* by Donald Hughes, Doubleday, 1959. **Pages 118, 121:** Illustrations from *The Contemporary Curtain Wall* by William Dudley Hunt, McGraw-Hill, 1958. **Pages 130, 131:** Illustrations from "Fire and Fire Protection," by Howard W. Emmons. Copyright © 1975 by Scientific American, Inc. All rights reserved. **Page 132:** Illustration from *Scientific American.* **Page 135:** Photograph courtesy of *WNYF* magazine. **Page 144:** Photograph courtesy of the *Journal of American Insurance.* **Page 160:** Photograph courtesy of National Foam System, Inc. **Pages 175, 176, 177, 180, 191:** Photographs courtesy of *WNYF* magazine. **Page 194:** Photograph courtesy of the New York Fire Department. **Page 213:** Photograph by John Bassano. **Page 214:** Photograph courtesy of *WNYF* magazine. **Page 222:** Photograph by Alexander Donchin. **Page 234:** Photograph courtesy of the New York Fire Department. **Page 261:** Photograph courtesy of *WNYF* magazine. **Page 262:** Photograph courtesy of the National Board of Fire Underwriters. **Page 270:** Photograph courtesy of the New York Fire Department. **Page 278:** Photograph courtesy of *WNYF* magazine. **Page 288:** Photograph courtesy of the New York Fire Department. **Page 291:** Photograph from "Remote Sensing of Natural Resources" by R. N. Colwell. Copyright © 1968 by Scientific American, Inc. All rights reserved. **Page 295:** Photograph by John Bassano. **Page 296:** Photograph courtesy of the New York Fire Department. **Page 305:** Photograph by John Bassano. **Page 316:** Photograph courtesy of the New York Fire Department. **Page 322:** Photograph by John Bassano. **Pages 326, 333, 344, 354:** Photographs courtesy of the New York Fire Department. **Page 359:** Photograph by John Bassano. **Page 360:** Photograph courtesy of the New York Fire Department. **Page 365:** Illustration from "Fire and Fire Protection" by Howard W. Emmons. Copyright © 1975 by Scientific American, Inc. All rights reserved. **Page 368:** Photographs from "Fire and Fire Protection," by Howard W. Emmons. Copyright © 1975 by Scientific American, Inc. All rights reserved.

CONTENTS

TO CATH

PREFACE

In the interval between the first edition and this revision, the need for standardized training in fire service has become increasingly evident. The Wingspread Conference on Fire Service Administration, Education, and Research, attended by the chiefs of the three largest fire departments and directors of the most prominent fire-service training institutions in the United States, and the report *America Burning,* prepared by the Presidential Commission on Fire Prevention and Control, both emphasized the need for standardized training in the fire service. The objective of this revision is to provide the type of training referred to.

It is as true to state now as it was fourteen years ago when the first edition of *Firefighting Strategy and Leadership* was published that firefighting or fire suppression is fundamentally a science founded upon principles and laws capable of quite definite statement. Just as fundamentally, it is an art demanding skill, ingenuity, experience, and judgment. This revision aims at helping firefighters and fire officers of all ranks, paid or volunteer, learn this science and art, and thereby develop the strategy of firefighting. In so doing, training that is both standardized and appropriate will be provided. The science and art to be taught will enable officers of all ranks to effectively use the same mental approach or action plan at all types of fires. Concomitantly, this training stresses leadership with a venturesome and progressive approach to applying principles of management—particularly those related to *modern* concepts of leadership, discipline, morale, and command.

We have not attempted to cover elementary knowledge or training in the use and care of equipment, mechanics involved in stretching and laying hose lines and raising ladders, and so on. This training is given by all fire-service organizations to their recruits. Instead, we have shown how these basic skills should be applied—to explain when and where to do what, and especially, why. We have discussed the relationship between factors in fire situations, the impact on firefighting of new materials, and the application of a fire plan to various types of structures. The goal is to develop in students decision-making skills that will result in more effective firefighting.

Certain features have been added, deleted, or rearranged in this edition to develop more fully the main theme of the original edition—the application of a standardized mental process at all types of fires. We have updated much of the information in recognition of the impact on the fire situation by high-rise buildings, increased use of plastics, new equipment and apparatus, increased emphasis on fire prevention, findings specified in the report *America Burning*, and the advent of a National Fire Academy with its implications on the fire-service training area. Recognition has also been given to legislation pertaining to the metric conversion, which is in the offing. To help the learning process for students, questions are offered at the end of each chapter for review purposes.

A prerequisite for the effective use of the action plan that can be used at all types of fires is an adequate knowledge of the science and art of firefighting. Part One explains how such a science and art can be developed and the premises on which the use of the action plan is predicated.

Part Two provides information for carrying out the first two steps in the action plan: To note and evaluate as accurately as possible the primary factors that are pertinent in the given situation, and to select objectives and activities on the basis of the evaluation made. Part Two covers eleven chapters and is extensive because relationships among primary factors conceivably pertinent at all types of fires have to be considered.

Part Three deals with the remaining steps in the action plan, in which decisions are implemented: The assigning and coordinating of activities by knowledgeable supervision and through effective communication. Derivative or minor objectives are introduced at this stage.

Part Four describes the practical application of the information provided about the science and art of firefighting, and thereby the use of firefighting strategy. It shows how the action plan is actually applied at fires, and how practice in using the action plan can be beneficial.

Part Five is concerned with the role of managerial principles in relation to the skill or art with which the science of fire suppression is applied. As these principles—in addition to those discussed earlier—are mastered, firefighting can achieve truly professional status.

Part Six deals with fire prevention. Recommendations in the report *America Burning* indicate that fire prevention may finally get proper attention. The system of managing by objectives is followed in covering this subject—that is to say the major and minor objectives of a fire-prevention program are specified. Explanations are then offered for achieving the minor or derivative objectives, thereby automatically attaining the major objective.

Grateful recognition is due to the many who have contributed, directly and indirectly, to this work. In this group are members of all ranks and many fire departments. Special thanks are due to Joseph M. Redden, Chief Engineer of the Newark Fire Department for his constructive observations and comments.

PART ONE

Managerial Principles and Firefighting Strategy

1

The Science and Art of Firefighting

A prerequisite for the effective use of the action plan to be recommended is an adequate knowledge of the science and art of firefighting.

THE SCIENCE OF FIREFIGHTING

Science can be defined as a systematized body of laws, principles, or knowledge pertaining to a specific field. The science of firefighting could therefore be defined as a systematized body of laws, principles, or knowledge pertaining to the field of firefighting. Arriving at this body of laws requires the use of the scientific method.

The Scientific Method

Scientific method involves observing facts, then testing the accuracy of these facts through continued observation. If the facts prove accurate, the scientist seeks some causal relationships between them and other happenings from which logical hypotheses can be deduced. Such hypotheses are in turn tested, and if they are found to be true—to explain some aspect of reality and therefore to have value in predicting what will happen in similar circumstances—they are called principles.

For decades, the fire service has been observing facts at fire situations. These facts have led to valid hypotheses, or principles. By putting these principles into some logical order—that is, by *systematizing* them, we can take a step toward developing a science of firefighting. We can take an even greater step if we accept the help of scientific researchers in experimentation and verification.

Recognized authorities maintain that science is said to be systematized not only because the underlying principles have been discovered, but also because the relationships between *variables* and *limits* have been ascertained. Accordingly, in this text, we will try to explain what variables and limits are, and how relationships between them have been ascertained. We will also try to explain how principles relating to firefighting or fire suppression have been derived.

Variables. In this text, variables associated with fire situations are classified as primary and secondary factors. This is done to indicate a time sequence for evaluation and not necessarily a degree of importance. Such factors are variables because they change from fire to fire. Primary factors are the conditions or elements that should be recognized and evaluated on arrival and during operations. Those that may have to be considerered at practically any type of fire are listed in column 1 of the primary-factor chart on page 20. Their relationships with other variables are indicated in columns 2 and 3 of the chart.

Secondary factors are the activities undertaken to achieve objectives; such activities include forcing entry, ventilating, using hose lines, overhauling, making decisions, establishing command posts, and so on. These variables have reciprocal relationships because of the inevitable effect of one activity on another. For example, effective ventilation facilitates the advance of hose lines, yet ineffective stretching or laying of hose lines nullifies the effectiveness of ventilation or even causes it to be harmful if it results in spreading the fire before a line is ready to operate. Therefore, secondary factors can also affect such primary factors as extent of fire after arrival, heat and smoke conditions, exposure hazards, duration of operation, requirements to operate, and so on.

Limits. Limits are specifications for acceptable solutions at fires (Chapter 2). For practical purposes, there are two particular limits, or specifications: (1) If there is a life hazard for occupants, risks to personnel ranging from merely unusual to extreme may be warranted, and (2) if there is no life hazard for occupants, personnel are never to be jeopardized unnecessarily.

In an effort to systematize a science of firefighting, the relationships between variables and limits and the effects of these relationships on fire operations will be indicated and explained in much of this book.

Underlying Principles. Underlying or basic principles are fundamental truths applicable to a given set of conditions or circumstances; they indicate what may be expected to happen under these conditions or circumstances. Scientists have already discovered some of these principles that concern combustion; extinguishment of fire; transfer of heat by conduction, convection, and radiation; and the flow of liquids and gases. Using these basic principles, the fire service has formulated more specific principles governing fire activities, such as ventilating and using hose lines.

Principles governing flow of fluids enable fire personnel to expect that when an obstruction reduces the velocity of convection currents rising through a vertical structural channel, pressure will increase and will be greatest immediately beneath the obstruction. This will cause the fire to spread toward areas of lower pressure—or least resistance—either mushrooming or moving horizontally. But if the obstruction is removed—for example, by making an opening in the roof directly above the involved vertical structural channel—the velocity of the rising gases will increase and the pressure will decrease. The decreased pressure will then minimize the possibility of horizontal spread of fire in the cockloft (the space between the top-floor ceiling and the underside of the roof). Similar reasoning applies to the formation of many other specific firefighting principles.

Firefighting principles are universal in application. That is, they are the same for all departments, large or small, paid or volunteer. However, they may have to be applied in different ways because the primary factors change from one community to the next. For example, the primary factor of water supply available varies in different communities, but when there is a life hazard for occupants the same principle applies to all: the first water available is used as quickly as possible and as long as necessary between the fire and the endangered occupants or their means of escape. It does not matter whether the water comes from a booster tank in a small community or a standpipe riser in a city high-rise building. It is this universal application of firefighting principles that makes standardized training both possible and practical.

Firefighting as an Inexact Science

In the chapter "Management as an Inexact Science," Koontz and O'Donnell, authors of *Principles of Management,* say much that applies with equal force to firefighting as an inexact science.

> It is often pointed out that the social sciences are "inexact" sciences, as compared to the "exact" physical sciences. It is also sometimes indicated that management is perhaps the most inexact of the social sciences. It is true that social sciences, and management in particular, deal with complex phenomena about which little is known. It is true likewise, that the structure and behavior of the atom are far less complex than the structure and behavior of groups of people.
>
> But we should not forget that even in the most exact of the "exact" sciences—physics—there are areas where scientific knowledge must be replaced with speculation and hypothesis. As much as is known of bridge

mechanics, there are still cases where bridges fail through such causes as vibrations set up from wind currents. And as we move from the longer-known areas of physics into the biological sciences, we find that areas of exactness tend to diminish.

Since virtually all areas of knowledge have tremendous expanses of the unknown, people working in the social sciences should not be defeatist. A scientific approach to management cannot wait until an exact science of management can be developed. Had the physical and biological sciences thus waited, man might have still been living in caves.

Certainly, the observations of perceptive managers must substitute for the desirable laboratory-proved facts of the management scientist, at least until such facts can be determined. Statistical proof of principles of management are desirable, but there is no use waiting for such proof before giving credence to principles derived from experience. After all, no one has been able to give statistical proof of the validity of the Golden Rule, but people of many religions have accepted this fundamental precept as a guide to behavior for centuries, and there are few who would doubt that its observance improves human conduct. . . .

To be sure, management is an inexact science. But the questions one must ask are these: Does the use of such theory as is available or postulated help us understand management and aid in improving management practice now? Are we better off using such theory now—for guidelines in research and practice—or waiting until that perhaps distant future when the science can be "proved"? Does such theory help in substituting rationality for confusion? Does it increase objectivity in the understanding and practice of management?[1]

In considering answers to the questions posed in the foregoing quote, assuming they are related to firefighting as an inexact science, keep in mind the overall fire situation so eloquently and graphically described in the report *America Burning,* prepared by the Presidential Commission on Fire Prevention and Control. After indicating that heads of fire departments know that the key to their performance, and the performance of those under them, lies in training, the report states:

At both State and local levels in this Nation, the quality of training ranges from excellence to total absence. Usually the quality of training is tied to economic circumstances. But poor training programs could be improved, at little cost, if they followed the example of outstanding programs. At present, however, there is no systematic interchange of information among educators in the fire services.

One possible remedy has almost unanimous support within the fire suppression and protection fields—namely, a National Fire Academy. What most experts envision is an institution that not only has advanced education

[1] Harold Koontz and Cyril O'Donnell, Principles of Management, McGraw-Hill, New York, 1964, pp. 6–7.

> programs of its own, but also lends help to States and local training and educational programs. In addition to conducting classes and seminars at its own facility, the Academy would serve as the hub of an educational network. The Academy system would use existing fire training school programs, fire science education programs in community colleges, and fire management and fire protection engineering programs at the college or university level in each State. The Academy would function as the core of the Nation's efforts in fire service education—feeding out model programs, curricula, and information, and at the same time receiving helpful advice from those schools and the fire services.[2]

From the foregoing, it is apparent that a current major weakness in the fire service is still the lack of standardized education and training. At this particular time, therefore, a strenuous effort to develop a *science* of fire suppression is warranted, because only by systematizing authoritative knowledge can standardized education and training be achieved. The present time seems to be especially propitious in view of the National Academy's mandate. However, the development of a firefighting science should not be a unilateral undertaking by the fire service. Obviously, for more satisfactory results, the assistance of fire-research scientists and fire-protection engineers is required, particularly in the areas of experimentation and verification.

THE ART OF FIREFIGHTING

The task of an art is to create useful applications of scientific knowledge. From a fire-service viewpoint, this means that the art of firefighting involves the use of skill and ingenuity in applying the science (knowledge) of firefighting. Science and art are not mutually exclusive but are complementary. As the science of firefighting improves, so should the art, as has happened in other fields. Strategy in firefighting, of course, encompasses both science and art. The action plan recommended in this book incorporates this concept of strategy; it provides the knowledge (science) required to make correct decisions at all types of fires so that every fire officer can implement such decisions with skill (art).

Knowledge of principles or theory will not, however, guarantee successful practice because there is no science in which everything is known and all relationships are proven. A thorough understanding of techniques is needed in order to utilize scientific theory. This text will therefore discuss the techniques involved in the assignment of activities and their coordination via supervision and communication. Consideration will also be given to the roles of experience, judgment, leadership, motivation, and

[2] National Commission on Fire Prevention and Control, *America Burning*, 1973, p. 41.

morale because they too affect the efficiency and skill (art) in applying knowledge (science).

A skillful officer uses what he has on hand (water, apparatus, equipment, and personnel) to maximum advantage and quickly anticipates additional needs. Obviously, much depends on the training of personnel as to the use of water, maintenance and placement of apparatus, and the maintenance and use of equipment.

It is obvious that an action plan is necessary to achieve the maximum level of success. Engaging in a random pattern of activity is unlikely to result in a successful operation. Leaders must lead, and in order to do so, the professional fire officer has an action plan.

REVIEW QUESTIONS

1. What is an acceptable definition of the science of firefighting?

2. What can the fire service do to develop a science of firefighting? How can this effort be supplemented by fire-research scientists?

3. What are variables (in this text) comprised of?

4. What do primary factors consist of? What is their role?

5. What do secondary factors consist of? Why do they have to be evaluated?

6. Define "limits." Indicate practical limits.

7. Define "principles." Indicate why firefighting principles are fundamentally based on scientific principles.

8. Give some reasons why an effort is justified and even mandatory in view of our huge fire losses, to develop a science of firefighting, however inexact it may be.

9. In what way is firefighting an art? What is the task of an art, from a fire service viewpoint? What is the relationship between the art and science of firefighting?

10. What is needed for the effective use of scientific theory?

2

The Action Plan

The action plan is a mental process that is practical for fire officers of all ranks at all types of fires. It is designed to standardize thinking and acting at fires and utilizes the scientific approach described in the Chapter 1.

THE DECISION-MAKING PROCESS

The action plan is based on a decision-making process used in modern business situations, as described by Peter Drucker in an article in *Our Nation's Business*. This decision-making process can be used in fire situations, since they also consist of conditions or circumstances that are the factors in making and implementing decisions. The steps in the process for making business decisions are: (1) Define the problem; (2) establish specifications for an acceptable solution; (3) define expectations about the objective and action taken; (4) develop alternate solutions; (5) make the decision; and (6) convert the decision into action.

Define the Problem

A tool used extensively to define the problem is critical-factor analysis. This tool employs the following principle. When analyzing a situation, attention should be focused on those factors that are limiting or strategic to the decision involved. At a fire situation, officers can employ critical-factor analysis by recognizing and evaluating those factors that are limiting or strategic to decisions about objectives and activities. In this context, only primary, and not secondary, factors qualify as limiting or strategic. It may be noted that limiting or strategic factors can change during a fire operation, thereby changing the problem and the objectives and activities to be undertaken. Hence, critical-factor analysis is a continuous process at fires.

Establish Specifications for an Acceptable Solution

Business managers are likely to think in terms of financial risks. Fire officers must think in terms of life hazard for occupants and personnel. Hence, specifications for acceptable solutions at fire operations are the limits specified in Chapter 1: (1) If there is a life hazard for occupants, risks for personnel ranging from merely unusual to extreme may be warranted, and (2) if there is no life hazard for occupants, personnel should never be jeopardized unnecessarily.

Define Expectations

Fire officers should have logical expectations about objectives sought and activities undertaken. Only in this way can they make an intelligent appraisal of what is being done. If things are going according to plans, fine; if not, the officer is mentally prepared to make necessary corrections. Expectations will be more logical if an officer is adequately trained to recognize and evaluate the effects of the factors that are pertinent in a given situation.

Develop Alternate Solutions

Offhand, developing alternate solutions at fires may seem somewhat impractical because of the pressure of time. However, in some cases, common sense may dictate that a minute or so be taken to consider possible alternatives, especially where the decision may involve a choice between initiating an interior or exterior attack. In other cases, there may be sufficient reason to consider alternatives such as evacuating some or all of the occupants, protecting a rear or side exposure first, using high expansion foam or cellar pipes at cellar fires, forcing entry at the rear or front first, backing up the first interior line with the second line or stretching the second line via the fire escape, and so on. Training officers to consider alternate solutions is an antidote against the one-track-mind approach.

Make the Decision

Business decisions are made by weighing the range of alternate solutions against the specifications for an acceptable solution. The action plan incorporates the same idea.

Convert the Decision into Action

Fire officers implement their decisions by assigning activities and coordinating them via adequate supervision and communication. All decisions, in the business world or elsewhere, are implemented in this way.

FIRE-SERVICE ADAPTATION

The action plan is an adaptation of the process for making business decisions. The plan consists of the following steps: (1) Note and evaluate as accurately as possible the primary factors that are pertinent in the given situation; (2) select objectives and activities on the basis of the evaluation made; (3) assign activities; and (4) coordinate activities by adequate supervision and communication. Each of these steps should be fully understood.

Note and Evaluate the Pertinent Primary Factors

The action plan, like the process for making business decisions, is concerned with defining the problem as soon as possible. At fires this can be done only after the pertinent primary factors have been recognized and evaluated, or in other words, after the critical factors have been analyzed.

Select Objectives and Activities

Business decisions are made by weighing the range of alternate solutions against the specifications for an acceptable solution. In the action plan, the relationships among the pertinent primary factors should suggest a range of alternatives. These are then weighed against the limits or specifications for an acceptable solution. (It may be noted that a range of alternatives exists because inevitably there is more than one way to operate at a fire. There is, however, only one best way and the firefighter must learn to recognize it.)

On the subject of objectives, Koontz and O'Donnell state "Planning can be a useful managerial function only if objectives are properly selected. Improper selection or faulty specifications of objectives will vitiate planning time and cost, result in frustration, and make the entire planning activity futile. . . . Every enterprise should spell out its own precise objectives. The cost of failure to do this will be a helpless undertaking, battered by chance events because it can make no concrete plans. On the other hand, identification of objectives permits plans to accomplish them."[1] Agreement on major objectives is essential for standardized training. Accordingly, major objectives for the fire service have been designated as (1) rescue, (2) extinguish, and (3) confine, control, and extinguish. These are determined upon arrival solely by the primary factors that are critical or pertinent. After the operation gets under way, secondary factors or activities also become pertinent. This is because the manner in which secondary factors affect primary factors and each other can be limiting or strategic to later decisions. An example is that ineffective

[1] Koontz and O'Donnell, *Principles of Management,* p. 100.

ventilation may make it difficult to advance hose lines. These circumstances would adversely affect such primary factors as extent of fire after arrival, heat and smoke conditions, exposure hazards and so on. In such cases, plans are not meeting the expectations of the commanding officer, who may then have to back units out of the fire building and change the major objective from extinguish to confine, control, and extinguish.

Rescue. When life hazard for occupants is the limiting or strategic factor, the major objective is rescue. A life hazard exists when the human element in the occupancy is endangered because escape from heat and/or smoke is impeded by the location or extent of the fire. Thus, only a few primary factors can create a life hazard, the foremost limiting or strategic factor in making decisions about objectives and activities at fires. However, the severity of the life hazard can be increased or decreased by any or all of the other primary factors.

Extinguish. If there is no life hazard for occupants and the fire is considered to be controllable by a direct and/or indirect attack extinguish is the major objective. Pertinent primary factors that would determine whether the fire were controllable in such cases could include location and extent of fire, heat and smoke conditions, construction, occupancy, and availability of adequate personnel, water, apparatus, and equipment.

Confine, Control, and Extinguish. If there is no life hazard for occupants and the situation is temporarily uncontrollable, confine, control, and extinguish is the major objective. Pertinent primary factors that might indicate the fire was uncontrollable could include location and extent of fire; severe heat and smoke conditions and exposure hazards; actual or possible structural collapse; construction; occupancy; unfavorable weather conditions; or inadequate personnel, water, apparatus, and equipment.

In some cases, confine, control, and extinguish tactics have to be employed even though rescue continues to be the major objective. This can happen when trapped occupants in a collapsed fire building cannot be reached in any other way. Another example would be a fire in a heavily involved, unoccupied structure. Here, the fire should first be confined by protecting occupied exposures, pending removal of the occupants. Then, the major objective reverts to confine, control, and extinguish.

Assign Activities

It is extremely important that all fire officers working at the fire know what the major objective is, since it influences the risks undertaken while performing the activities necessary to attain that objective.

Activities associated with rescue may include locating the fire; ascertaining its extent; forcing entry; searching for and removing occupants; using hose lines, ladders, elevated platforms, roof ropes, life nets, and masks as required; and rendering first aid and obtaining medical treatment as may be necessary.

Activities associated with extinguish may include locating the fire; ascertaining its extent; selecting the extinguishing agent (it may not be water); forcing entry; ventilating; using hose lines, apparatus, and equipment as required; checking for extension of fire; overhauling; and salvage.

Activities associated with "confine, control, and extinguish" may include locating the fire; ascertaining its extent; establishing an order of priority in covering exterior exposures by master streams; and supplying standpipe, sprinkler, or other available fixed systems until the first phase "confine" has been achieved. Simultaneous activities may include operating in exterior exposures to close windows; moving endangered stock; extinguishing incipient fires by hand lines or other means; shutting down sprinkler systems no longer needed; forcing entry; ventilating on the unexposed side of the structure if feasible; checking for any possible extension of fire; and overhauling.

To achieve the second phase "control," activities may include forcing entry; ventilating; and attacking the main body of fire with offensive lines supplemented by those previously used defensively. To achieve the final phase extinguish, activities would depend in large measure on the structural condition of the fire building. If an investigation warrants sending units into the fire building, activities may include forcing entry; ventilating; checking for extension of fire; using hose lines, masks, and lighting equipment as required; draining floors; overhauling; and salvage. It is assumed that gas and electricity have been shut off and that precautions are taken to avoid injuries that may be caused by structural conditions as a result of the fire and use of water. If interior work is not warranted because of a possible or imminent structural collapse, activities may include the use of streams operating from a safe distance and eventually the use of one or two "watch" lines similarly located as a precaution against rekindles.

A final activity may be to demolish the fire building. Note that activities associated with the final phase extinguish are quite different from activities undertaken when the major objective logically and solely is extinguish.

Care should be taken to assign activities according to the functions of the various units. Decisions can only be effectively implemented if appropriate activities are assigned to the appropriate company. Assigning activities such as laddering or even manning life nets to the engine company first to arrive can be a serious error when a hose line is required to protect a life hazard.

During inspection programs it has been found that efficiency, teamwork, and hence coordination were impaired when ladder company personnel responded to fires with pumpers, and engine company personnel responded with trucks. The former were not sufficiently adept in stretching and using hose lines, and the latter performed likewise in using truck company tools. Quite often, commanding officers found it advisable to make adjustments—engine company personnel handled hose lines and ladder company personnel handled tools.

It is true that during the probationary period firefighters are trained to carry out both ladder and engine company evolutions. Afterward, however, they become more familiar with the mechanical skills associated with the work of the engine or ladder company to which they belong. It is also true that in some departments units annually go to Company (or otherwise designated) school at which they are tested in performing ladder and engine company evolutions. Frequently, they perform creditably, mainly because the prevailing conditions are known beforehand. However, at actual fires, the conditions that will prevail cannot be anticipated. When ladder company functions are assigned to engine company personnel (and vice versa) the resulting performance leaves much to be desired.

Coordinate Activities

Supervision by company officers ensures that the principles governing unit activities are being properly applied. Supervision by the commanding officer ensures that activities are being supplemented or modified according to how they are measuring up against expectations. The establishment of a command post aids communication between subordinate and commanding officers. By keeping in constant contact, the commanding officer can keep abreast of as well as anticipate developments, insofar as they are affecting expectations. This accelerates any required action and improves coordination.

Establish Command Post. The command post is the position taken by the commanding officer at a fire operation. A command post should be established promptly so that the commanding officer (1) is readily visible and accessible to units reporting in for their assignnments, and (2) can effectively coordinate activities through communication with subordinate officers and the dispatcher.

Provided they are feasible, street intersections are favorable locations for command posts. If visibility is impaired by the time of the fire, smoke, snow, rain, or whatever, responding units can be informed by radio about the location of the command post. Lights on chiefs' cars are a help here.

Unless the commanding officer is readily visible and accessible, there will be a delay in getting and carrying out assignments. Worse still, incoming units may start to work on their own initiative, which can damage coordination and disorganize an operation. For example, in a fire operation where several lines are going to be required, the battalion or district chief who goes into the fire building and remains there with the first-to-arrive engine and ladder companies is derelict. The chief should hasten back to the street and establish a command post as soon as it is apparent that one line may not be adequate.

At one-line fires (about 90 percent of all fires require no more than one line), there is hardly any need for a command post, assuming there is no life hazard for occupants. In such cases, the officers of the first-to-arrive engine and ladder companies go in and put the fire out, without establishing a command post. However, if it is discovered that more lines may be needed and other units have not arrived, the commanding company officer should go out into the street and establish a command post to direct incoming units. This officer should maintain the command post until relieved by a responding superior officer. The other company officer should remain in the building to direct both of the first-to-arrive units. The quick establishment of a command post prevents such errors as stretching a second line up a fire escape when it should be backing up the first line, or backing up the first line with a second line, when it should have been stretched up a fire escape. Without proper direction, similar mistakes can be made about additional lines and the placement of incoming truck company apparatus.

The prompt establishment of a command post provides important contact between the commanding and subordinate officers. Reports can be quickly evaluated and activities supplemented or modified as required. From a command post, the commanding officer can supervise the overall operation more efficiently. Thus, the command post furnishes the two ingredients necessary for a coordinated effort: communication and supervision (Fig. 2-1).

REVIEW QUESTIONS

1. Why does the process for making decisions in business situations also apply in fire situations?

2. What tool is used extensively in defining a problem?

3. What kind of factors at fire situations are regarded as limiting or strategic? Why is critical-factor analysis an ongoing process at fires?

4. What does the expression "establish specifications for an acceptable solution at fires" mean to a fire officer?

FIG. 2-1. The command post plays an important role at the fire operation. From the command post, a chief can maintain contact with units and supervise and coordinate the activities assigned.

5. Why is it logical for officers to define their expectations about objectives and activities at fires?

6. Why is it practical and advisable to consider alternate solutions at fires?

7. How is the decision actually made in the process for making business decisions? Is this method practical at fires?

8. How does the action plan incorporate the ideas embodied in the process for making business decisions?

9. Why is the third step in the action plan—assign activities—particularly important?

10. What steps in the action plan implement decisions made?

11. How are activities coordinated at fires?

12. Compare supervision of company and chief officers.

13. Comment on the importance of the command post.

14. Why is it imperative for every enterprise, including the fire service, to spell out its own precise objectives?

15. When is rescue the major objective at fires? What activities are associated with its achievement?

16. When is extinguish the major objective? What activities are associated with its achievement?

17. When is confine, control, and extinguish the major objective? What activities are associated with its achievement?

18. Why is it important for all officers working at a fire to know what the major objective is?

19. When may confine, control, and extinguish tactics have to be employed when rescue is the major objective?

PART TWO

Evaluating Factors and Selecting Objectives

3

Primary Factors

It has been established that to systematize the science of firefighting, it is necessary to define variables and limits and to ascertain the relationships between them. Primary factors are considered variables because they change from fire to fire. The study of primary factors and of the relationships between them will help officers carry out the first two steps in the action plan. These two steps are quite important: to note and evaluate as accurately as possible the primary factors that are pertinent in the given situation, and to select objectives and activities on the the basis of the evaluation made.

The primary factor chart on page 20 facilitates the study of relationships between primary factors, and indicates the sequence of coverage. Extensive explanations are necessary since all primary and secondary factors that could conceivably be pertinent at all types of fires are considered. Lectures or discussions of past or structured fire situations, even when supplemented by motion pictures or diagrams, are of limited value. This is because the critical factors are specified, whereas at an actual fire critical factors must be recognized and evaluated under possible hectic conditions. In addition, such lectures or discussions may only provide helpful information about one set of circumstances in one given situation. Officers actually need helpful information about any set of circumstances in any fire situation. This chapter attempts to provide such information.

REVIEW QUESTIONS

1. Why is a study of the relationships among primary factors relevant to the development of firefighting strategy?

2. What is such a study designed to do?

3. What is the purpose of the primary-factor chart?

4. Why are explanations of relationships among factors so extensive?

5. What must be recognized if the training of officers is to be realistic?

6. Why are lectures or discussions of past or structured fires of limited value?

TABLE 3–1 Primary Factor Chart

	PRIMARY FACTORS	AFFECTS	IS AFFECTED BY
1	Life hazard for occupants	2–5, 8, 9, 15, 16, 18–20, 24–26, 31, 32, 34, 35	*
2	Life hazard for personnel	1, 5, 8, 15, 16, 24–26, 31, 32, 34, 35	*
3	Location of fire	1, 2, 4, 5, 8, 9, 15, 16, 18–20, 22–26, 28, 29, 31, 32, 34, 35, 37	1, 4, 6–25, 27–29, 32–34
4	Extent of fire on arrival	1–3, 5, 8, 9, 15, 16, 22–26, 28, 29, 31, 32, 34, 35, 37	1, 3, 6–30, 32–34, 36–38
5	Extent of fire after arrival	1, 2, 8, 9, 15, 16, 22–26, 28, 29, 31, 32, 34, 35, 37	*
6	Construction of fire building	1–5, 7–16, 22–26, 31, 32, 34, 35, 37	1, 7–14, 22, 34
7	Construction of exposures	1–6, 8–16, 22–26, 31, 32, 34, 35, 37	1, 6, 8–14, 23
8	Occupancy of fire building	1–7, 9–16, 18–26, 31, 32, 34–38	*
9	Occupancy of exposures	1–8, 10–16, 18–26, 31, 32, 34–38	*
10	Height of fire building	1–9, 11–16, 22–26, 31, 32, 34, 35, 37	1, 6–9, 11–15, 22, 32, 34
11	Height of exposures	1–10, 12–16, 23–26, 31, 32, 34, 35, 37	1, 6–10, 12–16, 23, 32, 34
12	Area of fire building	1–11, 13–16, 22, 24–26, 31, 32, 34, 35, 37	1, 6–11, 13, 14, 22, 34
13	Area of exposures	1–12, 14–16, 23–26, 31, 32, 34, 35, 37	1, 6–12, 14, 23, 34
14	Proximity of exposures	1–13, 15, 16, 22–26, 31, 32, 34, 35, 37	1, 6–13, 22, 23, 34
15	Structural collapse of fire building	1–5, 8–11, 14, 16, 22–26, 31, 32, 34–37	*
16	Structural collapse of exposures	1, 2, 5, 8–11, 14, 22–26, 31, 32, 34–37	*
17	Time fire started	1–5, 8, 9, 15, 16, 18–38	8, 21, 24, 25, 27, 32, 33
18	Time of discovery	1–5, 8, 9, 15, 16, 19–38	1–3, 6–14, 17, 19, 21, 22 24–30, 32
19	Time of alarm	1–5, 8, 9, 15, 16, 18, 20–38	1, 3, 6–14, 17, 18, 21, 22, 24–30, 32
20	Time of response	1–5, 8, 9, 15, 16, 22–38	1, 3, 19, 21, 25–27, 30, 31 33, 36, 37

21	Time of week or year	1–5, 8, 9, 15–20, 22–38	17
22	Auxiliary appliances in the fire building	1–6, 8, 10, 12, 14, 18–20, 24–26, 31, 32, 34, 35, 38	1–6, 8, 10, 12, 14, 15, 17–21, 24–29, 31, 32, 34–38
23	Auxiliary appliances in the exposures	1–5, 7, 9, 11, 13, 14, 24–26, 31, 32, 34, 35, 38	1–22, 24–29, 31, 32, 34–38
24	Heat conditions	1–5, 8, 9, 15–20, 22, 23, 25, 26, 28, 29, 31, 32, 34, 35, 38	*
25	Smoke conditions	1–5, 8, 9, 15–20, 22–24, 26, 31, 32, 34–36	*
26	Visibility	1–5, 8, 9, 15, 16, 18–20, 22–25, 31, 32, 34–36	*
27	Weather	1–5, 8, 9, 15–20, 22–26, 28–37	3, 17, 21; 28–30
28	Wind direction	1–5, 8, 9, 15, 16, 18–20, 22–27, 31, 32, 34, 35, 37	3–5, 17, 21, 24, 25, 27, 33, 37
29	Wind velocity	1–5, 8, 9, 15, 16, 18–20, 22–27, 31, 32, 34, 35, 37	4, 5, 21, 24, 27, 33, 37
30	Humidity	1–5, 8, 9, 15–20, 24–27, 31–37	3, 21, 27
31	Requirements to operate	1, 2, 5, 8, 9, 15, 16, 22–26, 34–37	*
32	Explosion—back draft	1–5, 8, 9, 15–20, 22–26, 31, 34–37	*
33	Topography	1–5, 8, 9, 20, 24–26, 28, 29, 31, 34, 35, 37	3, 17, 27, 30
34	Exposure hazard	1, 2, 4, 5, 8, 9, 15, 16, 22–26, 28, 29, 31, 32, 35, 37	*
35	Duration of operation	1, 2, 5, 8, 9, 15, 16, 22–26, 28, 29, 31, 32, 34, 37	*
36	Street conditions	1–5, 8, 9, 14–16, 20, 22–26, 31, 32, 34, 35	*
37	Simultaneous fires	1–5, 8, 9, 15, 16, 20, 22–26, 28, 29, 31, 34–36	*
38	Class of fire	1–5, 8, 9, 15, 16, 22–26, 31, 32, 34, 35	3, 6–9, 21–23, 31, 32

* Signifies all other primary factors.

4

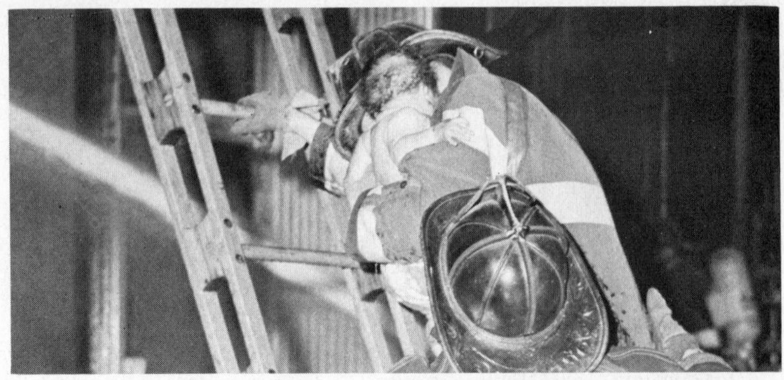

Life Hazard for Occupants and Personnel

The primary factors "life hazard for occupants" and "life hazard for personnel" rate top priority because they are associated with human life. Specifications for acceptable solutions at fires are: (1) If there is a life hazard for occupants, risks to personnel ranging from merely unusual to extreme may be warranted; and (2) if there is no life hazard for occupants, personnel should never be jeopardized unnecessarily.

If occupants are endangered, it is the responsibility of the commanding officer to make risks for personnel merely unusual rather than extreme, when possible and feasible. This can be accomplished by providing helpful ventilation, protective lines including back-up lines, proper masks, and effective coordination via adequate supervision and communication. At times, when there is only one company on the scene, help may not be available. In such situations, rescue efforts could involve extreme risks. However, if the fire building has collapsed and fire conditions are such that usual rescue measures would only uselessly increase the casualty list, there definitely would be limitations on the risks to be taken. This is why the words "may be warranted" instead of "are always warranted" are used in describing the limits.

The guideline that personnel should never be jeopardized unnecessarily when there is no life hazard for occupants should not be construed to mean that a safer exterior operation is always preferable to a perhaps riskier interior operation. The former tends to result in greater water, fire, and smoke damage, while the latter (admittedly more dangerous) is usually much more efficient. However, a major responsibility of the fire service is to protect property at fires and some reasonable risk is warranted in meeting that responsibility, without unnecessarily jeopardizing personnel.

The foregoing comments are substantiated by the "Creed for Firefighters Safety" formulated by the Uniformed Fire Officers Association of the New York Fire Department. This creed states

> When a man enters the Fire Department, he has the right to expect the Fire Department to continually make every effort to protect him from any predictable hazards of the firefighting profession.
>
> The union rejects the theory that injury and death is 'the unavoidable cost' of firefighting. While tragedies will continue to occur and firefighters will die protecting the public, we believe that firefighter deaths and injuries can be reduced by means of safety education and training.
>
> The traditional disregard of the firefighter for his own safety in order to rescue persons in imminent danger is commended and extolled; however, that which is praiseworthy where lives are in danger, becomes foolhardy and reckless where only property is involved, and must be discouraged.

LIFE HAZARD FOR OCCUPANTS

The potential life hazard for occupants in the event of fire is recognized by local and state legislation. This includes building and fire-prevention codes, zoning resolutions, and laws regulating such factors as construction, height, area, type of occupancies, distances between buildings, and presence of auxiliary appliances.

Effects on Other Primary Factors

It has been established that a life hazard for occupants can increase the life hazard for personnel because risks ranging from merely unusual to extreme may be warranted. In addition, a life hazard for occupants can adversely affect other primary factors. Such factors may include location and extent of fire on and after arrival, occupancy (human element and contents), heat and smoke conditions, visibility, exposure hazards, duration of operation, and requirements to operate effectively (personnel, water, apparatus, and equipment). For example, people trying to escape (and at times, civilians and police officers attempting rescues before the arrival of the fire department) may leave doors open in critical places. This allows an inflow of air into otherwise unventilated fire areas, which accelerates combustion and causes a back draft or smoke explosion that can collapse the structure.

Effects of Other Primary Factors

The life hazard for occupants (as indicated in the third column of the primary factor chart) can also be affected by each of the other primary factors. These factors are thoroughly discussed in the following chapters. Note carefully the effects of each primary factor upon the life hazard for occupants. Such knowledge is important when making a decision about the evacuation of a building. Remember, the chief responsibility of the fire department is to save lives at fires. Accordingly, the fire department is the only agency authorized to force entry, if necessary, and without a warrant, into a home or place of business and order occupants to move out. A sound knowledge of the effects of various factors upon the life hazard for occupants will ensure proper use of this unique authority to evacuate buildings.

Effects on Secondary Factors

Rescue Work. When a life hazard is present, the major objective is rescue and time is usually short. Forcible entry is made with less regard for structural damage, exposure hazards, or the availability of a hose stream to protect personnel. Ventilation unfavorable to controlling and extinguishing the fire may be needed to draw heat and smoke away from endangered occupants. Available hose lines are used as required to cover the life hazard. In short, a life hazard for occupants and the resultant rescue activities can delay efforts to control the fire, making extinguishment more difficult.

Covering Exposures. A life hazard may make the task of covering exposures more difficult and may also delay the attack on the fire itself. For example, aerial ladders used to remove occupants may have to be repositioned to use ladder pipes in protecting exterior exposures. This can entail a harmful delay, intensifying existing and possibly creating new exposure hazards. In addition, involvement in rescue work will delay the evaluation of factors to establish an order of priority in covering exterior exposures. Such coverage may have to wait for the arrival of greater-alarm units.

LIFE HAZARD FOR PERSONNEL

Risks incurred or assumed by fire personnel while performing their duties are referred to as the life hazard for personnel. These risks, however, may only be warranted if there is a life hazard for occupants. Lacking this condition, personnel should never be jeopardized unnecessarily. Despite practically universal agreement with these limits, more casualties occur in performing activities not associated with rescue. This fact presents a worthwhile subject for inquiry, especially in view of the logical demand in these times to increase productivity. There is no better way to do this than by reducing injuries among personnel, the most expensive item in every fire department's budget. Such factors as back draft or smoke explosion, structural collapse, and heat and smoke conditions are major causes of casualties not incurred in rescue work. Chapters 9, 10, and 13 discuss these factors and their associated hazards. A thorough study of those chapters will show that the hazards are predictable to a certain extent and that casualties can be reduced appreciably.

The problem of unnecessary casualties also suggests the need for further research into the state of the science and art of fire suppression. The lack of a satisfactory science and art might possibly be the reason for defective and harmful attitudes. There are firefighters who fatalistically maintain that "what will be, will be." Some chiefs advocate that fire personnel should try to keep hand lines in a building as long as there is a toehold, even if the building is already a total loss. Other chiefs express disdain for officers who give thought to possible structural collapse. Still other chiefs express concern over "killing the fire spirit" by undue caution. As a result, a company officer fearful of criticism if he starts outside operations, will try to make the fire before the chief arrives, and at times, this can be disastrous. Such attitudes are not in accord with the specifications for acceptable solutions at fires or with the "Creed for Firefighters Safety."

Tradition is a powerful force in the fire service, and in most respects, rightly so. Some units have a well-deserved reputation established and maintained by unfailing self-sacrifice. Newcomers to such units are promptly apprised about requirements to "measure up." Some traditions, however, are not beneficial to the fire department. An example is the tradition of not wanting to give up a line when ordered to do so. This has caused some unnecessary injuries and impaired the efficiency of some operations.

Effects on Other Primary Factors

Since risks cannot be completely eliminated in carrying out fire operations, the primary factor life hazard for personnel can have a bearing on many other primary factors. These include life hazard for occupants; extent of fire after arrival; occupancy; structural collapse; heat and smoke conditions; visibility; requirements to operate; explosions and back drafts; exposure hazards; and duration of operation.

Effects of Other Primary Factors

The life hazard for personnel can be increased or decreased according to whether any or all of the other primary factors worsen or better the fire situation.

Effects on Secondary Factors

Acceptance of *warranted* risks is essential for good results in carrying out fire activities.

Conclusion

So far, only relationships between the factors affecting a fire operation have been considered. There has been no reference to the many injuries incurred responding to and returning from alarms, or to those (less frequent) injuries incurred in quarters. However, unquestionably, the most intelligent and satisfying way to reduce casualties among personnel is to establish and maintain an effective fire-prevention program. Chapter 31 deals with this problem and suggests such a program.

Finally, there are those who maintain that a hazard control officer is a major solution to the problem of reducing personnel casualties. Such an officer would monitor the risks to which personnel are being subjected at fires. To do so, the officer would have to learn the effects of relationships between factors, the limits for acceptable solutions, and the principles to be applied in carrying out activities. Essentially, the officer would be trained to employ the plan. It is illogical to train only a staff hazard control officer in this way. All line officers should be trained to recognize and even anticipate hazards, thereby avoiding or minimizing their harmful effects.

REVIEW QUESTIONS

1. Why do the primary factors life hazard for occupants and life hazard for personnel rate top priority?

2. If there is a life hazard for occupants, what is the responsibility of the commanding officer insofar as risks to personnel are concerned? How can the commanding officer meet this responsibility?

3. Why are safer exterior operations not always preferable to perhaps riskier interior operations when there is no life hazard for occupants?

4. In what ways does the "Creed for Firefighters Safety" corroborate the views expressed in this chapter about life hazard for personnel?

5. In what ways does life hazard for occupants affect legislation?

6. How does life hazard for occupants affect other primary factors?

7. How is life hazard for occupants affected by other primary factors? Why is this knowledge so important?

8. How does life hazard for occupants affect rescue work and secondary factors?

9. What is meant by life hazard for personnel?

10. Why would a reduction in casualties be the best way to increase the productivity of personnel?

11. What can cause unnecessary casualties among fire personnel?

12. How can life hazard for personnel affect other primary factors?

13. How can life hazard for personnel be affected by other primary factors?

14. How can life hazard for personnel affect secondary factors?

15. What is debatable about the idea of having a hazard control officer to reduce personnel casualties?

5

Location and Extent of Fire on and after Arrival

LOCATION OF FIRE

The level of the fire, the specific section of the level involved, and the address of the fire building are referred to as the location of the fire. The section of the level involved is indicated as north, east, south, or west, unless the floor is fully involved. In some cases the terms rear or front are sufficient. In addition to the location of the fire, an officer must note where the fire is spreading—in the structure or to other structures—and surmise the extent of the fire. This must be reported to headquarters. Location and extent of fire are treated as separate factors because their relationships with other factors are not identical.

Effects on Other Primary Factors

Fires located between occupants and their means of escape create a life hazard. In addition, the location or level of the fire can increase or decrease the severity of the hazard. For example, a fire on a lower floor can jeopardize more people than the same fire on a higher floor. Likewise, a fire on an upper floor of a high rise building requires an interior operation, making rescues very difficult.

Serious but not always dangerous smoke conditions can be caused on top floors of fire-resistive construction. This can happen as high as ten floors above the fire, if it is located near the elevator shafts. Such heavy smoke conditions require prompt attention lest occupants of these floors overestimate their danger and become panic-stricken. The smoke conditions can be alleviated by adequate ventilation and panic avoided by assuring occupants that they are not in serious danger.

In unusual cases the type of occupancy affected produces a severe life hazard and rescue work must be started before the fire is definitely located. An example of this situation would be a subway fire. To contend with possible panic, the commanding officer may have to commit all available manpower to rescue work. The job of locating and extinguishing the fire may have to wait until more units arrive.

Fires located near vulnerable vertical and/or horizontal structural arteries tend to be extensive on arrival and to spread rapidly thereafter. Such arteries include air shafts, stairways, pipe recesses, and common cocklofts. Fires in these areas can intensify heat and smoke conditions, maximize life hazard, jeopardize interior and exterior exposures, and possibly cause structural collapse. A spark-and-ember hazard at such a fire could start other fires downwind, increasing duration of the operation and requirements to operate effectively.

A fire in an inaccessible location such as a cellar, subcellar, and or substructure of a pier might not be discovered until it had spread considerably. This could delay the transmission of alarm and response of apparatus. In addition, an inaccessible location could hamper ventilation and advancement of hose lines. Lack of proper ventilation can cause heavy smoke conditions, resulting in back drafts or smoke explosions.

The location of the fire can have a bearing on the effectiveness of sprinkler systems. For example, fires in materials on shelves or under workbenches limit the extinguishing capability of sprinklers.

The direction of the wind can also be affected by the location of the fire. This is because the fire heats the air, reducing its density and causing it to rise. A constant flow of air begins: Air is drawn towards the fire, then is heated and rises. At past wartime conflagrations, air flowed toward the fire at very high velocities. At structural fires, an inflow of air can result in a disastrous back draft if the fire area is not adequately ventilated.

The type of community in which the fire is located has a bearing on many other primary factors. These include: life hazard, construction, height, area, extent of fire on arrival, proximity of exposures, auxiliary appliances, occupancy, street conditions, and the availability of the requirements to operate.

Effects of Other Primary Factors

The location of the fire (level or section of the level involved on arrival) can be affected by a life hazard. The fire can spread through doors that are left open during efforts to escape or rescue prior to the arrival of the fire department. The location of the fire can be delimited by fire-resistive construction, height which determines the possible fire level, area which is effectively subdivided, auxiliary appliances such as alarm or sprinkler systems, and occupancies with normal fire loads.

The location of the fire can also be affected by explosions; structural collapses; heat, smoke, wind, and weather conditions that may cause windows to be open or closed; proximity of exposures; exposure hazards; street conditions affecting response, topography, especially at woodland fires, simultaneous fires set by arsonists, and class of fire (class B and C fires generally occur in certain places such as electric-generating plants and oil-storage facilities).

Effects on Secondary Factors

Forcible Entry. It is preferable to force entry near the location of the fire, especially when the area involved is large. This enables personnel to get water on the fire more quickly and minimizes the physical hardship entailed in advancing hose lines.

Ventilation. One of the main objectives of ventilation is to localize the fire—to stop its horizontal spread within the structure. This is usually achieved by opening up as directly over the fire as possible and feasible. For example, if a fire is extending into a cockloft via a pipe recess or similar channel, the roof should be opened. If this were done in the wrong place, it could be disastrous. Opening a roof in the front when the fire is coming up in the rear can involve the entire cockloft and turn a first-alarm into a major-alarm fire. It is not advisable, however, to open up directly over a fire on an intermediate floor, because this could involve the upper floor.

Another objective of ventilation is to protect occupants pending rescue. For example, if the fire has cut off the escape of occupants, the decision where to vent is determined by the need to draw heat and smoke away from them. This is done even if the required openings increase the intensity of the fire and the possibility of spread (not to occupied areas, of course). If only horizontal ventilation is required, the location of the fire will indicate the floor to be vented and the openings to be used.

The decision when to vent is determined by whether or not the location of the fire is creating a life hazard. If it is, ventilation may have to be started as soon as possible, even if hose lines are not ready or if an unoccupied exposure hazard may be created or intensified.

The location of the fire can affect how to vent. Fires below ground level may have to be vented by opening sidewalk deadlights or other covers or by using fog streams or smoke ejectors. Fires in stairways may have to be ventilated by opening the roof bulkhead. Fires in windowless buildings may have to be ventilated by utilizing air-conditioning systems. (Refer to Chapter 18 for a more thorough discussion of ventilation.)

Removal of Occupants. A fire on the first floor of a five-story residential building could endanger all occupants and necessitate their removal. However, if the same fire originated on the fourth floor, it may be better to move occupants of the fifth floor to the first or second floors. This is true especially if the fire occurs on a cold night and occupants are scantily clothed.

Checking for Fire Extension. A fire near a vertical or horizontal structural channel will spread readily. Officers assigned to check for fire extension should therefore note the location of the fire and keep in mind how heat travels by conduction, convection, and radiation via existing exposed channels.

Placement and Use of Hose Lines. The location of the fire determines the amount of hose to be stretched, and in some cases, the size. The minimum hose lines to provide adequate protection for fire personnel at below ground fires are $1\frac{3}{4}$ or $2\frac{1}{2}$ [44 or 63 mm]. Smaller hose lines may not provide adequate water to combat the heat. If the location of the fire has created a life hazard, hose lines should be placed to facilitate rescue, and should be operated as soon as possible and until rescue is completed. If there is no life hazard, the fire location will still govern the placement of lines.

Use of Special Equipment. High-level fires may require the use of standpipe systems, ladder-pipe, or other high-caliber streams. Perforated-pipe systems, high-expansion foam, or equipment to breach walls and apply water may be advisable at fires located in cellars or subcellars. The fire location may also influence the decision to use sprinkler systems, cellar pipes or fixed systems of various types.

Conclusion

Getting fire operations under way in the most efficient and effective manner depends upon accurate information about the location of the fire. Unfortunately, operations are occasionally begun with a mistaken idea of the fire location. This means delay, and it may increase the life hazard, make rescue work more complicated, and extinguishment more difficult. Fire officers who quickly and accurately determine the location of the fire are best able to make and implement correct decisions, because they can more clearly define the problems that the fire presents.

EXTENT OF FIRE

"Extent of fire" means the extent at the time the fire department arrives plus any subsequent extension. A fire of moderate extent on arrival may sometimes become much larger. This is apt to be true if, for example, a length of hose bursts, the water supply is inadequate, or the need for rescue work entails venting in a manner unfavorable to control and extinguishment. As the extent of a fire may vary during a fire operation, so may the decisions on how best to fight it.

In some structural fires, it may even be impossible to determine accurately the extent of the fire in its early stages. Just determining its location, however, generally produces some information about its extent. Areas of heavy involvement may be known, of course, but excessive heat and smoke, poor visibility, life hazard, and concealed channels by which the fire can spread may make exact determination difficult—certainly until all such channels and exposed areas have been checked. Actually, by that time extension will probably have stopped and confinement and control will be imminent.

Officers in command should, however, be reasonably conservative in reporting on fire extent. Premature reports that a fire is under control frequently lead to disastrous consequences. In many such cases, some companies are sent back to quarters too soon. If commanding officers then try to handle the situation with only the assignment still on the scene, personnel may have to pay the price. It is much better to promptly recall the units sent back to quarters and even to transmit an additional alarm if necessary.

Premature reports of control are most likely to be made at cellar or subcellar fires, at ground-floor fires that drop into the cellar, or at structural fires involving concealed spaces. Such spaces might include pipe recesses or other vertical channels through which the spread of fire is not easily detected.

Effects on Other Primary Factors

Generally, fires that are extensive on arrival have considerable momentum and potential for spreading after arrival. They can thus worsen heat, smoke, visibility, and exposure hazards. They are also apt to cause structural collapse and other fires, and to increase life hazard for occupants and personnel, as well as requirements to operate and duration of the operation.

The extent of fire can indicate whether sprinkler systems in the fire building or exposures should be shut down or supplied, and whether available standpipe or other fixed systems should be supplied. The extent can also affect the direction and velocity of the wind. It can thereby contribute to the development of a back draft and—obviously—affect the human element and content aspects of occupancies.

Effects of extent of fire on other primary factors—both on and after arrival—are approximately the same, as shown on the primary-factor chart.

Effects of Other Primary Factors

The extent of fire after arrival can be altered by all the other primary factors. The extent of fire on arrival has not yet time to be affected in this way, as can be seen from the primary-factor chart.

Effects on Secondary Factors

Forcible Entry. A light haze of smoke visible through heavy glass doors that feel cool to the touch usually indicates fire of small extent. In such circumstances, entry should be made in the manner least damaging to property, possibly even by getting and using a key. Where the extent of fire is obviously substantial, however, such consideration is not warranted. Speed in getting an efficient operation under way is more important.

Ventilation. The extent of fire should have a reasonable relationship to the amount of structural damage done in ventilating. That is, a hole should not be made in the roof when opening top-floor windows is sufficient. But if the extent of fire is great, proper ventilation is more important than any necessary structural damage.

Fire extent may determine whether roof ventilation should be attempted at all. For example, if two or more floors in an old loft building are fully involved in fire, one should not try to work either on the roof or in the structure unless rescue work makes it essential. If the fire is so great that roof or fire-escape venting is out of the question, heavy outside streams or aerial-ladder tips may be used to break windows that do not have shutters.

Other Effects

The extent of fire affects the amount and kind of overhauling needed and the nozzle type, size, and pressure; the hose size; the number of lines; and the kind of operation (interior or exterior). At extensive fires, notably in woodland areas, supervision can be impaired by large perimeters, smoke, and poor visibility. Communication can be hampered by hills that interfere with radio and television reception. And coordination, so dependent upon supervision and communication, can be adversely affected.

REVIEW QUESTIONS

1. In general, to what does location of fire refer? At a structural fire, to what does it refer?

2. How does location of fire affect other primary factors?

3. How does location of fire affect secondary factors?

4. Why is it important to quickly and accurately determine location of fire?

5. What additional information does location of fire convey apart from what is asked in question 1?

6. How can location of fire be affected by other primary factors?

7. What does extent of fire mean?

8. What can make it difficult to determine the extent of fire in its early stages?

9. At what types of fires are premature reports of control most likely to be made? How should such mistakes be rectified?

10. How can extent of fire on and after arrival affect other primary factors?

11. How does extent of fire on and after arrival affect secondary factors?

6

Construction

It is important for fire officers to know the type of structure that is burning. Such knowledge will help them determine the speed with which the fire may spread, whether it will spread vertically or horizontally or both, and how the objectives of rescue and extinguish can best be achieved.

Knowledge of construction is essential if officers are to operate efficiently at structural fires, which often provide the consummate test of their knowledge and skill (science and art). Officers who can more frequently check internal extension of fire without unduly jeopardizing their subordinates should be rated more highly than those who resort to exterior operations. These latter operations cause maximum instead of minimum damage and consequently more often "lose the building."

Officers are not expected to be personally familiar with the structural layout of every building in their districts or communities. It is not unreasonable, however, to expect them to be familiar with structural features of special significance in local *types* of construction. In addition, they should realize that a fire in a high rise or large industrial complex is no longer the sole responsibility of city firefighters. Hence, surburban as well as city firefighters should know how the construction associated with such occupancies can affect other related factors, and thereby an entire fire operation.

Fire resistiveness in buildings depends—among other things—on the manner in which floors, walls, partitions, ceilings, columns, and girders are constructed. It also depends upon floor areas, combustibility of the structural parts, roof conditions, and the degree to which horizontal and —especially—vertical channels are firestopped.

Horizontal channels could include hallways, corridors, ceiling spaces—particularly common cocklofts and plenums—floor spaces, doors, windows, and ducts. Fires can also travel horizontally when heat is conducted, for example, by metal beams, through intervening walls and partitions, or from wooden beam to wooden beam when they abut.

Vertical channels could include partitions, stairways, elevator shafts, dumbwaiter shafts, laundry chutes, ramps, escalators, air and light shafts, recesses enclosing pipes or electrical conduits, conveyors, and ducts associated with air-conditioning systems and large cooking ranges. Fires can also spread vertically by burning through floors or ceilings, or from floor to floor on the outside of the building.

Materials used in construction naturally affect the spread of fire. Some masonry materials with high fire-resistive ratings contain water in their makeup. This water slows down the heat transfer rate—it absorbs large amounts of heat and delays transmission until it has been evaporated. On the other hand, good insulating materials generally have low fire-resistive ratings. They, too, can slow down the heat transfer rate—partly by means of entrapped air which absorbs heat and delays transmission—but not as effectively as high fire-resistive materials.

Column 2 of the primary-factor chart indicates that construction of the fire building affects such other factors as its height, area, occupancy, proximity of exposures, exposure hazard, and auxiliary appliances. Building and fire-prevention codes, zoning resolutions, and other laws govern these relationships in many communities. Column 3 of the chart indicates that, reciprocally, construction is affected by these other factors, in accordance with the laws referred to.

The National Commission on Fire Prevention and Control report *America Burning* reveals that there are about 14,000 local building codes in the United States. Necessarily, only a few types of construction and their relationships with other factors can be discussed here.

BRICK-JOIST CONSTRUCTION IN RESIDENTIAL BUILDINGS

This construction is common in congested areas of many large cities. It features many combustible structural members, fire escapes with gated windows, and numerous inadequately firestopped horizontal and vertical structural arteries, both open and concealed. It has one advantage, however: it is not tight enough to prevent the escape of considerable heat by convection and radiation, and it is therefore less likely to cause a back draft or smoke explosion. Roof bulkheads, doors, and plain glass windows allow for ordinary ventilation. Fire escapes are covered separately, below.

Effects on Other Primary Factors

The disadvantages cited for this type of construction can cause extensive fires on and after arrival, severe heat and smoke conditions, poor visibility, interior exposure hazards, and a life hazard for occupants. The open construction allows much heat to escape, but this can become a disadvantage if it creates or worsens an exterior exposure hazard. Such a hazard can increase requirements to operate effectively, prolong the operation, and increase hazards for personnel.

Effects on Secondary Factors

Forcible Entry. In recent years it has taken longer to gain entry because of the increase in crime and the resulting increase in locks on hallway doors and extended locked gates inside fire escape windows.

Ventilation. Horizontal ventilation is achieved by opening doors and windows. Vertical ventilation is achieved by opening roof bulkheads and making openings in the roof as conditions warrant. Usually both kinds of ventilation are required.

Placement and Use of Hose Lines. Usually, hose lines are stretched via the interior stairs, and less frequently via fire escapes, ladders, and ladder-tower platforms.

Overhauling. There is likely to be more overhauling than usual if combustible structural parts are involved and concealed spaces have to be checked out.

Removal of Occupants. Interior stairs are the preferred means of removing occupants, unless they are above the fire. In the latter case, fire escapes are usually used, but aerial ladders or tower-ladder platforms may sometimes be necessary. Occupants trapped in the rear without access to fire escapes or not reachable from ladders may be removed by firefighters who are lowered to them by ropes. In some instances, occupants taken to the roof via rear fire escapes can be brought down interior stairways of adjoining buildings.

BRICK-JOIST CONSTRUCTION IN COMMERCIAL BUILDINGS

These brick-joist structures have combustible structural members that burn readily. They also lack firestopping material, thus enabling fire to spread quickly both horizontally and vertically. In addition, they are frequently old, which aggravates structural defects. Older loft buildings have still other unfavorable characteristics: subcellars, unusual depth (in some instances, 200 ft [61 meter (m)], unprotected metal columns, wide floor spans, and iron shutters (Fig. 6-1).

Effects on Other Primary Factors

The disadvantages cited above can—because of the building codes and other laws that permitted them—unfavorably affect location and extent of fire on and after arrival. These disadvantages can in turn unfavorably affect heat and smoke conditions, visibility, exposure hazards, back drafts or smoke explosions, requirements to operate, duration of the operation, and consequently the life hazard for occupants and personnel.

Effects on Secondary Factors

Forcible Entry. In commercial buildings, entry is seldom a problem during business hours. At other times, entry is hindered by iron shutters and door locks that are often intricate and difficult to force.

Ventilation. Ventilation is greatly hampered by iron shutters. Sometimes there are conditions conducive to a smoke explosion. In such cases, it is of utmost importance to vent at the roof, side, rear, or front of the fire

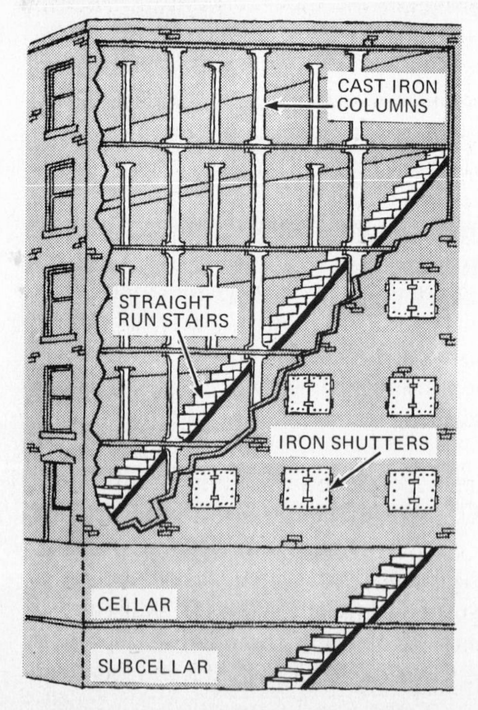

FIG. 6-1. Construction of a brick-joist commercial building.

area before opening up at lower levels or the fire floor for entry. Fires in cellars or subcellars have limited means of ventilation: deadlights in sidewalks, sidewalk covering of entrances into cellars, and, under definitely controlled conditions, openings made in floors above the fire. In some situations, fog lines or smoke ejectors can be helpful.

Rescue Work. Rescue work can be impeded by a number of factors. High temperatures are common; severe smoke conditions may impair visibility; floor space may be crowded by workbenches, machines, and other materials—in some instances only a 3-ft [.9 m] aisle space is required by law. Frequently there are many employees, and exits may be unfavorably located, particularly above the second floor. Some exits, for example, are about 40 ft from the street front because the stairs run straight back to the third-floor landing. In addition, these exits may be in the heart of the fire. In such cases a ladder pipe or other appliance can be operated through the street-front windows so that entry can be made with a hand line (two lines if warranted) via ladders or even the interior stairway.

Ventilation, a potent weapon for minimizing the life hazard, cannot always be used effectively unless the fire is located favorably or a roof opening can be quickly made to draw heat and smoke away from the life-hazard area. Occupants may be found conscious but so panic-stricken that rescuing them becomes unusually complicated. Some, unless strongly urged and guided, seek safety blindly, even disastrously.

Some of these buildings have two interior stairways, located at a distance from each other; others have one stairway and a fire escape. Occupants can also be rescued by means of ladders, tower-ladder platforms, and interior stairways of adjoining buildings when such buildings are accessible from the roof of the fire building. In extreme cases, life nets have been used.

Placement and Use of Hose Lines. Where a life hazard is present, the first hose line is stretched and operated as quickly as possible between the fire and the endangered occupants or between the fire and the means of escape. Outside streams may be needed at times to help with the positioning of hand lines.

Where no life hazard exists, lines are placed and used according to principles discussed in Chapter 17. However, in these loft buildings, the first line is more often stretched up the interior stairway and used to execute a holding action—to confine the fire to the involved occupancy—while the second line is brought up the fire escape, if one is available, to put out the fire. Because of the unusual depth of some buildings, and if the location of the fire is favorable for using such a technique, this is an acceptable variation of the usual procedure of operating the line from inside. Any auxiliary appliances (sprinkler, perforated pipe systems) must be supplied as conditions dictate. The correct inlet must be supplied, or unwarranted and severe water damage may result. At low-level fires the correct inlet may be chosen by feeling for heat conducted from the fire area by connecting piping. Even the location of the fire may be found this way if all inlets are properly marked.

Supervision. The unusually risky nature of operations at loft fires requires extremely careful supervision. Entire fire companies have been injured or killed by falling through collapsed roofs and floors. Effects of relationships among pertinent factors at these fires must be weighed with exceptional care before operations are initiated. All officers should be keenly alert to signs indicating possible structural collapse so that firefighters can be removed in time, and communication should be promptly established so that all units can be quickly contacted. Some authorities recommend an exterior operation when two or more floors in such buildings are fully involved. This is a sound recommendation, but an exterior operation may be advisable even before the fire reaches that extent if there are conditions such as wide floor spans, unusual depth, combustibility of structural parts, unprotected metal columns susceptible to failure when heated and struck by cold water, excessive age of the building, long duration of the fire, and/or the presence of heavy machinery or stock that absorbs water.

Overhauling. Since many parts of these structures are combustible, more than unusual amounts of structural overhauling can be anticipated.

Overall Effects of Fire Escapes

The primary purpose of a fire escape, as its name says, is to provide a way by which occupants can escape from fire. The annals of the fire service abound with reports of rescues made by firefighters entering exposed floors from fire escapes and removing endangered occupants. In some cases, entry, search, and removal are achieved even without the protection of hose streams, which may not be operating.

A secondary purpose of the fire escape, when the safety of occupants is not jeopardized thereby, is to provide the fire department with a means of advancing into involved buildings to extinguish fires and protect exposures. Fire stairs fulfill the primary purpose better than a fire escape only when they are equally accessible. In many modern apartments without fire escapes, for example, fire stairs are useless if the occupant is trapped in an apartment.

As a rule, forcible entry is faster and easier in structures with fire escapes, since strong locks and doors may be opened more easily from within. In addition, much less property damage is done.

At times, an entire building can be ventilated fairly quickly by means of fire escapes. By contrast, when there are no fire escapes ventilation may be difficult and time-consuming, especially when the fire and exposed floors are beyond the reach of, and otherwise inaccessible to, aerial ladders. (This is even true of fires on lower floors of many modern residential projects where aerials or ladder-tower platforms are practically useless because the trucks cannot be positioned effectively.)

Fire escapes are a great asset in stretching and operating hose lines. Often a line advanced from the fire escape can put out the fire when lines from the interior cannot be moved in. It is also helpful to have the option of attacking the fire from either side instead of from only one. Where there are no fire escapes, an inability to advance lines from the inside may compel officers to use heavy streams from the outside, which invariably results in greater property damage. In such an event, the door to the fire area is closed if possible and personnel are moved to a safe position.

Frequently, at greater-alarm fires, fire escapes have been used as vantage points from which to operate lines. Such vantage points remove the need to force entry into surrounding structures.

There are a few cautions to be observed when using fire escapes. (1) Officers should not overload fire escapes when they are stretching hose lines. (2) They should be aware of the direction of the wind when ventilating from fire escapes or when using aerials or ladder-tower platforms to remove occupants from above the fire. (3) If a plaster hook or pike pole is used from the fire escape to break a window first on the windward side, positions on the balcony can become so untenable that other windows cannot be broken or opened. (In removing occupants from above the fire, the windward side of the balcony may be subjected to less heat.) (4) Operating on a fire escape above the fire can be very dangerous but, if deemed necessary, protective lines should be made available in time.

FRAME CONSTRUCTION

This type of construction is usually referred to as wood frame construction. There are many variations of the exterior wall: wood shingles, clapboard, matched boards, brick veneer, stucco, metal-clad over wood sheathing, and so forth. Private, one-, or two-story dwellings feature such construction.

Some authorities maintain that large multistory frame buildings can be made reasonably safe if proper attention is given to protection against the horizontal and vertical spread of fire, against exposure fires, and against fire conditions which may be anticipated on the basis of expected fire loads. However, the same authorities also point out that conflagrations may occur at fires in primarily residential sections due to closely built combustible construction and wood shingle roofs. Conflagrations are considered possible where certain construction practices are allowed, and where protection forces are weak and water supplies inadequate. As a matter of fact, such fires can occur even where protection is strong and water supply is adequate, as attested by a fire involving many beach bungalows within the limits of a large city.

Statistics show that there is a large loss of life in rural and urban dwellings, presumably of frame construction. The lack of a prompt and adequate response by firefighters has much to do with these statistics, but structural features also play an important role.

FIRE-RESISTIVE CONSTRUCTION

In fire-resistive construction, walls and structural members are made of noncombustible materials or assemblies with the following minimum fire-resistive ratings: four hours for exterior walls, fire walls, party walls, piers, columns, and interior structural members which carry walls; three hours for other girders, fire partitions, floors (including their beams and girders), roofs, and floor fillings and required stairway enclosures. Such construction does not feature central air-conditioning systems. The Empire State Building fire is a good example of how fire-resistive construction affects primary and secondary factors.

The Empire State Building is of steel skeleton construction and, although it differs somewhat from the specifications mentioned above, it fully qualifies as fire-resistive. The fire occurred after a United States Air Force bomber crashed squarely into the upper part of the building, spraying approximately 800 gal [3 meter^3 (m^3)] of gasoline over where it struck. Parts of the seventy-eighth and seventy-ninth floors caught fire and burned furiously. Gasoline also descended one elevator shaft and caused a shaft fire all the way down to the basement level. In their report of this fire, NYBFU states, "Too much cannot be said of the sturdy, well constructed and fire resistive nature of the Empire State Building. Structural damage is comparatively negligible. The fire did not spread to other floors or portions of the building."[1]

At the same time, it must be admitted that the fire is not always confined to one floor in the type of construction referred to as truly fire-resistive. The Woolworth Building had a grease-duct fire that extended from the basement to the roof, and the Empire State Building had a

[1] The New York Board of Fire Underwriters, *Report on the Collision of U.S. Army Air Force Bomber with the Empire State Building*, 1945, p. 1.

water-pipe insulation fire in a shaft that reached from the thirty-first to the sixty-sixth floor. Fires in shafts enclosing electrical cables, as well as elevator shafts, contribute to the spread of heat and smoke in any construction. However, the records show that spectacular fires were very few and the loss of life was minimal in such structures as compared with their successors. The exception, of course, was the plane crash that sprayed 800 gal of gasoline on two floors of the Empire State Building, although it still did not jeopardize the strength of the building.

Fire towers in fire-resistive buildings are in all likelihood the best means devised for the escape of occupants at fires. With some exceptions, older building codes required at least one such tower for public and business buildings that were 75 ft [22.8 m] or more in height. Enclosing walls have a four-hour fire-resistive rating. Outside balconies or fireproof vestibules connect the fire tower and the structure. Such balconies or vestibules are separated from the structure and the stairs by self-closing fire doors which can be opened from both sides without a key. They open on a street or yard, or on a vertical open court, which has a minimum net area of 105 ft^2 [9.8 square meters (m^2)] and a minimum dimension of 7 ft [2.1 m]. It is practically impossible for heat and smoke to get into such fire towers because when doors are open between an involved occupancy and the balcony—to advance a line or for other reasons—emerging heat and smoke rise vertically through the open court rather than travel horizontally through the fire door into the tower. Stringent regulations governing openings in court walls minimize the hazard created by rising heat and smoke.

Effects of Construction on Other Primary Factors

Fire-resistive structural features (apart from occupancy contents, area, location of fire, and so on) tend to delimit the location and extent of the fire on and after arrival, and to minimize interior and exterior exposure hazards. They also minimize duration of the operation, requirements to operate effectively, and life hazard for occupants, particularly if a fire tower is available. However, much heat can be absorbed and retained by masonry materials, and this greatly increases hardships and the life hazards for personnel. This hazard is intensified by smoke, which impairs visibility, and which, along with the heat, is difficult to alleviate by ventilation. Paint-work, venetian blinds, and floor and ceiling covering can contribute to the development of troublesome heat and smoke conditions in the involved occupancy as well as in long corridors leading to the fire area. Smoke and heat rising in elevator shafts can create a hazard for occupants on the top floor served by the elevators; smoke can be the cause if the fire is at a low level; smoke and heat can be causes if the fire is at an upper level. In either case, the affected floor should be promptly ventilated.

Effects on Secondary Factors

Forcible Entry. In commercial structures, the watchman may cause some delay if he waits for the fire department at the fire floor instead of at the street level from where he can direct the firefighters. In residential fire-resistive structures, such as hotels, one should use great care in opening obviously hot doors to unventilated involved guest rooms; this can cause a back draft with drastic results for personnel. In such cases, it is suggested that the door be kept closed and indirect attack used by injecting fog through a small opening made in the partition to the fire room. The fire room can then be entered.

Ventilation. Doors and windows are used for cross-ventilation. Make openings first on the leeward and then on the windward side. Elevator shafts are not recommended because they only transfer the ventilating problem to an upper floor, endangering occupants who may be using the elevators and personnel working near the open shaft. In addition, they may cause unnecessary damage to the elevator mechanism. Use of fire stairs for ventilating is not recommended because the rising heat and smoke could endanger occupants trying to come downstairs. Also, if the stairs are not needed by the occupants, they could be used to alleviate heat and smoke conditions on the top floor.

Placement and Use of Hose Lines. Fire departments have their own regulations about supplying and using hose lines from standpipe systems. Usually, it is advisable to stretch the first line from the outlet on the floor below the fire and the second line (if needed) from the outlet on the fire floor. Also, lines are usually advanced from the windward side of the fire, especially down long corridors. Other applicable principles are discussed in Chapter 17.

Overhauling. Overhauling is likely to be confined to contents rather than to structure, even though a major purpose of overhauling after control is established is to check contents and structural parts for any lingering fire. Since structural channels have to be checked out in order to declare a fire under control, comparatively little remains to be done about such channels thereafter, although smoldering contents may require much overhauling.

Removal of Occupants. In buildings featuring fully fire-resistive construction without central air-conditioning systems, the fire department rarely has occasion to vacate occupants above the fire floor. As a matter of fact, at fires in hotels of this type, it is preferable to leave occupants of the fire floor in their own rooms rather than take them out of smoke- and heat-free areas into hot and smoky corridors toward fire stairs.

In very unusual cases, however, fires can extend vertically in the best fire-resistive construction due to explosions or via exterior windows or shafts enclosing electrical conduits or insulated water pipes. In these instances, occupants above the fire floor must be removed, preferably by means of fire towers. Other ways include using a stairway that does not enclose the standpipe riser used to supply hose lines—because such a stairway is open at the fire floor and allows smoke and heat to enter and rise. Where removal of occupants from above the fire is necessary, particularly at high levels, it can be accelerated by leaving them just a floor or two below the fire unless the need for medical attention dictates otherwise. Elevators exposed to heat and/or smoke, or affected by call buttons responsive to heat, smoke, or flames should not be used.

MODERN HIGH-RISE BUILDINGS

These structures have generally been built in the international style of steel and glass, with open floors, service cores, sealed windows, air-conditioning systems, and plenums (the space between the ceiling and the floor above). The cores are of reinforced concrete and contain stairs, elevators, utilities, and air-conditioning equipment. Plenums contain air-supply ducts, lighting fixtures, power lines in conduit, telephone cables, and communication cables. Careful study of the World Trade Center fire shows how modern high-rise construction affects other primary factors.

The building is subdivided into many vertical components so that the possibility of total involvement in fire is almost impossible. There are only three vertical shafts (elevator) that travel the height of the building, and only one of these has openings on every floor and is designated for fire department use. The other two shafts open only at the ground floor, the sky lobbies (forty-fourth and seventy-eighth floors), and in the upper third of the building. The chimney effect so often mentioned in high-rise buildings is not 110 stories in effect, but is divided into four components by the action of the air-conditioning systems. None of the stairways runs straight from the top to the bottom of the building. Stair towers are offset at various floors where the size of the core changes or the number of elevators serving a floor is reduced. At each of these points, horizontal passageways lead to the new shaft location and fire doors are provided in the passageway. These doors would prevent smoke from contaminating a stairway from top to bottom. The arrangement of elevators is such that they could not carry fire throughout the building but could be a factor in only a limited number of floors. It might even be feasible to use most elevators for evacuation—all except those that served the seven or eight floors that included the fire floor.

Construction of the Trade Center (Figs. 6-2 and 6-3) is unique in that the exterior walls are load-supporting. Large steel trusses connected to the interior core are structurally a part of the 4 in [101.6 mm] concrete floors they support. The core, which contains elevators, stairs, restrooms, and building services, consist of numerous steel columns with concrete floors. Concrete floors butt against the exterior steel spandrel beam so no vertical flues are formed. In addition, the interior face of the exterior columns is protected by plaster so that no vertical flue exists at this point. All vertical shafts are enclosed by 2-hour fire-rated walls. In discussing design faults of building codes in general, the NYBFU report states, "Unfortunately, no provision seems to be made for protection of openings in floors or walls. Consequently, some of the holes are not filled or others are filled with materials that disappear in the first seconds of fire. It is ridiculous to spend time and money to prove that a floor or wall can withstand a two-hour fire and then allow holes to be cut in it that destroys the fire resistance."[2] The fire in question did spread through such holes.

An item of particular interest to the fire service is the fact that the air-conditioning system can be placed in the purge mode after a fire alarm is received. This means that fresh air is blown into the core area to keep it free of smoke, and that air is drawn out of all the tenant areas on the affected floor to prevent smoke from spreading throughout the building. By supplying fresh air to the core and shutting down its normal vents, elevators as well as stairs can be pressurized and exit corridors kept free of smoke. To draw air out of the tenant areas, only the return air fans operate and discharge to the outside of the building. Proponents of this system apparently feel that a normal temperature would exist in the return

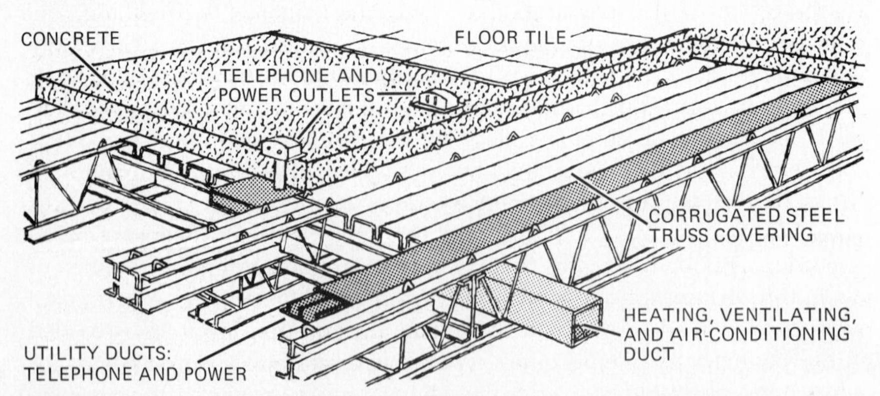

FIG. 6-2. Floor construction of One World Trade Center.

[2] The New York Board of Fire Underwriters, Report on One World Trade Center Fire, 1975.

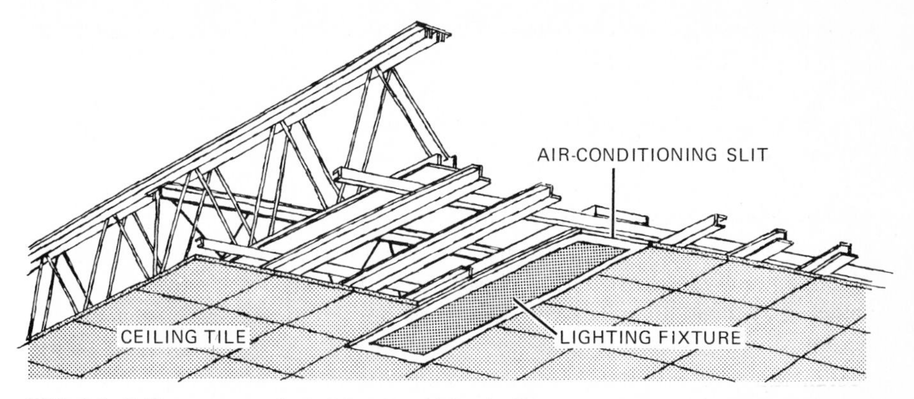

FIG. 6-3. Ceiling construction of One World Trade Center.

air shaft because of the large volume of cool air being drawn in from other floors. In the meanwhile, the supply air fans are shut down, and a question arises about the overall effects of such tactics on occupants in the tenant areas affected, especially if the fire operation is prolonged and the weather is hot.

In any case, it is believed that the fire on the date in question originated from an unknown source, in an executive office on the eleventh floor, at about 11:30 P.M. The office contained a wooden desk, lamp tables, sofa, and four chairs. Cushioning for the sofa and chairs was foamed-polyurethane with dacron covering. The overstuffed furniture burned fiercely and the fire spread along the corridors toward the open office area, involving the file room in the process. A gallon can of methyl-alcohol duplicating fluid in the file room was set aflame by the heat.

It appears that a flashover, apparently with the effects of a smoke explosion, occurred as fire-security personnel reached the office door, driving the fire out of the office of origin, and breaking seven windows in the east wall. An inflow of air through the broken windows accelerated spread of the fire toward the core of the building. Fire also spread through the horizontal channel in the peripheral heat diffuser units to adjoining offices. At about this time, according to the report, the air conditioning (otherwise inactive at night) was placed in the purge mode and the exhaust fans pulled air into the plenum to the return air duct. Telephone cables in the plenum were ignited by the fire. Because of openings around the cables where they passed through the telephone closet walls (in the outer section of the core) and because of louvers in the telephone closet door, fire penetrated the closet to ignite telephone panel blocks and cables. The polyethylene and polyvinyl chloride cable insulation and plastic panel blocks burned readily. Fire spread to upper and lower levels through openings about 12 × 18 in [304.8 × 457.2 mm] in the floors of these closets and emerged from the telephone closets on the twelfth and thirteenth floors in the same manner it had entered on the eleventh floor. The openings mentioned were the ones referred to in discussing construction of this building.

Effects on Other Primary Factors

For practical purposes, fire officers want to know what the effects are *as of now,* because since the fire the troublesome openings mentioned above have been firestopped. In addition, the louvers in the door to the telephone closet will be sealed with a steel plate. Evaluations will be affected by the building arrangement, prevailing laws governing departmentation or the installation of sprinkler systems, and the assertion in the NYBFU report that stair and elevator pressurization worked very well during the last fire. Based on these premises, construction will tend to delimit the location and extent of fire on and after arrival, and to reduce exposure hazards, requirements to operate effectively, duration of the operation, and the life hazard for all concerned. These tendencies may not be apparent, however, unless (1) combustibility of contents and furnishings are strictly limited by law, (2) fires can be quickly discovered and reported by alarm systems, (3) occupants can be promptly advised of fires and told what they should do, (4) well-trained fire-security personnel are always available, (5) prefire plans, including drills with fire-department units, are diligently formulated, and (6) recommendations of the NYBFU are implemented as promptly and fully as possible.

The NYBFU report indicates that there are bad points about the World Trade Center towers: fireproofing of the steel may be missing in places; fire rating of the shafts is just above minimum requirements; wiring ducts under the floor (as in many other buildings) have questionable fire resistance; and construction hazards and deficiencies due to incomplete construction are still present. Two unsatisfactory conditions associated with the telephone installation are also mentioned. First, cables going from one closet to another on the same floor pass through the plenum, exposing the combustible insulation to any fire that may be drawn into the plenum. Second, large groups of plastic-insulated telephone cables pass through the mechanical-equipment levels, such as one of the air-handling rooms for the air-conditioning system which is located where fire in the cables could spread smoke to as many as 32 floors. Smoke detectors at the filters would prevent this smoke from being recirculated but they would not prevent it from traveling through the shafts when the system was shut down.

Recommendations have been offered in the NYBFU report for eliminating or at least minimizing the foregoing hazardous conditions. The report concludes that the good points in the World Trade towers outweigh the bad and, in this connection, mention—among other items—two unique fire safety features: their limits on contents combustibility and their fire safety plan. Obviously, evaluations must be properly qualified. Our own evaluations of the construction factor, for example, specifically in relation to the World Trade Center towers, would be drastically different if the occupancy factor were unfavorable or the purge mode proved ineffective.

SPECIAL STRUCTURAL FEATURES

Air-Conditioning Systems

In one of the new types of high-rise construction, the air-conditioning system serves the core only, and occupancies around the perimeter of the building are provided with individual units. Thus smoke and heat cannot be conveyed into these occupancies by air-conditioning-system ducts. In case of fire, the main system can be shut down temporarily and the individual units can be operated on exhaust, thereby creating a favorable flow of smoke and heat, facilitating the advance of hose lines, and expediting extinguishment. At the same time, smoke and heat are being driven away from rather than toward the main air-conditioning system, thus lessening the likelihood of the spread of smoke to other floors via the ducts in such systems. If necessary, smoke and heat in the plenum area over an involved occupancy can be effectively dissipated by operating the return-air fans only and dumping outside the building. Or low-velocity fog injected into involved plenums could minimize heat conditions. In addition, it is possible that individual air-conditioning units, operating on intake on the fire floor—except in the involved occupancy—may abet the flow of heat and smoke out of the structure. They may also make it unnecessary to break windows to get air.

In the best fire-resistive construction, mechanical failure of controls, inadequate control over flammable contents, and structural defects which negate the fire-resistive ratings of floors and partitions, increase the possibility of both vertical and horizontal spread of heat and especially smoke at high-rise fires. Aside from such possibilities, however, construction that features an air-conditioning system that serves the core of the building only and individual units for occupancies around the periphery of the building is a major improvement in fire protection. Moreover, fire operations can be carried out more safely and effectively. Finally, such construction suggests a favorable alternative to the use of pressurized stairways and elevators. Pressurized systems (see Chapter 9) are not features of the construction discussed here.

Although the air-conditioning systems in new high-rise buildings present some hazards, older systems present even more. Many systems in use today were installed before effective laws governing their installation came into being. At the least, this lack resulted in little standardization, and generally it resulted in many shortcomings. A few examples are the following. Combustible materials, such as cotton, paper, steel wool, and felt were used for filters, and many were also coated with high-flash-point oil to catch the dust. Some portions of the ducts were lined to reduce transmission of noise or heat through the duct walls, and sometimes the lining was combustible. In addition, ducts could be dangerously near other combustible materials which would eventually be susceptible to ignition. Sometimes, too, coils containing a toxic or flammable refrigerant were inserted in air passages. In case of a leak or a fire, the refrigerant gas could then be distributed throughout the ducts, intensifying the life hazard for all concerned.

Dampers may help to check the spread of heat in both new and old systems, but to date there is no evidence that they can satisfactorily control the spread of smoke and gas. Another common feature is supply inlets in exterior walls that create an exposure hazard from fires in other buildings.

Other Air Ducts

Of special significance are air ducts with roof fan housing that are designed to remove gas and heat from large cooking ranges in restaurants, night clubs, hospitals, and so forth. With improper maintenance, grease can accumulate in the ducts and ignite—usually in peak hours in restaurants and night clubs. Rescue work can be difficult when owners are reluctant to let customers go without paying their bills and customers are reluctant to leave without their coats and hats.

Smoke may obliterate exit signs and occupants may try to get out by the way they entered. Or a sudden worsening of heat and smoke conditions caused, for example, by prematurely shutting down the fan, can result in panic and havoc. The life hazard can be worse if the involved occupancy is below ground or inaccessible from ladders or tower-ladder platforms.

Regulations governing access to exits are frequently violated. Difficulties are complicated by the fact that protective hose lines have to be stretched without interfering with the egress of occupants. In such cases, fans should be kept operating to alleviate smoke and heat conditions pending rescue, assuming a worse hazard is not thereby created on the lee side of the roof fan housing.

While operating in an involved kitchen area, personnel should be aware of some unusual hazards: gas valves still in the "on" position after the flame is out, large cooking pots containing very hot contents which can be overturned, and floors made slippy by melted grease. Extinguishment can usually be achieved in the kitchen area by one or two fog lines. Extension can be reduced by sweeping the exterior surface of exposed ducts with fog, or injecting fog into the duct if there is sufficient heat to cause vaporization. A line to the roof fan housing may also be needed to extinguish and prevent extension of fire.

The hazards presented by these fires can be greatly reduced by the installation and proper maintenance of auxiliary appliances such as approved steam extinguishing, carbon dioxide, dry powder, fine water spray, or the newly developed combination fan-and-grease collector systems described in the *WNYF* magazine article "Coolin' It," by A. J. Kranz, Jr.[3] However, such protection is not provided everywhere and, in addition, there is always the possibility of mechanical failure or human shortcomings relative to maintenance.

Ducts are present in many forms in various occupancies, and in many respects their effects resemble those created by horizontal and vertical structural channels. Fog can be helpful in coping with fires in such ducts and channels, provided there is sufficient heat to vaporize the fog so that it can exert an effective smothering and cooling effect. In vertical ducts and channels it is preferable to inject the fog at fire level. Experimentation by fire reserach scientists in this area is desirable.

Solar Heating and Cooling Systems

In the not too distant future many structures will have solar heating systems, and perhaps in the more remote future, solar cooling systems. Energetic business groups are presently engaged in promoting such developments, which are a natural outgrowth of the fuel and energy shortage. Solar heating systems will supplement rather than replace existing systems. Sizable panels are required for collecting radiant heat. Even larger panels are required in the colder areas. In some cases heated air will be transferred throughout the structure by ducts. How much the advent of such systems will contribute to the fire hazard is unknown at this time. It is known, however, that more structures will have ducts, and two heating systems instead of one. It is also known that more energy will have to be collected for cooling than for heating a house. Hopefully, these facts will be given due consideration in formulating the appropriate regulations for installation and maintenance of solar heating and cooling systems.

[3] A. J. Kranz, "Coolin' It," *WNYF,* magazine, vol. 31, no. 1, pp. 11, 14.

ALTERATIONS IN BUILDING CONSTRUCTION

Alterations are not always effected in accordance with the law or with recommendations in nationally recognized codes, nor are defects of many years standing always fully rectified by retroactive laws. In addition, structural changes required at times by the nature of a new occupancy are not always made. As a result, in some cases areas remain excessive, metal columns are inadequately protected, and door assemblies in dividing walls lack the required fire resistiveness. In other cases, installed dropped ceilings (Fig. 6-4) cover structural defects, and frequently alterations increase the number of concealed spaces by various kinds of falsework, double or triple flooring, and so on. Before explaining how alterations in construction can affect other primary and secondary factors, let us review an actual fire in a building that had undergone alterations.

The fire originated about noon on a Sunday in September. It started in the kitchen area at the rear of a restaurant in a seven-story residential brick-and-joist building erected in 1887. Originally a brick and concrete roof served as a terrace but it was subsequently covered with wood roofing, seven layers of tar paper, and tar finish (Fig. 6-5). Many years later, in 1967, extensive alterations were made to increase the number of apartments per floor. In the process, ceilings of bathrooms and the public hall were dropped $1\frac{1}{2}$ ft [.5 m], and $\frac{5}{8}$ in [about 9.6 millimeter (mm)] sheetrock was used between apartments and public halls. The sheetrock was omitted, however, on the inner side of public hall partitions next to furred-

FIG. 6-4. Suspended ceiling of an altered building, through which fire can spread.

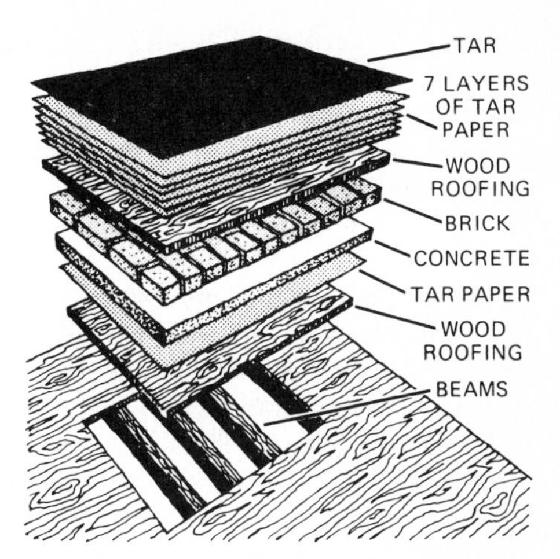

FIG. 6-5. Many different materials are found in the various layers of roofing in altered buildings. These materials, of course, affect the methods of ventilation used.

out pipe spaces, and these pipe spaces were not adequately firestopped (Fig. 6-6). The required sheetrock had also been omitted on one or both sides of public hall partitions, but this omission was covered by the hung ceiling. Fiber glass bats used between studs for firestopping and soundproofing proved to be completely ineffective as a firestop under the heavy fire conditions. As a result, the fire entered an L-shaped 5 × 5 ft [1.5 × 1.5 m] pipe recess at the second floor and raced unimpeded into the cockloft. Openings punched through walls for wiring and plumbing lines caused horizontal spread of the fire, but were also covered by the hung

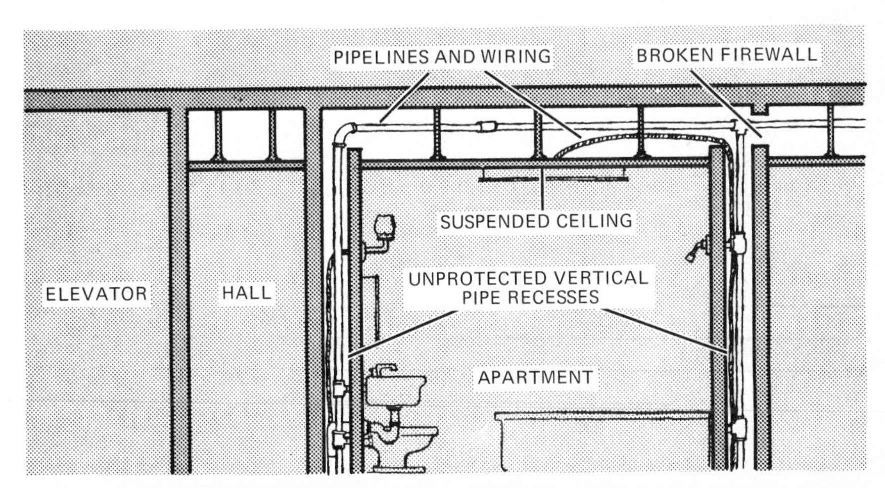

FIG. 6-6. Construction in an altered building often includes holes punched through firewalls to accommodate wiring and plumbing; inadequately firestopped vertical pipe recesses; and suspended ceilings that hide these defects and also provide a protected channel through which fire can spread horizontally.

ceiling. Fortunately, the time of origin minimized the life hazard for occupants who could be readily alerted and removed. Fortunately also, fire escapes had been provided during the alterations and were helpful in the operation.

The major objective was to confine, control, and extinguish. The greatest difficulties were in trying to confine and control the fire in the cockloft. The fire there had to be attacked from below, and the severe heat and smoke conditions and poor visibility could not be effectively alleviated by roof ventilation. Control was finally achieved by a third-alarm assignment and the use of eleven hand lines, aided by the work of ladder and rescue companies. Thirty-seven injuries were reported as a result of this fire and forty-four air cylinders were expended, attesting to the severity of the smoke conditions and the absence of effective roof ventilation.

Effects on Other Primary Factors

Alterations in this case represented errors of omission as well as comission and adversely affected location and extent of fire on and after arrival. They also worsened heat, smoke, and visibility conditions and the exposure hazard; increased the duration of operation and requirements to operate, especially rescue company equipment; intensified the life hazard for personnel, and could have created a very serious life hazard for occupants if the fire had occurred at night.

Effects on Secondary Factors

Alterations abetting horizontal spread of the fire and the pipe recess abetting vertical spread were not visible and made it difficult to determine the location and extent of the fire on and immediately after arrival. Concrete and brick construction on the roof prevented ventilation that could have localized the fire and alleviated smoke and heat conditions. This made it extremely difficult to operate lines used to attack the fire in the cockloft from the floor below. Roof construction and other alterations increased the amount of structural overhauling needed both before and after control was established.

Other Effects

If required structural alterations are not made when an occupancy changes from noncombustible to combustible, highly undesirable results can accrue. Such alterations may necessitate new partitions to subdivide areas, protection of metal columns, a higher fire-resistive rating for doors in subdividing walls, and so forth. In one case, however, failure to comply with these requirements resulted in total involvement and collapse of the fire building, and a $2 million fire loss.

BUILDINGS UNDER CONSTRUCTION

During the day, fires in these structures usually do not present serious life hazards because they are discovered quickly and the average worker can readily get out of harm's way. At night, however, one or more watchmen may have to be searched for by the Fire Department, and the hazards in general, already great, are intensified.

Fires in buildings under construction, particularly at high levels, can quickly reach major proportions. There are many reasons for this, some of which follow. Construction is wide open, providing ample oxygen. Much fuel is supplied by combustible debris, wooden interior scaffolding, chutes, sheds, shanties, and possibly concrete forms. Paints, oakum, excelsior, tarpaulins, and the like, are an additional source of fuel. Tanks containing flammable gases for use in cutting torches, dangerous gases for heating purposes, and cartridges used for riveting, may be present and exposed. Winds prevailing at high levels are generally strong. And there may be, and often are, abnormal delays in getting water to the fire floor. In addition, openings in floors and the absence of windows, doors, and completed walls and partitions abet horizontal and vertical extension of the fire. (See Fig. 6-7.)

There is always the danger of timbers falling from topsides. The exposed steelwork on top may buckle, weakening the structure. The concrete beams and slabs on upper stories may not be set; if the wooden supporting forms burn, the floors may drop. Fires at high levels beyond the range of high-caliber outside streams present the most serious problems, especially if standpipe systems are inoperative. In such cases, a fire that actually requires only one stream for extinguishment may necessitate the use of a full first-alarm assignment because then the line may have to be stretched up the outside of the structure and this is a laborious process, requiring much manpower.

Danger from the use of explosives for blasting in the very early stages of construction is somewhat alleviated by strict regulations, careful surveillance, and competent watchmen. Some fire departments, as a precautionary measure, prohibit the use of radio transmitters on department vehicles within 150 ft [45.7 m] of magazines containing explosive caps; at close quarters radio waves may energize the detonating mechanism.

Effects on Other Primary Factors

To be realistic, structural features are considered in conjunction with occupancy contents (such as combustible materials and flammable gases) and an assumed height factor (12 stories). Such total situations can have adverse effects on the location and extent of the fire on and after arrival. They can create smoke conditions that develop problems in exterior

FIG. 6-7. Hazards of a building under construction: (1) flammable gas for heating; (2) hoists which are dangerous to use; (3) lumber; (4) salamanders; (5) tanks of flammable gas; (6) gas heaters for drying; (7) tarpaulins; (8) wooden forms; (9) wooden bridging.

exposures, heat conditions that can buckle exposed steelwork, and a spark-and-ember hazard that can worsen the exterior exposure hazard. They can also result in greater-alarm requirements to operate and a prolonged operation, with considerable danger for personnel—especially if the fire is at night—and even more danger if tanks containing flammable gases explode.

Effects on Secondary Factors

An oddity about this type of structural fire is the minimal need for forcible entry and ventilation. Stress is on the placement and use of lines. When lines are being advanced via stairways and on floors, supervision is exceptionally important because stairs at the upper levels may be unfinished and floor openings may be protected by inadequate guard rails. Operative standpipe systems should be supplied and used, or else lines may have to be stretched up the outside of the structure, in which case the need for sufficient personnel must be anticipated.

A minor fire in buildings under construction can sometimes be kept minor by the prompt use of a deck pipe or similar equipment. For example, in one case in which the standpipe system was inoperative a deck pipe was used to extinguish a fire in debris on a setback on the twelfth floor while a line was being stretched up the outside of the building.

If the fire is within reach of streams from ladder pipes or tower-ladder platforms, fog from the windward side can be effective. Solid streams would be advisable, however, if greater penetration is needed.

The exterior exposure hazard in buildings under construction can present multiple problems because, besides the danger to nearby buildings from radiation and convection of heat, a spark-and-ember hazard may exist. Sparks can start other fires at surprising distances from the original fire building. At times, they are drawn into buildings by fans in exterior wall openings on the lee side of the fire.

Officers should remember that hoists for materials are not intended to transport people. If possible and feasible, however, they can be used to convey rolled-up hose and other equipment to upper floors. Where elevators designated for fire-department use are required and provided, it is preferable to use them if they are not affected by the fire. When such elevators have been installed, guards should be provided to operate them.

In some localities, standpipe systems are required under certain conditions—for example, when floors are in place above the seventh story, or more than 75 ft [22.8 m] above the curb level. Quite often, however, these systems are not dependable at night because of carelessness about closing valves that have been opened during the day.

New high-rise building construction has presented serious problems for the fire service, but in some respects it has one advantage: The buildings are erected more quickly because of the curtain wall construction, thereby reducing the period of fire hazards that are inherent in buildings under construction.

BUILDINGS UNDER DEMOLITION

Much that has been said about buildings under construction applies to those under demolition. For example, the fire department may have trouble getting water to high fires if the standpipe system has already been put out of service but the structure is still 25 stories tall. Some contractors, using modern techniques, just peel off exterior walls, remove undesirable partitions, and erect a new building on the metal framework of the old.

Dismantling of sprinkler systems can also have disastrous results, as demonstrated on several occasions. A notable example was the Wanamaker fire which originated in the building's subcellar on July 14, 1956, in New York City. The building was in the process of demolition. The fire injured more than 200 fire personnel, did extensive damage to the subway system, and took several days to extinguish. A major cause of this disaster was the fact that the dismantled sprinkler system in the subcellar and cellar prevented the department from discharging water upon fire in inaccessible areas.

Effects on Other Factors

The effects on other primary and secondary factors are similar to those of buildings under construction, assuming comparable heights.

DEMOLITION IN SLUM-CLEARANCE AREAS

Buildings being demolished are usually of brick-joist or frame construction, and piles of combustible debris may accumulate to a dangerous degree. In some communities official permission is granted to burn this debris under specified and controlled conditions. Unless stringent surveillance is exercised, contractors may ignore controls and try to burn the material in larger volumes than the law allows, which may result in extensive fires that are at times of greater-alarm proportions. Some fires, started legally to burn debris, are buried under tons of brick and are evidently thought to be extinguished when the contractor leaves at the end of the day, only to flare up later. To avoid all doubt, it is advisable to use bulldozers at these fires so that they can be more readily and adequately exposed for complete extinguishment. This technique minimizes the hazard to personnel. It also lightens their workload, which is otherwise magnified by the high incidence of fires in slum-clearance projects, which may include several blocks of buildings.

Some buildings marked for demolition in slum-clearance areas are still occupied. Some tenants, mostly older persons, are uncooperative about moving away from old friends and neighborhoods, and remain as long as possible. Their reluctance is understandable, but their isolated presence is hard to detect since large areas and numerous houses may be involved in the project. In addition, vacated buildings become havens for vagrants and, in some cases, hangouts for teenagers. Both these groups may create an unsuspected life hazard. In partly vacated buildings, an additional hazard may be created by gas that is still being supplied. Such buildings are characterized by the worst features of brick-joist or frame construction, plus holes in floors, weakened or missing stairways, windows that are boarded up or covered with sheet metal, large accumulations of combustible rubbish, repeated fires, and old age.

Effects on Other Primary Factors

Fires are more frequent than usual in buildings being demolished and can be extensive on arrival, creating much heat and smoke and many interior as well as exterior exposures, prolonging operations, and increasing requirements to operate. Unless the fire is already coming out of the structure on arrival, blocked windows make it difficult to alleviate heat and smoke conditions, and also reduce visibility and jeopardize any occupants and personnel trying to perform rescue work.

Effects on Secondary Factors

At fires in these structures, special attention should be given to the limit (or specification) that where there is no life hazard for occupants, personnel should never be jeopardized unnecessarily. Remember, these structures and their contents are practically worthless and are only awaiting demolition. Fire personnel should not be jeopardized unnecessarily by an interior operation. It is preferable to make the major objective one of confine, control, and extinguish and to achieve it by an exterior operation, assuming it has been ascertained that there is no life hazard for occupants.

REVIEW QUESTIONS

1. Generally speaking, how does a knowledge of construction help fire officers?

2. What can officers reasonably be expected to know about construction?

3. What does the fire-resistiveness of construction depend upon?

4. What do horizontal structural channels include? Vertical channels?

5. Why do some masonry materials have a higher fire-resistive rating than materials regarded as good insulators?

6. How does ordinary or brick-joist construction in residential buildings affect other primary factors? Secondary factors? What are some features of this construction?

7. How does ordinary or brick-joist construction in commercial buildings affect other primary factors? Secondary factors? What are some features of this construction?

8. What are the overall effects of fire escapes on fire operations, especially on secondary factors?

9. What are the effects of fire-resistive construction *without* central air-conditioning—on other primary factors? On secondary factors?

10. In very unusual cases, fire can spread vertically in fire-resistive construction. What causes such a spread?

11. At the great majority of fires in hotels (assuming fire-resistive construction), why is it advisable to leave most occupants in their own rooms rather than take them to exits?

12. In the New York Board of Fire Underwriter (NYBFU) report on the fire at One World Trade Center, what was meant by "placing the air-conditioning system on the purge mode"?

13. What are the advantages of the building arrangement at One World Trade Center?

14. Why is the trade center construction unique and advantageous? What negated the effectiveness of this construction at the fire in question?

15. What faults were pointed out in the NYBFU report? Recommendations?

16. How could the building arrangement and construction of One World Trade Center affect other primary factors? How and why would this evaluation have to be qualified?

17. Why is vertical extension of fire possible in the best fire-resistive construction?

18. In one of the newer types of high-rise buildings, the construction features an air-conditioning system which serves the core area only and individual air-conditioning units which serve the occupancies around the periphery of the building. What are some possible advantages of this construction?

19. What are some effects of air-conditioning systems in other than new high-rise buildings?

20. What are the effects of ducts with roof fan housings?

21. Why do alterations in buildings frequently create structural defects from a fire-service point of view? How did alterations in the case described affect other primary factors? Secondary factors?

22. In a building under construction, what are the effects of structural features on other primary factors, assuming a fire at the twelfth-floor level? On secondary factors? Consider the usual occupancy contents in estimating effects.

23. What are the effects of vacated or deteriorated buildings on primary factors? On secondary factors?

7

Occupancy

Occupancy has two aspects: the human element present; and the contents and their associated fire hazards. Much of what is said about the occupancy of the fire building applies also to the occupancies of exposures.

Occupancies are generally classified as residential, mercantile, commercial, public, and so on, but at fires they are more specifically described as one-family or multiple residences or hotels, department stores or supermarkets, factories, hospitals, schools, institutions for the blind or aged, and the like. In some communities local expressions are used, such as "old- or new-law tenement," or "taxpayer" (store). The description conveys information about the human element that may be present and the content that may be burning or exposed to the fire. In this context, an occupancy could be a brush or forest area in which a human element might be endangered. The content aspect would be the vegetation. Likewise, a crashed plane could be the occupancy. Obviously, if such fires spread, other occupancies can become involved.

HUMAN OCCUPANCY

An awareness of the mental, physical, emotional, or other relevant condition of the human element in various types of occupancies helps officers gage the severity of the life hazard and anticipate problems in rescue work.

In nightclubs, churches, theaters, and so forth, the allowable occupancy (maximum legal number that can be accommodated) may be so large and of such density as to induce panic in the event of fire or smoke. Schools can present somewhat similar problems.

In hospitals and institutions for the care of infants, the elderly, the blind, the deaf and dumb, or other physically handicapped, occupants may have to be carried or led out of the building with unusual care.

In jails or mental hospitals, rescuers may have to contend with uncooperative, hostile, or generally difficult occupants, as well as cope with heavy, locked doors, or cut through bars on windows.

In multiple-residential structures, particularly those of old, brick-joist construction in tenement or ghetto districts, occupancies are frequently overcrowded—at times by tenants who speak mostly foreign languages and therefore have difficulty in describing where other occupants are trapped.

In assessing the effects of the human element on the life hazard, fire officers must always consider the *time* of the fire. The human element greatly increases the life hazard at night fires in residential occupancies, workdays in commercial occupancies, classtimes in schools, showtime in theaters, holiday seasons in churches and department stores, and rush hours in subways and tunnels.

Effects on Other Factors

When occupants are endangered, the effects of the human element on other primary factors and on secondary factors are similar to the effects of a life hazard for occupants (Chapter 4).

CONTENTS

Some authorities refer to the contents of an occupancy in terms of "fire loads"—which are equal to the pounds of combustible material present, times the average calorific value of those materials in Btu per pound, dividided by the floor area in square feet. Others maintain that the overall fire load encompasses both the structural parts and the contents. From a practical point of view, however, the fire officer's concern about contents revolves around such questions as "How readily do they ignite?" "How much heat do they give off?" "Do they give off explosive or toxic gases or both?" "What is the proper extinguishing agent if water should not be used?" "How will the storage or placement of contents affect the effectiveness of streams?" and "How will the type of gases given off affect ventilation?"

Effects on Primary Factors

The answers to the above questions determine how occupancy contents affect other primary factors.

Effects on Secondary Factors

These same answers also indicate how occupancy contents would affect such secondary factors as ventilating, advancing hose lines, and selecting the proper extinguishing agent. These and other secondary factors are discussed below.

Ventilation. Oils, fats, rubber, wax, tar, and some plastics produce large volumes of smoke, which may be largely unburned vapors. The heat of this type of smoke is low, as is its buoyancy. Visibility is therefore impaired, and ventilation is slowed down. Some materials give off gases that are toxic or injurious to the eyes or skin. Burning silks and woolens, for example, give off carbon dioxide and hydrogen cyanide gases. Both are toxic, and the latter can be absorbed by the skin. Ammonia is also given off and injures the eyes, lungs, and damp skin areas. PVC (polyvinyl chloride) gives off chlorine gas and forms hydrochloric acid with water in eyes, armpits, groins, and wherever the human body perspires.

Ventilation is achieved more slowly in such cases because firefighters must take time to don masks and protective clothing and are hampered by poor visibility.

Where the presence of explosive mixtures or substances is suspected, exterior ventilation measures should be taken to prevent an explosion or minimize its results.

Placement and Use of Hose Lines. Difficulties in ventilating can reduce the effectiveness of hose lines. Effectiveness may also be adversely affected by an excessive amount of contents and by the manner in which they are stored. Stock may be piled so high that it reduces the effectiveness of sprinklers and streams. Such a situation makes it hazardous for units to advance lines.

Where contents are combustible and plentiful, as in a lumberyard, fires are characterized by rapid spread, high temperatures, and a spark-and-ember hazard. To extinguish the main body of fire, heavy-caliber and high-pressure streams are in order. Lighter mobile lines can cover the spark-and-ember hazard and finish up the job.

In some cases, two occupied structures can be equally distant from and endangered by a fire in an unoccupied building. *Equally endangered* implies similar construction, height, area, and so on. In such an event, the human element in the occupancy presenting the greater life hazard would be covered first. Thus, a hospital would be given priority over a factory because many occupants might not be able to walk. With two unoccupied structures, the nature of the contents decides the placement and use of lines. Thus, a commercial occupancy would be covered before an abandoned building. Covering exposures is discussed in Chapter 8.

Selecting an Extinguishing Agent. In some cases water will spread the fire. For example, gasoline, kerosene, and similar materials are lighter than water, will float on top of it, and thus spread the fire. Calcium carbide with water gives off acetylene gas and may cause an explosion. Some flammable liquids are miscible with water, and unless they can be diluted to a point at which flammability is no longer possible, the fire may spread. Water used improperly in the presence of combustible dusts, such as wood, flour, zinc, or magnesium, may throw them into suspension and develop an explosive mixture. The use of water near acid in carboys, such as in a wholesale drug occupancy, may cause failure of the carboys by sudden chilling or impact of stream, permitting spread of the acids. The resulting release of gases may intensify and abet extension of the fire. Extinguishing agents for class B fires are discussed in Chapter 15.

In some cases, contents react dangerously with water. Calcium carbide may cause an explosion. Calcium oxide (quicklime or unslacked lime) gets hot on contact with water and could conceivably ignite nearby combustible materials. When potassium and sodium nitrates in large quantities are involved, they may fuse in the heat. Applying water may cause extensive scattering of the molten material. Nitrates are soluble in water, and where such solutions come in contact with ordinary combustible material they may affect it so that it becomes highly explosive when dry. Sodium and potassium liberate hydrogen in contact with water, with sufficient heat to ignite the released hydrogen. Sodium and potassium peroxides react vigorously with water and release oxygen and heat.

These examples are far from exhaustive. Fire officers should carry reference material on all major pieces of apparatus in order to help them evaluate the occupancy factor in case unusual hazardous flammables or chemicals require a special extinguishing agent. It is dangerous to depend on memory or the availability of competent advice, as is indicated by the following incident.

At about 2:45 P.M. on September 11, 1969, near Glendora, Mississippi, there was a railway wreck in which 15 cars were derailed. Eight of these were tank cars containing vinyl chloride, one of which was ruptured and, at 8:00 P.M., was reported to be leaking—but there was no fire. At about 9:00 P.M. the vinyl chloride became ignited from sparks. At about 6:45 A.M. the next day one car of vinyl chloride exploded, causing another car to be ruptured and ignited. Due to some erroneous information (that burning vinyl chloride would produce phosgene) approximately 30,000 people were evacuated from their homes unnecessarily. Mississippi National Guard troops were called in and martial law was put in force. Then a special army team from Fort McClelland Chemical School in Alabama reported there was no phosgene in the area.

Of the eight vinyl chloride cars involved or exposed, one exploded, two burned, and five remained intact and were recovered. Returning residents found dead livestock near the wreck, but a report issued later indicated that these deaths were not connected with phosgene. About four months later, after an intensive laboratory investigation, the Manufacturing Chemists Association issued a report to the effect that no detectable amount of phosgene is produced when vinyl chloride monomer burns, except under very specific conditions which involves premixing vinyl chloride monomer with oxygen.[1]

Occasionally, occupancy contents include extremely valuable or irreplaceable books or works of art. In such cases, if feasible, low-velocity fog is preferable to high-pressure fog or solid streams.

Where radioactive materials are present, low-velocity fog is advocated so as not to disturb and possibly expose them.

Molten materials that react dangerously with water sometimes start burning in elevated factory kettles if temperature control fails, and the fire may extend to surrounding objects. The judicious use of low-velocity fog applied on and off beneath and around the kettle has been able to cool the contents of the kettle below its ignition temperature and put out the surrounding fire. This technique has been used based on the assumption that the fog would vaporize at the prevailing temperature and the resulting steam would be carried away from the kettle by convection currents. In any event, there was no reaction whatsoever between water and the kettle contents. However, this case is not a criterion for disregarding the dangers of using water when a harmful reaction is possible.

Overhauling. Overhauling is affected by the quantity of material involved, the manner in which it is stored, its nature, and the degree to which it has been subjected to the fire. The quantity of baled cotton, rags, and paper, and crated material can complicate and prolong overhauling. Baled material may have to be removed from the premises, and crated materials may have to be opened for examination. Other materials, such as mattresses and upholstered furniture, particularly the foam-rubber type, produce much smoke and may also have to be removed unless a wetting-agent solution is available. In addition, the degree to which contents have been subjected to the fire and heat affects the amount of overhauling required.

Supervision. Officers must be alert when contents include acid carboys, explosive dusts, heavy machinery, and electrical equipment. The dangers of acid carboys and explosive dusts have been explained above. Heavy machinery can be a menace to personnel working below if there is any

[1] Daniel R. Stull, review in *Fire Research* Abstracts and Reviews, vol. 13, no. 3, 1971, pp. 164, 178.

question at all about the stability of the structure. Electrical equipment can cause casualties by charging accumulations of water, especially in cellars.

The human element and the content aspect of occupancies can create "target" hazards if serious inherent hazards are intensified by other factors, such as height (as in high-rise buildings), area (as in lumberyards, shipyards, oil-storage plants), construction (as from age, lack of fire-resistiveness, combustible structural parts, inadequate departmentation). Fire departments in many cities are placing greater emphasis on these target hazards by inspecting them more frequently and carefully, and focusing attention on them in prefire planning.

METALLURGICAL INDUSTRIES

Some modern industries create special problems in firefighting because of certain materials in use or certain circumstances of operation. An example of this problem is the metallurgical industry.

Metals can be classified into the following groups:

1. The alkali group, which includes sodium, potassium, and lithium.
2. The alkali-earth group, which includes barium, strontium, calcium, and magnesium.
3. The iron group, which includes iron, aluminum, chromium, manganese, zinc, cobalt, nickel, and titanium.
4. The heavy-metal group, which includes copper, antimony, tin, tungsten, lead, mercury, platinum, and gold.

Alkali Group

Metals of the alkali group may cause serious dust explosions. They also have other dangerous characteristics. On contact with water they release hydrogen, with the possibility of ignition and explosion. Alkali metals are so highly oxidizable that they are generally stored under hydrocarbons. They react spontaneously in the presence of chlorine or sulfuric acid, and violently with common extinguishing agents such as water, foam, carbon tetrachloride, and halogenated hydrocarbons.

Sodium, in fine particles, can ignite spontaneously in moist air at room temperature. It reacts violently with halogenated hydrocarbons, such as carbon tetrachloride, and other extinguishing agents, such as foam and water from sulfuric acid extinguishers. Carbon dioxide will not extinguish sodium fires.

In some occupancies, such as scientific institutions and Atomic Energy Commission (AEC) installations, sodium or some other alkali metal may be used as a heat-transfer agent to carry heat from an atomic pile to water. The water is converted to steam and then used in standard equipment to create electrical energy. In these circumstances, two types of sodium are used. One, referred to as *primary sodium,* is radioactive and is kept behind the biological shield. The other, *secondary sodium,* is not radioactive but could be involved in an accidental water-sodium explosion in the boilers. Liquid sodium, exposed to the air, may burn spontaneously, giving off sodium oxide which combines with moisture in the air to form sodium hydroxide (lye).

Potassium is in many ways similar to sodium. In some cases, however, potassium reacts more violently: with bromine, it detonates; with sulfuric acid, it explodes. Sodium-potassium alloys compound the fire hazard because they react more vigorously than the individual component metals alone.

Lithium resembles sodium and potassium in many ways. However, the lithium-water reaction does not generate enough heat to ignite the hydrogen released.

Alkali-Earth Group

Barium, strontium, and calcium decompose in water to yield hydrogen which may or may not burn, depending on the mass of the metal involved. In the massive state, decomposition may not be vigorous enough to create sufficient heat to ignite the hydrogen. In finely divided form, ignition will be spontaneous.

The alkali-earth metal magnesium merits separate consideration because the increased use of magnesium, especially in alloyed form, has concomitantly increased the frequency with which the problems of magnesium fires must be met.

The ignition temperature of magnesium is about the same as its melting point (1202°F, or 650°C). However, magnesium ribbons and chips or fine shavings can be ignited at about 950°F [510°C], and finely divided magnesium powder ignites at temperatures below 900°F [482°C]. Under certain conditions some alloys even ignite at 800°F [427°C], depending on the form and size of the alloy and the duration of exposure to the heat.

The high thermal conductivity of magnesium makes it difficult to ignite large pieces of the metal. Ignition may not occur until the temperature of the entire mass is raised to the proper level. The difficulty or ease of ignition therefore depends on the size and shape of the magnesium.

Magnesium has a great affinity for chlorine and, when burning, reacts violently with carbon tetrachloride. Water applied to molten magnesium or burning chips may decompose, releasing hydrogen and thereby adding to the intensity of the fire. When magnesium is burning intensely, the actinic light produced is very harmful to the eyes.

Magnesium dust is explosive and generates high pressures very quickly. If a structure lacks adequate ventilation to dissipate these sudden pressures, the walls or roof may collapse.

Iron Group

In the massive state, metals of the iron group are not combustible in the ordinary sense of the word. However, at high temperatures they may be oxidized in some cases, yielding light and heat. In the finely divided state they are more combustible.

Iron and steel are obviously important metals from the point of view of the fire service since they are used so extensively in building construction. Cast iron may break when heated and then suddenly cooled (as when heated by fire and suddenly cooled by a hose stream). Of the strength of cast iron at 70°F [21°C], 76 percent is retained at 930°F [499°C] and 42 percent at 1100°F [593°C]. The strength of cast steel at 930°F [499°C] is 57 percent of its strength at 70°F [21°C]. (See Figs. 7-1 and 7-2.) Steel expands 8 parts in a million for each degree of rise in temperature. Therefore, when heated to 1000°F [556°C] above normal (that is, to 1070°F, or 577°C), a 100-ft [30.5 m] length of steel will expand about 9 in [22.8 cm]. Such expansion may push out bearing walls and cause collapse of the structure. (See Fig. 7-3 and Table 7-1.)

Titanium is used as a structural metal in atomic reactors because of its superior radiation-resistance characteristics. Titanium and titanium alloys are also used extensively in aircraft; titanium is as strong as many varieties of steel but only about 56 percent as heavy. But titanium presents a serious fire hazard during production of the raw sponge, melting of the sponge, and subsequent operations that create fine turnings and chips—which burn rapidly when involved. Fine-powder samples have ignited in air as a cloud at 896°F [480°C]. Combustion can also occur in atmosphere other than air. For example, above 1475°F titanium dust burns readily in atmospheres of pure nitrogen, and a layer of titanium powder can be ignited in carbon dioxide at 1256°F. Unlike zirconium, plutonium, ura-

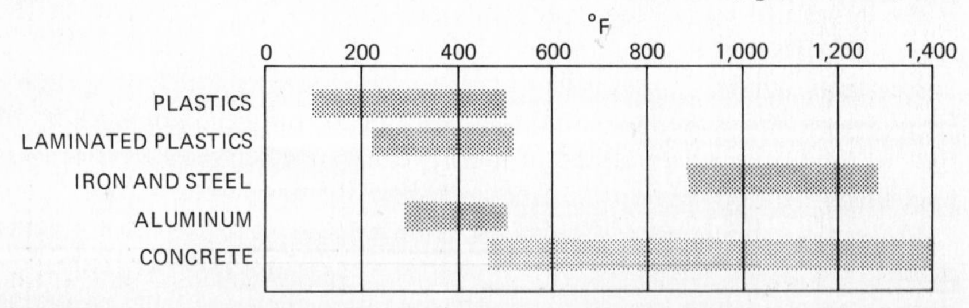

Fig. 7-1. Upper temperature limits for various materials (commonly accepted ranges). Ranges for concrete and metals are assumed to be those in which the materials have lost 50 percent of their original strength.

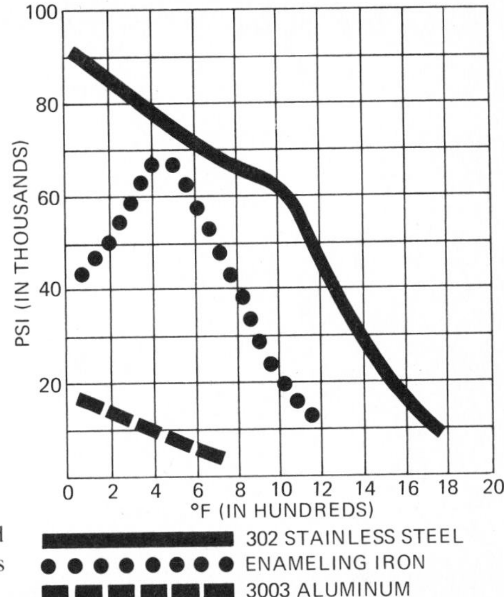

FIG. 7-2. Tensile strengths at elevated temperatures of aluminum, stainless steel, and enameling iron.

nium, thorium, and some other metals, titanium in dust or layer form does not ignite spontaneously.

Other Metals

Zirconium, hafnium, and atomic-fuel metals also present unusual fire hazards.

Zirconium, too, is used as a structural metal in atomic reactors because of its superior radiation resistance. In massive form, zirconium can withstand very high temperatures without igniting, but in dry-powder form it may ignite at low temperature. Layers may ignite below 400°F [204°C].

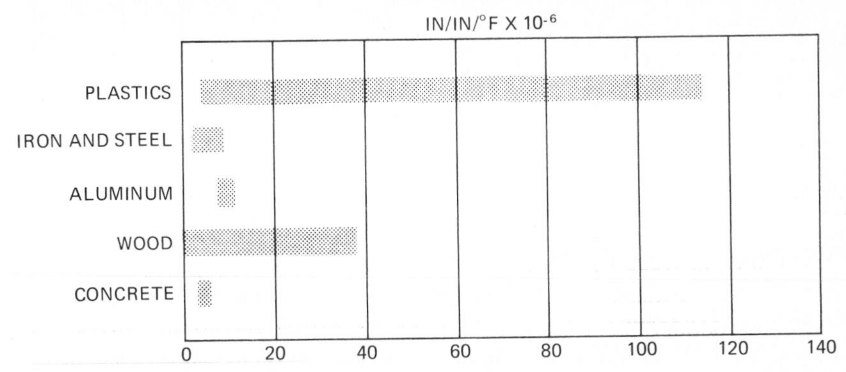

FIG. 7-3. Thermal expansion range for various materials. This expansion is commonly expressed in inches of expansion per inch of original material size per range of temperature in degrees Fahrenheit.

TABLE 7–1. Dimensional Changes of Materials Caused by 100°F Temperature Change*

MATERIAL	APPROXIMATE COEFFICIENT OF EXPANSION, IN. PER IN. PER DEG F $\times 10^{-6}$	DIMENSIONAL CHANGE, IN. LENGTH OF MATERIAL		
		5 FT	10 FT	15 FT
Aluminum	13.0	0.078	0.156	0.234
Stainless steel	9.6	0.058	0.115	0.173
Carbon steel	7.0	0.042	0.084	0.126
Glass	5.0	0.030	0.060	0.090
Concrete	8.0	0.048	0.096	0.144
Stone	4.0	0.024	0.048	0.072
Reinforced plastic	16.0	0.096	0.194	0.286

* Note that measurements given here are in USCS units and may be converted to metric units if necessary.

Dispersed into the air as a cloud, zirconium dust explodes spontaneously, ignited by the static electric charge generated in the dispersal of the cloud or other source of ignition. As a fine powder, it ignites in nitrogen at 986°F [530°C] and in carbon dioxide at 1040°F [560°C]. Zirconium powder is safer to handle when it is wet because it is more difficult to ignite, but once ignited it burns violently, decomposes the water, and uses the oxygen released for its own combustion. Spontaneous explosions have occurred during handling of very finely divided, contaminated zirconium scrap that is only moist rather than wet. Powder containing 5 to 10 percent water is the most dangerous in this respect.

Hafnium is similar to other metals in that its size and shape determine the degree of combustibility. Large pieces withstand high temperatures without ignition, but small pieces ignite easily. If a source of ignition is present, dust clouds explode in air, nitrogen, or carbon dioxide. Hafnium burns with little flame but much heat. It reacts with water to release hydrogen, with a possibility of its ignition, and in general is considered to be more reactive than titanium or zirconium under comparable conditions.

Atomic-fuel Metals

This group includes uranium, plutonium, and thorium. All are radioactive. Since radioactivity is not influenced by fire, radiation will continue during the fire and even spread if the radioactive materials are disturbed and moved about. Smoke and water in contact with these materials do not become radioactive but can become radioactively contaminated, and spread radioactive materials around.

Uranium in finely divided form, such as chips, shavings, or turnings, may ignite spontaneously unless it is kept under water. Very fine uranium dust has even been known to ignite under water if it is allowed to accumulate. Massive forms of uranium ignite only after prolonged exposure to high temperatures and then burn very slowly. Burning uranium reacts violently with carbon tetrachloride. A nuclear accident is possible when enriched uranium (uranium 235) burns, unless water is used with great care (see below, Industries Using Radioactive Materials).

Plutonium is a radioactive, combustible metal somewhat more susceptible to ignition than uranium. In finely divided form, the metal can ignite spontaneously in moist air. Massive plutonium generally burns without visible flame. Fires are characterized by comparatively slow combustion accompanied by intense heat and brilliant white light. The light may be partly masked by an oxide coating over the burning metal surface. These fires are extremely difficult to extinguish and, wherever possible, are handled by isolating the burning metal rapidly—by covering it with extinguishant and placing it in a covered metal container to minimize dispersal of plutonium oxide contamination. The extreme toxicity of the oxides generated and dispersed is a severe hazard to firefighters and complicates control and extinguishment. Decontamination requires more time, money, and personnel, and hence is a much more expensive process than cleaning up after a small fire in other metals. For these and other reasons, it is best to limit the quality of plutonium kept at one location, thus minimizing the size of a fire in this metal.

Thorium is also a radioactive combustible metal. In powdered form it should be handled carefully because of its low ignition point. When pouring the powder into a glass container, for example, ignition may be caused by friction in the falling particles or by static electricity that may be generated. Powdered thorium is usually compacted into and stored as 1-oz [.028 Kg] pellets, and may then be converted into alloys with other metals.

Extinguishing agents to be used at metal fires are discussed in Chapter 15.

General Observations on Metals

Some metals are referred to as combustible; others as not normally combustible. Nevertheless, the latter, specifically aluminum and steel, may—in finely divided form—ignite and burn. Aluminum and iron dusts can be explosive.

Metal powders have a greater surface area on which to generate heat, a smaller mass in which to dissipate this heat by conduction, and a smaller percentage of total surface area available for loss of heat by radiation to external surroundings. By contrast, large masses of metal have comparatively less surface area on which to generate heat, greater potential for dissipating heat by conduction, and a larger percentage of total surface area available for loss of heat by radiation to external surroundings. Since heat-radiation losses take place at a rate proportional to the fourth power of the absolute temperature, it is not surprising that large masses of metal can withstand high temperatures for prolonged periods before igniting. Conversely, powdered metal ignites much more readily.

Some metals oxidize more rapidly than others in air and moisture and reach ignition temperatures more quickly. Some are more likely than others to explode. All can be made radioactive, thus adding to the hazards of the subject metal. On arrival at a fire involving metals, it is essential to recognize the hazard that exists and to select the correct extinguishing agent. In some cases water should not be used. In others, it may be used with discretion.

THE AVIATION INDUSTRY

The severity of plane-crash fires and the problems they create for the fire service depend on several primary factors.

Effects of Primary Factors

Life Hazard. Due to the fuel shortage situation, the number of plane flights has been curtailed. This has reduced collision incidence but has increased the number of passengers per flight, thereby increasing the potential life hazard per flight. In a serious crash fire the survival time for passengers has been estimated to range from 50 to 300 seconds—or it can be as short as 30 seconds if flame enters the cabin. The severity of the crash depends in large measure on the forces involved at impact. Where rescue is possible, the life hazard for personnel is magnified by the severe heat and smoke conditions that usually develop.

Location of Fire. Statistics indicate that most plane accidents occur at or within 5 miles of airports, which is favorable since the response of fire apparatus, equipment, and personnel is likely to be substantial. However, FAA (Federal Aviation Administration) statistics indicate that about 25 percent of accidents occur en route, and therefore perhaps in areas lacking adequate fire protection and difficult to approach, making a successful operation unlikely. See Proximity of Exposures and Exposure Hazards below.

Extent of Fire on and after Arrival. Impact forces may result in fuel spillage and atomization. Fuel in atomized or mist form may ignite and burn rapidly. It is usually consumed in seconds, in all probability before the fire department arrives. If the mist explosion propagates flame over large areas of spilled fuel, the extent of fire on and after arrival will be affected accordingly. Often, of course, there is no fuel spill and ignition. In any case, extent of fire after arrival can also be affected by weather and wind conditions, exposure hazards, and, obviously, requirements available to operate.

Construction. With the advent of jet turbine aircraft, fuselage skin and structure have had to become heavier because of higher-altitude flying and concomitantly greater cabin pressurization. In addition, the fuselage is reinforced by structural features to prevent explosive depressurization that might be caused by minor leaks in the skin. As a result, forcible entry is practically impossible, even with power tools. This is a critical matter if normal means of egress are jammed on impact. A further complication is that exit door locations and operating mechanisms are not standardized, even on American planes. And in some instances, instructions on the outside of the hull are not readable in the smoke.

This handicap is somewhat mitigated by the current firefighting technique which calls for an attack along the entire length of the fuselage. The attack presses outward in an attempt to drive the fire far enough away so that it no longer threatens either the fuselage or the lives involved. If such an attack is successsful, it is then more possible to locate and use the available exits. Much depends on the type of impact, fuel spillage, heat conditions, requirements available to operate, and the like.

Magnesium castings in aircraft engines, and titanium and titanium alloys in sheet and casting forms, both discussed earlier, may ignite and necessitate special extinguishing agents. These agents are further discussed in Chapter 15.

Occupancy. The human element consists of passengers and crew and can, of course, cause a life hazard. Contents that can contribute to the severity of plane-crash fires include aviation fuel (some planes carry more than 15,000 gal [57 m^3]), flammables in lubricating and hydraulic systems, oxygen supplies, cargo, and plastics in passenger cabins. The FAA has promulgated regulations governing air transportation of hazardous materials which may be radioactive, flammable, explosive, or give off toxic fumes if exposed. Such cargoes can add considerably to the fire hazard in a plane crash. Plastics in passenger cabins are sometimes used as foam padding covered by fiberglass. If the covering material is destroyed, the padding burns rapidly and gives off toxic gases. An Air Force officer who survived the recent disastrous ''Viet-orphan'' jet crash was quoted in a newspaper account as saying that he saw plastic-lined pillows explode. Cargo on military aircraft add a special dimension to the fire hazard in the event of a crash.

Height and Area. Aircraft have become taller, more spacious, heavier, and faster, all of which can increase the impact forces in an emergency or other landing. In "power-on" emergency landings the impact forces are so high that in most cases survival time is practically nil and rescue improbable. If impact is low, rescue is much more probable. See Effects on Placement and Use of Lines below.

Proximity of Exposures and Exposure Hazards. The effects of proximity of exposures and exposure hazards are assayed according to the location of the fire. If a plane crashes in a populated area, these factors become critical.

Time of Origin and Response. The time of the fire (day or night) can affect visibility and, therefore, fire activities. The time of response is of paramount importance to survival time and has a bearing on extent of fire on arrival and other related factors.

Heat and Smoke Conditions. High temperatures are developed in part because 1 lb [0.5 kg] of aviation fuel has a heat of combustion of about 19,000 Btu [2,945,000 joules]—and some planes carry 15,000 gal [57 m^3] of such fuel. Such heat greatly endangers both the passengers trapped in the cabin and the fire personnel trying to approach for rescue. "Proximity clothing," designed to reflect radiant heat and generally insulate the wearer, can help fire personnel to cope with high temperatures. But smoke conditions can impair visibility, making it difficult to find exit doors and to read the necessary instructions beside them.

Weather and Wind Conditions. Despite the fact that highly sophisticated instrumentation has eliminated much of the danger in blind flying, plane-crash fires sometimes occur when visibility is impaired by snow, rain, or fog. The effects of weather on fire operations are described in Chapter 11.

Water, Personnel, Apparatus, Equipment, and Special Extinguishing Agents Available. In or near airfields, requirements for handling plane-crash fires are likely to be adequate. In rural communities, however, the initial response may consist of a single pumper provided with booster tank equipment. The effects of these varying responses on such factors as extent of fire after arrival, heat and smoke conditions, exposure hazards, and, above all, the life hazard for passengers, can be gaged accordingly.

Topography. If a plane crashes in a mountainous, wooded area, the response of the fire service may be delayed by the lack of usable roads. In addition, response may otherwise be minimal in such areas. The adverse effects of these factors are described above.

Class of Fire. A plane-crash fire is unique in that it may present problems created by class A fires involving ordinary combustibles, class B fires involving flammable liquids, class D fires involving metals (magnesium and titanium), and, more remotely, class C fires involving electrical equipment. Initially, the fire department has to cope with a class B fire, and subsequently—if the magnesium and titanium are ignited—with class A and D fires. Class C fires are less likely because electrical systems are supposed to be deenergized before impact. Extinguishing agents are discussed in Chapter 15.

Effects on Secondary Factors

Where there is a life hazard, rescue is the major objective at any fire, and risks ranging from merely unusual to extreme may be warranted. The officers first to arrive cannot be certain that the crash involved high impact forces caused by a "power-on" landing, although they can make a fairly accurate guess by noting the condition of the plane and the heat and smoke conditions. Nor can they be sure that the survival time has run out, except in the most obvious situations.

Placement and Use of Lines. When there is a life hazard at any fire, the first line or lines available are placed to operate between the fire and the trapped occupants or their means of escape (Fig. 7-4). This principle ap-

FIG. 7-4. Fire operation at an airplane crash.

plies regardless of the extinguishing agent used, but here bearing in mind that a class B fire is in progress. An approach to the fuselage, and escape for the passengers and crew, must be provided (see above). If only fog lines from a booster tank are available, they should be used without waiting for a supplemental supply that might arrive after the survival time has expired. If, unfortunately, the survival time has already expired, these tactics may still help to confine and control the fire.

At such fires it is important, too, to advance from the windward side. If at all feasible, it is also important to work from an upward position (unlike working at brush fires) if this will help to drive the burning fuel downward and away from the fuselage—and will not create or worsen other hazards.

If a plane crashes into an occupied building or buildings, or the fire threatens to involve occupied buildings, the situation becomes extremely complicated. The severity of the life hazard determines the objective and activities to be undertaken and also determines the order of priority in covering the involved and/or exposed structures. Here again, confine, control, and extinguish tactics must be used even though rescue continues to be the major objective.

SUPERMARKETS

A fire in a supermarket is unusually troublesome when it is difficult to ventilate the structure so as to localize the fire and prevent horizontal spread. This observation could apply to other fires as well, but it applies particularly to supermarket fires. The difficulties referred to are caused by many primary factors but especially by construction, area, and occupancy.

An interior attack on a supermarket fire may be ineffective for several reasons. First, the extent of the fire is apt to be great since there is considerable fuel on hand and areas are open and spacious, thus abetting the spread of fire and the development of high temperatures. In addition, since supermarkets are often located in places remote from the major firehouses of more congested areas, frequently only small assignments are available. Water mains, too, are small and hydrants farther apart. Thus, only a limited attack can be initiated. Yet two or even three hose streams may be required to advance successfully because of the width of the store and the consequent danger of being outflanked by the fire. By the time lines adequate for such a maneuver are ready, an interior operation may be out of the question because of heavy involvement. Some supermarkets also have solid exterior walls, both side and rear, making it practically useless to attempt horizontal ventilation and magnifying the importance of roof ventilation, without which the interior attack is foredoomed. But roof ventilation over a heavily involved supermarket is an unusually hazardous undertaking and is therefore not advisable unless it aids in rescue work.

At less-extensive fires with no life hazard, usual tactics may be successfully resorted to. These include venting vertically and horizontally as effectively as possible; forcing entry in the manner that best facilitates the operation (for example, at the rear if the fire is in that area); operating the line or lines to achieve extinguishment while at the same time minimizing the possibility of extension. This means using the lines to drive the heat and smoke out of the store and not into it while putting the fire out.

Covering the life hazard in supermarket fires can be difficult. Many customers may be on hand, some of them accompanied by children. Turnstiles permit entry but may block egress. Some checkout counters may be chained off. Emergency exit doors may be provided in the rear and midway at the side, but they may open into other store areas rather than directly to the outside. These areas may also be filled with smoke and heat when the emergency doors are opened, and frequently access to the doors is blocked by stock. Usually there are two doors at the street front, an entrance and an exit. The average customer scarcely notices the emergency doors and is likely to head for these street-front doors. Panic can therefore develop readily at even the slightest sign of fire because of the congestion at the front of the store. In addition, rescue work will be impeded by the probable absence, at least initially, of helpful ventilation as well as the delay in stretching hose lines. Extensive parking areas in front of supermarkets can be congested during business hours.

Moreover, when hose lines are finally stretched, care must be taken to avoid interfering with the escape of the occupants. Of course, when there is a life hazard, roof ventilation must be attempted—to draw heat and smoke away from the exit area—but rescue work cannot wait until such ventilation is effective.

An exterior attack is also ineffective in terms of property loss, again largely due to the primary factors, occupancy, area, and construction. Blank walls make it difficult to reach the fire with exterior lines, especially if the cellar is involved. Lines onto the roof are of little help and can add considerably to the water damage. The operation of heavy-caliber streams through the broken street-front windows may be hampered by cars in the parking lot or gated protective devices currently in vogue. The history of these fires reveals that often the wooden or unprotected metal trusses beneath the wide roof spans fail and the structure collapses. Therefore, in anticipation of this development, lines should be set up to prevent spread of fire to nearby structures (unlikely with suburban supermarkets) while other heavy lines operate on the fire structure until hand lines can approach to finish up the job. In this way, the major objective of confine, control, and extinguish can be attained.

INDUSTRIES USING RADIOACTIVE MATERIALS

The hazards of radioactivity have been publicized mainly in connection with the threat of nuclear warfare. There is also a more common hazard because radioactive materials have become relatively inexpensive to produce and are increasingly available for use in more and more peacetime industries. Today, these materials can be encountered in universities, industrial laboratories and plants, research laboratories, doctors' offices, hospitals, atomic-power plants, and transportation facilities. Atomic-powered submarines are now a reality and a forerunner of what can be expected in future transportation by ocean liner, aircraft, truck, and automobiles. Much remains to be learned about radioactivity and its dangers by the scientific professions in general and by the fire service in particular.

The Nature of Radiation

Radioactivity may be defined as the spontaneous emission of radiation, generally alpha or beta particles—often accompanied by gamma rays—from the nuclei of an unstable, and therefore radioactive, isotope. Isotopes are forms of the same element that have identical chemical properties but different atomic mass (due to the different number of neutrons in their nuclei) and different nuclear properties—for example, radioactivity and fission. Not all isotopes are radioactive. Hydrogen has three isotopes, with atomic mass 1 (hydrogen), 2 (deuterium), and 3 (tritium). The first two are stable and hence not radioactive; the third is unstable and radioactive.

As a result of this spontaneous emission, the isotope is converted, or decays, into an isotope of a different element, which may or may not also be unstable. For example, the first step in the decay process of the uranium 238 atom is the emission of an alpha particle. Originally, this atom had 92 protons and 146 neutrons. After emission of the alpha particle, thorium 234, and its 90 protons and 144 neutrons, remains. (The alpha particle consists of 2 protons and 2 neutrons.) Thorium 234 is radioactive; hence, the decay process continues to form protactinium 234 and then uranium 234. The nuclei of some elements may have the same mass but a different ratio of neutrons to protons. It is because of this difference that thorium 234, with 90 protons and 144 neutrons, emits a beta particle and uranium 234, with 92 protons and 142 neutrons, emits an alpha particle.

The final step in the decay of the uranium 238 atom is conversion of polonium 210 to lead 206, which is stable and not radioactive.

Gamma radiation originates when the discharge of an alpha or beta particle does not take away sufficient energy to leave the nucleus in a contented, or stable, state. The remaining undesirable energy is given off as gamma radiation.

It is not possible to predict when disintegration will take place if only one atom is involved. However, if many atoms are involved, a statistical determination can be made that half the atoms will disintegrate in a given period of time. This is the so-called half-life of radioactive materials, and it may range from a fraction of a second up to thousands of years. The shorter the half-life, the more highly radioactive the material is. Fortunately, isotopes with very short half-lives are not widely used because they decay so rapidly, and this reduces the likelihood of the fire service encountering such materials.

The use of radioisotopes is based on the penetrating radiations that are emitted, which allow their location and movement to be traced with considerable precision. Radioisotopes function principally as "tracers." In this capacity, they are used in science, medicine, engineering, industry, and agriculture to show such things as how fertilizers move through the roots of plants and through the stem and leaves, how food is taken into the body and how it eventually gets into the tissues, and how certain components operate in an industrial process. Expanded use of radioisotopes can be anticipated: recently the Atomic Energy Commission estimated that in industry alone savings of half a billion dollars yearly already have been gained through use of isotopes. It is fortunate that fantastically small amounts of isotopes will suffice for tracing, since the potential hazard is thereby reduced.

Radiation Terminology

The standard terms for comparing radiation effects include roentgen, rad, RBE, and rem.

The *roentgen* measures only gamma (or X) radiation in the air. It must always be associated with a unit of time and is generally expressed as roentgens per hour.

The *rad* measures the absorbed dose of any type of radiation. A rad of one type of radiation may have more effect on the human body than a rad of another type. This difference is expressed by the *RBE* (relative biological effectiveness). For example, the RBE of gamma radiation is 1, whereas that of some alpha radiation is 20. This means that 1 rad of alpha radiation can have approximately 20 times the effect of 1 rad of gamma radiation.

The *rem* (roentgen equivalent man) is a measure of the effect of radiation in the body. It is the unit of radiation dose that makes it possible to express all types of radiation exposures in one term. A dose in roentgen equivalents man is equal to the dose in rads multiplied by the relative biological effectiveness of the type of radiation involved.

Amounts of radioactive material are measured in *curies, millicuries* (one-thousandth of a curie), and even *microcuries* (one-millionth of a curie). A curie is that amount of radioactive material which is disintegrating at the rate of 37 billion atoms per second. The curie bears no relationship to the weight of the material involved. If a material is very slightly radioactive, several thousand pounds might be required to give 1 curie of radioactivity; if it is very highly radioactive, a fraction of an ounce might give 1 curie of radioactivity. For example, 1 curie of cobalt 60 weighs 880 micrograms, but 1 curie of thorium 232 weighs 10 tons.

The curie does not measure radiation hazard; it simply indicates how many disintegrations are taking place every second. The hazard depends upon the quantity and type of radiation emitted.

The More Common Radioactive Isotopes

Isotopes of sodium, cobalt, iridium, potassium, iodine, iron, and gold are used. Sodium 24 gives the most roentgens per curie at 1 ft [0.3 m], but it is used only in highly specialized situations because of its very short half-life. Cobalt 60 is a popular source of gamma radiation because it gives the most roentgens (except sodium 24) among the isotopes readily available. Other isotopes give off comparatively less than half the number of roentgens given off by cobalt 60. Iridium 192, potassium 42, iodine 131, iron 59, and gold 198 give off, respectively, 5.5, 6.84, 2.7, 6.8, and 2.7 roentgens per hour at 1 ft [0.3 m] per curie. Atomic piles, cyclotrons, betatrons, and similar machines make it possible to radioactivate almost anything (materials made radioactive in this way are called *by-product radioisotopes*). Therefore, the fire service can expect to meet radioactive materials in many other forms than those specified above.

Firefighters opening the roof about 10 ft [3 m] above the radioactive source of 100 curies of cobalt 60 will be subjected to a radiation level of 15 roentgens per hour, since 1 curie of cobalt 60 gives off 15 roentgens per hour at 1 ft from the source. The 100 curies will give off 1,500 roentgens per hour at 1 ft, but by the law of inverse squares the roentgens per hour received at 10 ft will be 100 times less than that at 1 ft, or 15. If the firefighters on the roof take 10 minutes to do their job, their exposure will be about 2.5 ($\frac{1}{6}$ of 15) roentgens. The amount of exposure to which firefighters should be subjected depends to a large extent upon the objective sought. If the situation is urgent, greater risks will be taken; if it is not, minimum risk should be taken. Knowledge of the biological effects of gamma radiation is essential.

Fortunately, the very high-level multicurie sources require massive shielding and special remote control for normal handling. They are invariably installed in facilities built specifically for this purpose and in most cases are built into the ground. When in use, the source is raised out of the ground into a structure. In the event of fire, a heavy-duty fusible link melts and automatically lowers the source back into the shield; then there is no radiation problem at all. In many occupancies where medical work and research is carried on, the amount of radioactive material on hand is so small that the hazard is comparatively insignificant. Of course, it cannot be disregarded. Trained personnel in such occupancies should make certain that radioactive sources are restored to their shield in case of fire or, if necessary and possible, moved to a place of safety. Information on this point should be relayed to the responding firefighters.

Instruments and Dosimetry

Monitoring devices include radiation-detection instruments and devices to measure the radiation exposure of individuals. The best-known radiation-detection instrument is the *geiger counter* (Fig. 7-5), a gas-filled electrical device that detects the presence of radioactivity by counting the formation of ions. It is essentially a low-level instrument, since the maximum reading on most geiger counters is 40 or 50 milliroentgens per hour (a milliroentgen is one-thousandth of a roentgen). Many register both gamma and beta radiation. By closing a shield that covers the Geiger-Müller tube, beta radiation, which cannot penetrate the shield, is screened out and only the gamma radiation gets through. Beta radiation is then computed by taking the open-window reading (beta and gamma) and subtracting the closed-window reading (gamma only).

Ionization chambers, roughly similar to geiger counters and used in civil defense, can read higher levels of beta and gamma radiation, as high as 500 roentgens per hour.

The *scintillation counter* is a very precise radiation-measuring instrument that detects minute quantities of gamma radioactivity. It counts atomic particles by means of tiny flashes of light (scintillations) produced when the particles strike certain crystals.

Alpha radiation, with its very short range in air and little or no penetrating ability, must be detected on special *alpha-measuring meters,*

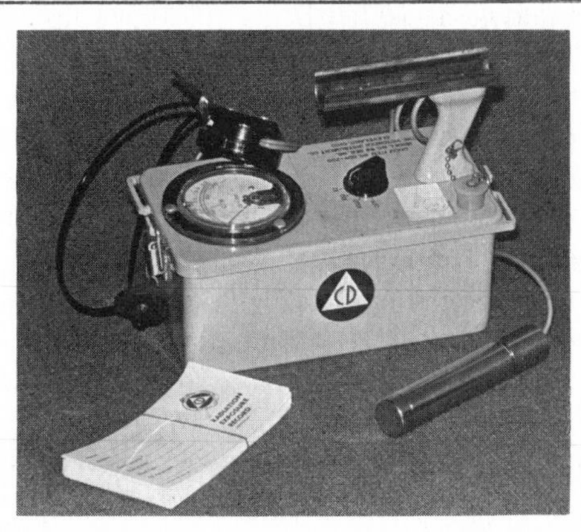

FIG. 7-5. The geiger counter.

whose measuring area must be brought directly to the contaminated surface. The surface must be checked thoroughly to detect the contamination. If a surface is irregular and the flat detecting part of the meter cannot be brought to bear, a piece of cloth or tissue is used to wipe a measured area of the surface. This paper or tissue is then checked to estimate the contamination level, which is generally stated as so many disintegrations per minute per 100 square centimeters (sq cm) of surface. The same wiping technique may also be used for beta- or gamma-contaminated surfaces to ascertain how much of the contamination is removable or fixed to the surface.

Radiation detectors are, in effect, rate meters, since they measure the rate of radiation received by the instrument at a particular time. When removed from the radiation field, they show no reading and past readings will not be reflected.

Devices designed to measure the accumulated radiation exposure of individuals are of two types: film badges and pencil dosimeters. *Film badges* are bits of dental x-ray film worn in special holders; they provide a permanent record of radiation exposure, but not an immediate one, since time is required to develop the film. A *pencil dosimeter* is a device that records the radiation received from zero to the limit of the scale. Some dosimeters can be read directly by holding them up to the light; others must be read in a special reading device. These devices must be worn at all times when a person is in a radiation area; when not being worn, they should be protected from radiation exposure.

These instruments serve little purpose unless they have been selected by informed persons, since developments in this field are rapid. In addition, they require skilled maintenance and calibration for effective use. Every branch of the fire service should set up a program, if one is not already in existence, to ensure that appropriate instruments will be available and in good operating condition when needed.

Transportation Accidents Involving Radioactivity

The transportation of radioactive materials and atomic weapons presents serious problems if an accident occurs, some of which apply to all transportation accidents and others of which derive from the nature of radioactivity.

The location of a transportation accident is unpredictable. This minimizes the important advantages that can be gained by preplanning. The problems can be anticipated only in a most general way compared with those in a static industrial situation. An accident may occur in a location that creates a severe life hazard, for example, in the vicinity of an occupied church or theater. It may cause a severe exposure fire hazard depending on the presence of vulnerable structures and the direction of the prevailing wind. It may occur in an area where the fire service is inadequately equipped to handle it or where the water supply is very poor. The transient radiation hazard may be associated with ships, planes, trains, trucks, transit sheds, truck depots, express depots, or transit warehouses. The individual problem may have complex aspects.

When radioactive materials are involved in transportation accidents, there is always the danger of external radiation to the surroundings. Protection against external radiation should be fully utilized; it depends on a combination of distance, time, and shielding. To be effective, these factors must be used in the proper combination.

Distance is an important protection because the radiation intensity varies inversely as the square of the distance. Therefore, if the radiation rate 1 ft [0.3 m] from the source is 1,000 roentgens equivalent man (rem) per hour, at 10 ft [3 m] it will be 100 times less, or 10 rem per hour (Fig. 7-6). This is the variation that can be expected in the immediate vicinity of radiation sources used in industry for penetrating radiation, because such sources are generally quite small in size. Where the source is large, such as the side of a reactor that covers a considerable area or such as might

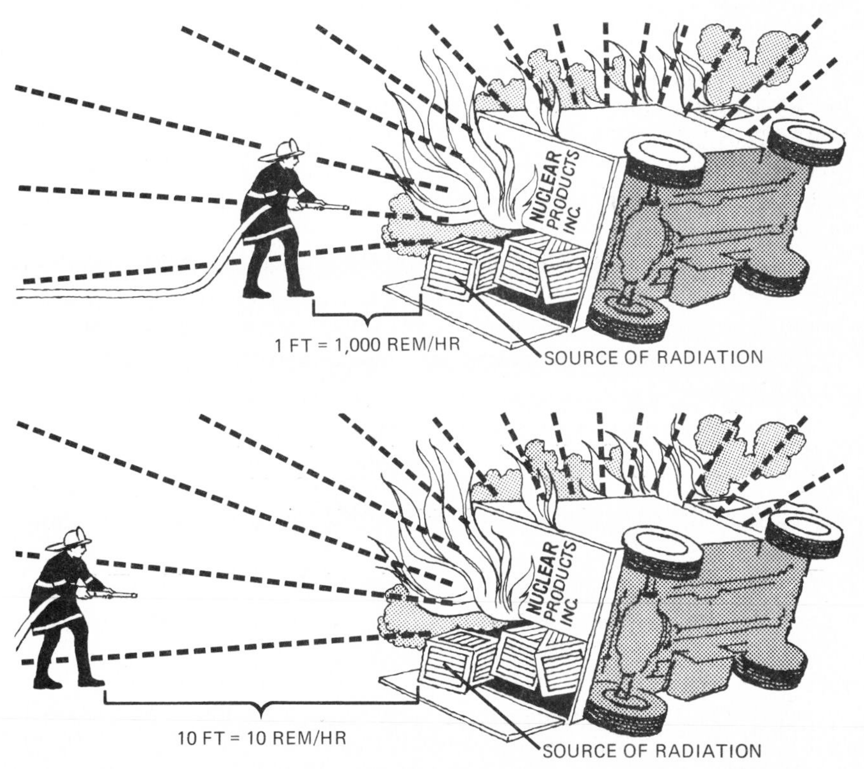

FIG. 7-6. Effects of distance on radiation intensity.

be the case if the radioactive material were dispersed, the inverse-square law does not apply until the distances involved are large in relation to the size of the source. To utilize the factor of distance in a transportation accident, the fire should be fought from upwind, thereby increasing the range of the stream of extinguishing agent and making possible an attack from maximum distance. A solid stream designed to break into a spray at the fire area would be effective and some fog streams would also suffice.

The protection of time should be used to keep single and overall exposures of personnel to a minimum. At ordinary fires, in which radioactivity is not a factor, officers know by experience when firefighters should be relieved because of existing heat and smoke conditions. If conditions are severe, relief should be provided more often. Where radiation is involved, the officer may not be as qualified to know when relief is needed. Special technical advisors may have to be called upon.

The most effective shielding protection from radioactivity is the preservation of the original shielding container, if that is possible.

To fight effectively a fire involving radioactive material, it is important to have some information about the material. The personnel in charge of transporting it may have been killed in the accident, and there may be no information on the contents and their nature except for required ICC red or blue labels (see below, Warning Signs). This is a serious situation if monitoring devices are not on hand or if fire personnel are not properly trained in their use. The radioactive materials being shipped may range from a tiny fraction of a curie of a short-half-life isotope to literally millions of curies of mixed fission products in the form of spent-fuel elements taken from reactors.

Water may react violently with some radioactive materials. Hence, it is important to find out, if possible, the chemical nature of the material (flammable, soluble, or liquid) as well as its physical state (solid, powder, or liquid) so that the most effective extinguishing agent can be selected and used with the least risk of worsening the radiation hazard.

When it is necessary to use water as the extinguishing agent, it is important to keep in mind that the water may be radioactively contaminated even though it is not made radioactive. Therefore, it is highly desirable to extinguish with a minimum amount of water (preferably in the form of low-velocity fog) and to confine the contamination as closely as possible to the point of origin. Standard fire-service salvage techniques should be used to confine the water close to the fire. As the water drains away, radioactive materials may be deposited and contamination may become widespread.

The problem of contamination as a consequence of these fires is a very serious one and usually requires aid or advice from specially trained persons. In addition to contamination of the water used to extinguish the fire, contamination may occur when materials involved in the fire are reduced to ashes and blow around the area. In this connection, it must be remembered that radioactivity is unaffected by chemical change by fire. For example, radioactive carbon will burn just as ordinary carbon burns, with the important exception that the resulting carbon monoxide, carbon dioxide, and smoke will be radioactively contaminated.

If it is necessary to carry out an emergency operation in a contaminated area, personnel should adapt their clothing to prevent the entry of dust. Such a precaution is essential to minimize the possibility of "beta burns" in the event of direct skin contact. It should be remembered, of course, that clothing offers virtually no protection against gamma radiation.

For dry operations, heavy pants and shoes are recommended, as well as cotton or canvas work gloves and a tight-fitting cap. In dusty areas it is advisable to tie the bottom of the pants and the ends of the sleeves (over the gloves) to prevent entry of contaminated material. A scarf around the neck also helps. It is advisable to avoid inhaling dust, and consequently a self-contained mask should be worn until it is determined that no airborne hazard exists. [Self-contained masks are oxygen and air-demand masks or oxygen-generating masks. Such masks provide a man with his own independent air supply and prevent him from breathing the contaminated air (see Chapter 12).]

For wet emergency or decontamination operations, water-repellent clothing, rubber boots, and rubber gloves are required. These can be cleaned with a stream of water and may be used several times, provided there are no breaks or tears. In addition to wearing the recommended protective clothing and equipment, personnel must be thoroughly checked for contamination after operations are concluded. Careful decontamination of apparatus, equipment, clothing, and persons is always in order. An unusually thorough shower and washing is recommended, with particular attention to those parts of the body that may have caught radioactive material, such as the hair, under the fingernails, and the crevices of the skin, especially around the face and ears.

At times a radiation survey of the area may be necessary. In such a case, technically qualified assistance must be obtained. In many states and municipalities, such aid is available from local agencies; it is also available through the AEC Radiological Assistance Plan. The assistance consists of a team of (1) persons trained and equipped to evaluate radiation hazards and recommend measures for control and (2) medical personnel who may be called upon to advise in the treatment of injuries believed to be contaminated by, or to have been caused by, exposure to radiation or radioactive material. If necessary to save life or to minimize injury, and if no other help is available, AEC personnel will themselves perform rescue work.

The AEC has designated eight of its twelve operations offices as regional offices for its Radiological Assistance Plan. The offices have the direct responsibility of providing appropriate AEC assistance upon request, when radiological accidents occur within their respective regions. The AEC team dispatched to the scene to preserve or recover the radioactive material or to assist in operations is under the direction of the regional operations office. If further AEC assistance is needed, the regional operations office will request it from the appropriate operations office or from the AEC–DOD (Department of Defense) Joint Nuclear Accident Coordinating Center. Appropriate AEC Washington headquarters divisions and operations offices are responsible for disseminating information about the Radiological Assistance Plan. The office of the general manager in Washington is responsible for maintaining in readiness a radiological incident center, which may be activated by appropriate headquarters division directors or by the general manager.

The dangers of shipping radioactive materials are modified by regulations (currently being revised) of the Interstate Commerce Commission, which specify the amount of radioactive material permitted to be shipped in a single container, the permitted radiation level at the surface of the container and at a distance of 39 in [1 m] from the center, and the assembly of a number of packages of external-radiation emitters that individually might be of low hazard but collectively might build up a high radiation level if assembled in one location. Other ICC limitations provide that not more than 2 curies of radium, polonium, or other members of the radium family or 2.7 curies of other radioactive materials may be packed in one shipping container. The exception to this limitation is a "specification container" for shipping up to 300 curies of solid cesium 137, cobalt 60, and iridium 192. The highest radiation reading on any accessible surface on the outside of this container must not exceed 200 milliroentgens per hour of gamma radiation, and the radiation level at 39 in [1 m] from the center of the container must not exceed 10 milliroentgens per hour of gamma radiation. FAA regulations govern transportation of radioactive materials by plane.

Occupancies with Radiation Machinery

Radiation machinery includes x-ray machines, cyclotrons, linear accelerators, and betatrons. Fires in such occupancies are complicated by the following factors: a high level of radiation is given off by the machine while it is operating; there may be large quantities of cooling oil present, either in the machinery or in associated electrical transformers; high-energy electrical equipment is present—a cyclotron produces energies of the order of several million volts, the particles to which this energy is impaired having energies of several million electron volts (MeV), and the betatron accelerator produces particles of a billion electron volts (GeV) energy; a radioactive target is located deep within the machine itself; large quantities of paraffin may be used for shielding, and the vapors given off when it is subjected to heat have a low autoignition point; ventilation may be difficult due to solid and windowless construction in some cases; interior attack may be delayed until radiation is stopped or reduced; many cyclotrons are located in rural areas, and the fire-service response may be limited.

Preplanning is emphasized so that unusual developments may be anticipated, adequate countermeasures decided upon in advance, and maximum coordination achieved between fire-service personnel and appropriate technicians. Preplanning is also indicated because the contents of these occupancies are unusually valuable. Determined, well-calculated efforts to extinguish the fire are required.

Supervision of fire personnel to reduce or eliminate the life hazard is very difficult when these occupancies take fire. All electric power must be shut down promptly to eliminate the hazards of electrocution and to stop the generation of radiation by the machine. Total-flooding carbon dioxide systems should be used if they are available. If they are not satisfactory, fog (low velocity, preferably) or foam may suffice. If entry with fog lines is necessary, overhead cranes should be utilized if available to remove ceiling sections and provide helpful ventilation.

After the fire is extinguished, technical help may be required to retrieve the radioactive target deep within the machine so that it will not present a life hazard during the overhauling phase. Other radioactive materials may be present, necessitating a survey by trained persons.

The fire hazard is not the same in all cyclotrons. Some are water-cooled and some are oil-cooled. Water-cooled machines present few oil-fire hazards, since the amount of oil on hand is generally inconsequential, and many fires can be readily extinguished by dry-chemical or carbon dioxide extinguishers. In oil-cooled cyclotrons, there is a much greater possibility of a serious class B fire, and smoke damage results because the oils are heavily sulfur-bearing. The sulfur and oil smoke hydrolyze the water in the air to form sulfuric acid, which when deposited on electrical contacts, renders them useless and makes it necessary to tear down all the equipment and rebuild it before power is restored.

Warning Signs

A purple-on-yellow propeller sign indicates a radiation hazard in an occupancy. Precautions are stipulated on the lower part of the sign (Fig. 7-7). Where radioactive materials are being transported, their presence is indicated by ICC red or blue radiation labels.

The blue label (Fig. 7-8) indicates that the materials in transit present a

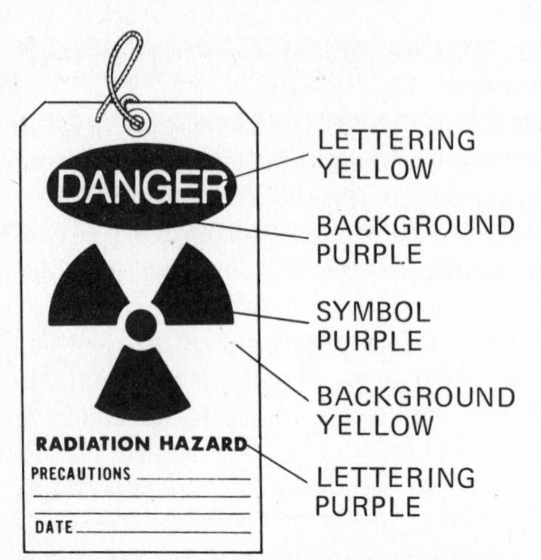

FIG. 7-7. Sign indicating presence of radioactive material.

radiation hazard only if the package is ruptured. The red label (Fig. 7-9) indicates that the materials present an external hazard from gamma rays or neutron emitters. It also indicates principal radioactive content, the activity of the contents, and the number of radiation units in each package. A radiation unit is defined as one milliroentgen per hour at a distance of 39 in [1 m] from the center of the container, and ICC regulations prescribe that not more than 40 units shall be loaded in one aircraft or held at one point or location. Trucks carrying any amount of "red-label" material must be placarded on each side and in the rear with a sign reading "Dangerous—Radioactive Materials."

Hazards of Radioactivity

Radioisotopes make the air electrically conductive, and the resultant accumulation of static electricity is a serious danger where explosive

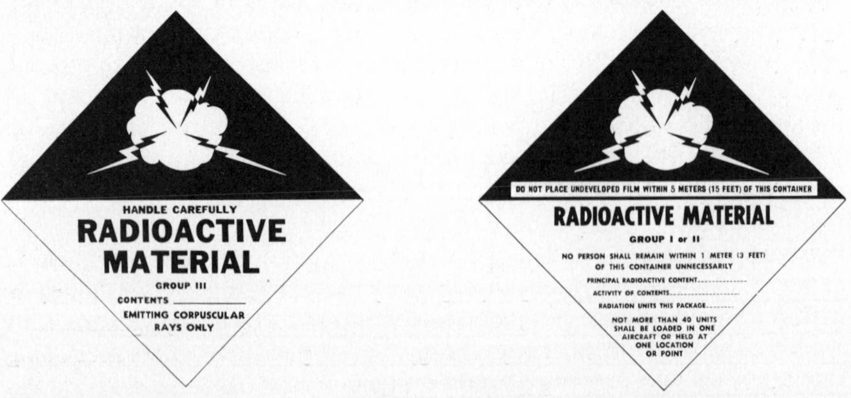

FIG. 7-8. The ICC blue label.

FIG. 7-9. The ICC red label.

vapor-air concentrations exist. The emission of alpha and beta particles and gamma rays may cause harmful biological effects.

An alpha particle is identical with the nucleus of the element helium. When it has lost most of its (kinetic) energy, it captures two electrons and becomes a harmless neutral helium atom. It is neutral because the two negatively charged electrons are matched by two positively charged protons. The range of an alpha particle depends upon its initial energy, but even those from plutonium, which have fairly high energy, have an average range of just over 1 in [2.5 cm] in air; in more dense media, such as water or body tissue, the range is much less. These particles can be stopped by a sheet of paper, and they are unable to penetrate even the outer layer of the skin (epidermis).

The great danger from alpha particles occurs when such elements as uranium or plutonium enter the body through the lungs, the digestive system, or breaks in the skin. The ionization caused by alpha particles from these elements may cause severe tissue, bone, or organ damage. Plutonium, for example, preferentially concentrates in bone, where it may cause serious harm by prolonged action (plutonium 239 has a long radioactive half-life, 24,000 years, as well as a long biological half-life, over 100 years, and its amount and activity decrease at a very slow rate).

Beta particles are charged particles of very small mass emitted spontaneously from the nuclei of certain radioactive elements, and like alpha particles, they are able to cause direct ionization as they pass through matter. But beta particles dissipate their energy less rapidly than alpha particles and so have a greater range in air and other media; many emitted from the fission products in a nuclear explosion traverse a distance of 10 ft [3 m] (or more) in the air before they are absorbed. However, beta particles are easily deflected in their course, and so their effective range is often less.

Injury to the body from beta particles can arise in two ways after a nuclear explosion: most, if not all, of the fission fragments emit (negative) beta particles, and if they come into contact with the skin and remain for an appreciable time, "beta burns" result. In addition, if the fallout area is extensive, the whole surface of the body is exposed to beta particles coming from all directions. Clothing attenuates this radiation to a considerable extent, but if outer layers of the skin receive a large dose, serious burns may result.

Gamma rays are electromagnetic radiations of high energy, originating in atomic nuclei and accompanying many nuclear reactions, for example, fission, radioactive decay, and neutron capture. Physically, gamma rays are identical with x-rays of high energy, the only essential difference being that the x-rays do not originate from atomic nuclei. Gamma rays are the most common form of radiation and the most penetrating: they have the same effects as x-rays. A dose of 25 roentgens of total body exposure within 24 hours produces no detectable clinical effects, but larger doses have increasingly more serious consequences; a dose of 450 roentgens within 24 hours would be fatal to about half the persons affected. A whole-body exposure of 700 or more roentgens in 36 hours would probably be fatal in all cases. A third-degree burn on a limited area of the body might not very serious, but a second-degree burn over a large area might prove fatal. Similarly, a dose of 1,000 or more roentgens of radiation on a small area would cause local damage but probably few overall consequences. If most of the body were exposed to the same dose in 24 hours or less, death would undoubtedly result.

A neutron is a neutral particle—that is, it has no electric charge—and is present in all atomic nuclei, except those of ordinary (or light) hydrogen. Neutrons are needed to initiate the fission process, and they are present in large numbers after the fission and fusion reactions in nuclear explosions. Neutrons, being electrically neutral, do not produce ionization directly in their passage through matter. They can, however, cause ionization to occur indirectly as a result of interaction with certain light nuclei. When a fast neutron collides with the nucleus of a hydrogen atom, for example, the neutron may transfer a large part of its energy to that nucleus. As a result, the hydrogen nucleus is freed from its associated electron and moves off as a high-energy proton. Such a proton is capable of producing a considerable number of ion pairs in its passage through a gas. By a similar mechanism, indirect ionization results, although to less extent, from collisions of fast neutrons with other light nuclei, for example, oxygen, and nitrogen. The ionization resulting from the interaction of fast neutrons with hydrogen and nitrogen in the tissue is the main cause of biological injury by neutrons.

The harmful effects of neutrons on the body are similar in character to gamma rays and can cause cataracts more readily than other nuclear radiations. At such distances from a nuclear explosion that they represent a hazard, nearly all the neutrons are received within a second of the explosion. As with other types of radiation, the neutron dose decreases with distance, partly because the neutrons spread over a large area and partly because of absorption and scattering.

Ionization is the dissociation of an atom into cations (with a positive charge) and anions (with a negative charge). This change is effected by the removal of one or more electrons from an atom, thereby disrupting the normal electric neutrality of the atom, which previously had an equal number of negatively charged electrons and positively charged protons. The removed electrons are negatively charged, and the remainder of the atom is positively charged. Each part is then an ion, and the atom is said to be ionized. Ionization can be achieved by chemical or electrical methods or by radiation. The phenomenon of ionization by radiation is utilized in monitoring devices, which indicate that their contents are being ionized when radioactive materials are within range.

It has been stated by some authorities that the radiation exposure of fire personnel should be determined by the fire chief. Theoretically, this is possible, but in those situations where immediate action is required, it is rarely possible to predict or evaluate accurately the radiation dose being received. This emphasizes the need for monitoring devices and for the training of fire personnel in their use. The National Committee on Radiation Protection has stated that a single, once-in-a-lifetime 25-rem emergency dose should have no effect in the radiation status of the individual. Radiation status is an accumulation of 5 rem per year starting at age eighteen. Fire officers should be guided by this knowledge and should rely on monitoring devices as well as on help from specially trained persons.

Conclusions

Preplanning plays a predominant role in fighting fires that involve radioactive materials. With the information in this chapter, fire officers can preplan more effectively and obtain better results.

It is important for fire officers to familiarize themselves with the hazards of radioactive materials. However, it should be noted that if a radiation survey of an area is necessary, technically qualified assistance must be obtained. Such technicians are trained and equipped to evaluate radiation hazards and recommend measures for control. Fire officers should be guided by these evaluations and recommendations accordingly.

PLASTIC INDUSTRY

In the past decade, the public and the fire service have become acutely aware of the effects of plastic products on human life and safety. In 1974, the Federal Trade Commission compelled 25 of the major companies in the plastic industry to sign a consent decree regarding advertising claims. As a result, plastic can no longer be advertised as "slow-burning," "noncombustible," or "self-extinguishing." Instead, promotional material must give the actual test information on the burning characteristics of the material. In addition, the Trade Commission insisted that the companies referred to fund a $5-million research program designed to test plastics on a scale comparable to its use in construction. One result of comparatively recent research is discovery of the fact that many plastics formerly considered self-extinguishing can be burned continuously under conditions that develop quite frequently at fires.

Manufacturers are presently concerned with flammability tests. However, tests that prove plastics to be nonflammable are not sufficiently meaningful. Toxicity tests are needed as well because the pyrolitic products of plastics exposed to fire include highly toxic gases. At present, the plastic industry has commercially available materials which are quite competitive with some common metals. Hopefully, the flammability and toxicity tests could result in a safer and more selective use of such materials.

The chemical and physical modifications of plastics and the incorporation of additives along with the thousands of trade names make it exceedingly difficult for fire officers to accurately evaluate the effects of the pyrolysis and combustion of plastics. However, there is one decision they can safely make and that is to use self-contained breathing appartus promptly, even before it is known that plastics are involved. Subsequent decisions depend, as always, on the primary factors that become pertinent, notably heat and smoke conditions, whether plastics are or are not involved—assuming here that the building is unoccupied. Fire officers can note the effects of the plastic involved even though they may not recognize the type.

The manufacture of basic plastic has the hazards of large quantities of flammable liquids or gases. Plastics are then converted into useful articles by molding, extrusion, or calendering, all of which involve the use of heat. Finally, the materials are fabricated by bending, machining, cementing, and polishing. These processes may be carried out in the same or in different plants, affecting hazards accordingly. It is suggested that plastic plants should be regarded as target hazards so that prefire plans made in conjunction with the management can be more constructive.

Problems with the finished product have increased in number and difficulty at fires in recent times. This trend is likely to continue, since the annual plastic tonnage is getting progressively larger. The most serious problems have been caused by the use of plastics in construction (about 20 percent of the total output), and as insultation for electrical wiring. The presence of plastic furnishing, such as flammable foamed cushioning, adds to the problems.

Effects of Other Primary Factors

Primary factors likely to be pertinent are construction, location and extent of fire on and after arrival, visibility, requirements to operate, exposure hazards, duration of operation, and life hazard for fire personnel. One factor that is especially important at fires where plastics are involved is heat and smoke conditions.

Authoritative reports on past fires reveal that foamed polystyrene (used as insulation) can spread fire via flaming droplets or gases. The reports also reveal that as the heat involved overstuffed furniture, its progress accelerated because of the amount of combustibles and the flammable gas given off by the foamed polyurethane. The report of the NYBFU on the fire at One New York Plaza indicates that postfire tests showed that polyurethane foam gave off flammable gases at 212°F [100°C].[2]

Smoke conditions at fires involving plastics are frequently dense. Gases are often toxic. For example, when polyvinyl chloride, or PVC, is decomposed at fires it gives off hydrogen chloride which, if inhaled, is converted into hydrochloric acid inside moist respiratory passages. PVC is widely used for wire and cable insulation, building materials, conduit, interior furnishings, rainwear, and so forth. It is one of the most heavily loaded plastics on the market, if not the most. Thus, fire officers can expect a worsening of smoke and gas conditions along with the greatly increased use of plastics, especially at fires in high-rise buildings.

Cellulose nitrate plastics require careful attention because of their flammability and instability to heat—even sunlight. These plastics are no longer used to make motion picture and photographic film but, temporarily at least, they are still used to make artificial leather, lacquer, enamel, and other articles. Special care must be taken when persons are exposed to the highly toxic effects of the oxides of nitrogen produced by the decomposition of cellulose nitrate because fatalities may occur hours or even days after exposure.

Methods suitable for extinguishing class A fires are also appropriate for plastics fires which, except for cellulose nitrate, are classified as ordinary combustibles. Large amounts of water are recommended for cellulose nitrate fires. Personnel should, if possible, work from the windward side to avoid exposure to the toxic gases given off. Self-contained breathing apparatus should always be worn at such fires.

The effects of plastics on a fire operation were amply demonstrated in New York City on February 27, 1975. The fire was believed to have started in electrical equipment in the subbasement shortly after midnight, and it required more than 15 hours to control. Over 1,000 tanks of compressed fresh air were used in masks by the nearly 400 firefighters at the scene. Fortunately, employees had escaped from the building, which was a major telephone switching center. Visibility, which was practically zero because of the smoke, made it very difficult to locate the fire. Firefighters also had trouble in breaking through the plastic substance behind the first- and second-floor windows which was provided to protect against vandalism. Hence, in their effort to alleviate the heat and smoke conditions, fire personnel had to endure abnormally high temperatures until the fire was located and could be attacked with streams.

These difficulties were compounded by the ceramic casings around the

[2] New York Board of Fire Underwriters, *Report of Fire at One New York Plaza,* 1970, p. 10.

involved cables, which were insulated by polyvinyl chloride material. The casings were unusually resistant to the equipment used. Progress in achieving control was therefore slow, and in the meantime, dense, acrid smoke permeated the neighborhood. At last, around midday (twelve hours after the alarm) one exterior exposure was evacuated.

REVIEW QUESTIONS

1. What helps officers assess the severity of a life hazard that is created when the human element in an occupancy is endangered at fires? Why must the nature of certain occupancies be considered in making this assessment?

2. From a practical point of view, what is the fire officer's concern relative to the content aspect of occupancies? Why is this concern practical in determining the effects of contents on other primary factors?

3. What are the effects of occupancy contents on secondary factors?

4. Why is it advisable for every apparatus to carry authoritative information about hazardous flammables and chemicals?

5. What is the role of the occupancy factor in creating target hazards? What emphasis should be placed on these hazards? How?

6. How are metals grouped in this chapter? What metals are included in the alkali group? What are some of their dangerous characteristics?

7. What metals are included in the alkali-earth group? Why does magnesium merit special consideration?

8. What should officers know about the characteristics of iron and steel?

9. Where can fires involving titanium be expected? Zirconium? Sodium?

10. What are the atomic-fuel metals? What are their collective and individual hazards? Why do plutonium fires present special problems? How should they be handled? (Methods and agents for handling fires in other metals are covered in Chapter 15.)

11. How can primary factors worsen the overall situation at plane-crash fires?

12. What must be considered in selecting major objectives at plane-crash fires? How would the objective of rescue affect the placement

and use of lines? What must be considered if a plane crashes into an occupied building or the fire threatens an occupied building?

13. Why can fires in supermarkets be unusually troublesome? Why can it be difficult to mount an effective interior operation? What tactics may be successful at the less-extensive fires? What complicates the problem of covering the life hazard at fires in supermarkets? Why are exterior operations often ineffective (in terms of property loss) at these fires?

14. What makes it difficult for officers to accurately evaluate the effects of the pyrolysis and combustion of plastics?

15. What hazards are associated with the manufacture of plastics? Why are problems with the finished product likely to continue and increase? What causes the most serious problems with the finished product?

16. What are the effects on heat and smoke conditions when plastics are involved at fires? What hazards are associated with cellulose nitrate plastics? What extinguishing methods are suggested for fires in plastics?

17. At the fire in the telephone building, how did the involvement of plastics affect the operation (primary and secondary factors)?

18. On a separate sheet of paper, match items from Column 1 with items from Column 2.

COLUMN 1	COLUMN 2
1. The spontaneous emission of radiation, generally alpha or beta particles, often accompanied by gamma rays, from the nucleus of an unstable, and therefore radioactive isotope	A. Not so widely used because they decay so rapidly
2. The undesirable energy remaining after the discharge of an alpha or beta particle	B. Forms of the same element with identical chemical properties but different atomic mass (due to the difference in the number of neutrons in their nuclei) and different nuclear properties
3. The time it takes for half the atoms in a radioactive isotope to disintegrate	C. Measures the absorbed dose of any type of radiation
4. Roentgen	D. Equals the dose in rads multiplied by the RBE of the type of radiation involved
5. Rad	E. Depends to a large extent upon the objective sought
6. RBE (relative biological effectiveness)	F. By-product radioisotopes
7. A dose of rems (roentgen equivalent man)	G. Require massive shielding and special remote control for normal handling
8. The amount of radioactive material that is disintegrating at the rate of 37 billion atoms per second	H. The more highly radioactive the material is
9. Isotopes	I. Half-life of radioactive material
10. A popular source of gamma radiation because it gives the most roentgens per hour (except sodium 24) among the isotopes readily available	J. Definition of radioactivity
11. The amount of radiation exposure to which personnel should be subjected	K. Is given off as gamma radiation
12. Materials made radioactive in betatrons and cyclotrons	L. Measures only gamma or X radiation in the air
13. Very-high-level multicurie sources	M. Expresses the difference between rads of different types of radiation relative to the effect on the human body
14. The shorter the half-life	N. A curie
15. Isotopes with very short half-lives	O. Cobalt 60

19. On a separate sheet of paper, match items from Column 1 with items from Column 2.

COLUMN 1	COLUMN 2
1. The period of time it takes for half of the radioactive material to be excreted from the body	A. Harmful biological effects
2. A combination of the radiological half-life with the biological half-life	B. Ionization, caused by alpha particles within the body
3. Radioisotopes	C. No detectable clinical effects
4. Alpha and beta particles and gamma rays may cause	D. 50 percent fatalities within a month
5. Alpha particles	E. Is present in all atomic nuclei except ordinary (light) hydrogen
6. When the great danger from alpha particles occurs	F. Is the ionization resulting from the interaction of fast neutrons with hydrogen and nitrogen in the tissue
7. May cause several tissue, bone, or organ damage	G. The biological half-life of a material
8. Plutonium	H. The disassociation of an atom into cations (with a positive charge) and anions (with a negative charge)
9. Beta particles	I. By chemical or electrical methods or radiation
10. Gamma rays	J. Cataracts
11. What a dose of 25 roentgens total body exposure within 24 hours produces	K. An accumulation of 5 rem per year starting at age eighteen
*12. What a dose of 450 roentgens total body exposure within 24 hours would cause	L. Can detect minute quantities of gamma radiation
13. What the harmful effects of neutrons can cause more readily than other nuclear radiation	M. When such elements as uranium and plutonium enter the body
14. The main cause of biological injury by neutrons	N. The effective half-life of the radioactive material in the body
15. Ionization	O. Cannot penetrate the outer layer of the human skin
16. How ionization can be achieved	P. Within the body, it preferentially concentrates in bone and has a long biological half-life
17. Why geiger counters detect the presence of ionization	Q. Make the air electrically conductive
18. A neutron	R. Cause serious burns if outer layers of the skin receive a large dose of them
19. The radiation status of an individual	S. The most penetrating form of radiation
20. The scintillation counter	T. Because their contents are being ionized by the radiation

* Authorities differ here

8

Height, Area, and Proximity of Exposures

HEIGHT

Effects on Other Primary Factors
The height at which people are trapped can intensify the life hazard for both occupants and personnel if rescue efforts entail the use of a scaling ladder in conjunction with an aerial ladder or, in a high-rise building, depend solely on an interior operation.

The height of the fire building establishes the potential location (level) and extent of a fire. For example, other things being equal, a one-story building more definitely limits the location and extent of a fire on and after arrival than a similar fire in a five-story building. In the latter case, fires of greater extent are more possible, and are more likely to increase requirements to operate and the duration of the operation. Ordinarily, since heat rises, higher (in relation to the fire level) rather than lower buildings aggravate the exposure hazard. In addition, other things being equal, higher structures provide more fuel than lower ones and hence have greater potential for developing heat and smoke conditions and impairing visibility.

The height from a floor to a ceiling, or a ceiling to the underside of a roof (the cockloft or attic area), is a measuring dimension that affects the volumetric area. This area in turn affects the time it would take for an explosion mixture to develop in inadequately ventilated fire areas. Volumetric areas and occupancy contents could affect the type and amount of extinguishing agent recommended as well as the method of application.

In some communities, in accordance with existing building and fire-prevention codes and other laws, the height of a building affects (1) its construction, area, occupancy, and proximity to other structures, and (2) the auxiliary appliances that may be required—such as a sprinkler or standpipe system or an elevator specifically designated for fire-department use.

Buildings limited in height because of inferior construction generally lack effective firestopping and therefore are more susceptible to total fire collapse than higher buildings of superior construction. Structural collapse can create a spark-and-ember hazard and other fires downwind.

Effects on Secondary Factors

Height can affect activities at fires in high-rise buildings as described in Chapter 6. At lower-level fires, roof or window ventilation may be possible, thus facilitating the advance of lines from either side of the fire as well as the search for and removal of occupants. Exterior lines may also be used. In either case, control is likely to be established more quickly than at a similar but higher fire which can be attacked only from the interior. This earlier control tends to minimize the overhauling required at lower-level fires.

AREA

The factor of area refers to the area of the building involved or exposed—not necessarily the area involved in the extent of fire.

Effects on Other Primary Factors

Large-area structures, such as churches, supermarkets, bowling centers, and so forth, particularly when not effectively subdivided, may provide enough oxygen to speed combustion, extent of fire on and after arrival, and involvement conducive to structural collapse. This was demonstrated at the Livonia and McCormack Place fires. Such possibilities can worsen the life hazard for occupants and make rescue efforts difficult and unusually dangerous. They can also create a spark-and-ember hazard that could start other fires and increase duration of the operation and requirements to operate. Effectively subdivided areas tend to offset these disadvantages.

Fires occurring in large-area structures or woodland occupancies can feature heavy smoke conditions that seriously impair visibility.

Area, like height, is a measuring dimension that affects the volumetric area. This area in turn affects the time it would take for an explosive mixture to develop and cause a back draft or smoke explosion, as after fresh air starts to flow into ineffectively ventilated fire areas.

In general, area can influence many fire-protection features, such as departmentation; the number, types, and location of means of egress;

and so on. In some communities, building and fire-prevention codes and other laws specify (1) the relationships of areas to height, promixity to other buildings, and allowable occupancies, and (2) how area can affect the fixed systems and/or portable extinguishers required.

Effects on Secondary Factors

If the fire can be confined to a small room, the fact that the total floor area is 200 by 200 ft [about 60 by 60 m] hardly matters. However, if such an area is not effectively subdivided and there is no small room, the extent of the fire can sooner or later correspond to the total floor area. If there is no life hazard present, this development could make it logical to select confine, control, and extinguish as the major objective. The decision could be to operate from the exterior but not to try to ventilate the roof, especially at supermarkets with their wide roof spans.

The extent of the fire—rather than only the large area of an occupancy, such as a lumberyard—influences the placement and use of hose lines. If the fire is not too extensive, heavy-caliber streams may be used to darken down the fire, allowing hand lines to advance, finish up the job, and readily cover the spark-and-ember hazard. However, if the fire is so extensive that exterior exposures are endangered, hose streams must first be used to confine the fire by protecting the exposures, then to darken down and control the fire, and finally to extinguish it.

The manner in which large fire perimeters at woodland fires affect supervision, communication, and coordination, is described in Chapter 22.

Proximity of Exposures

Proximity alone does not make an exposure vulnerable. To evaluate the effects of proximity in selecting objectives and activities, it must be considered in conjunction with other contributing factors—such as construction, location of fire, occupancy, and wind direction and velocity. Proximity is hardly a problem if the construction of both the fire and adjoining buildings features exterior windowless walls with four-hour fire-resistive ratings—assuming no inlets to air-conditioning systems are exposed. On the other hand, inferior construction, with inadequately protected openings in intervening shafts or narrow courtyards, can intensify proximity hazards.

Location of the fire can signify the degree to which proximity can be pertinent. Remember, *location* indicates the fire level as well as the section of the level involved. Proximity would scarcely be pertinent at a fire on a fifth floor if nearby exposures were only two stories high—unless the fire building collapsed, of course. But, due to the natural rise of heat and smoke, it would definitely be pertinent if the adjoining exposures were seven stories high.

The nature of occupancies in exposures naturally has an important impact on proximity. The human element and content aspects of an occupancy influence the placement and use of hose streams when other primary factors, including proximity, are equal. But proximity is the next deciding factor. Also, as previously noted, the proximity of occupied exposures can create a life hazard and establish rescue as the major objective even though the original fire building is unoccupied. In that case, the need to concentrate initially on rescue may prove detrimental to control of the main body of fire. In this way, proximity and occupancy factors can have an adverse effect on extent of fire after arrival; heat, smoke, and visibility conditions; and exposure hazards—with remote possibilities of structural collapse, and so on.

In evaluating proximity of exposures, special consideration must always be given to the factors of direction and velocity of the wind. These factors can minimize the effects of proximity on the windward side of the fire and maximize those on the leeward side to such a degree that the building nearest the fire is not necessarily the one most severely exposed.

Effects on Secondary Factors

The manner in which secondary factors are affected depends on the degree to which proximity and other factors maximize or minimize hazards in exposures and thereby determine objectives and activities.

REVIEW QUESTIONS

1. How does the primary factor of height affect other primary factors?

2. How does height affect secondary factors?

3. To what does the primary factor of area refer? How does it affect other primary factors?

4. How does area affect secondary factors if the fire can be confined to a small room and the floor area is large? If the floor area is not effectively subdivided? How would extent of fire on and after arrival—rather than the large area of an occupancy such as a lumberyard—influence the placement and use of hose lines?

5. The primary factor of proximity of exposures does not by itself make an exposure vulnerable. To evaluate its effects, it must be considered in conjunction with other primary factors. What are these primary factors and how do they make proximity of exposures pertinent in selecting objectives and activities?

6. How does proximity of exposures affect secondary factors?

9

Structural Collapse, Time Elements, and Auxiliary Appliances

STRUCTURAL COLLAPSE

Structural collapses that have occurred before the arrival of the fire department have usually been caused by gas explosions; they have also been caused by planes crashing into buildings. Only structural collapse after arrival of the fire department is being considered here. Explosions and back drafts are discussed in Chapter 13.

Effects on Other Factors

Effects of structural collapse are quite apparent, particularly if caused by a plane crashing into brick-joist or frame dwellings. The fire is likely to be extensive on and after arrival, thus creating a severe life hazard and much fire damage. It also worsens every aspect of the fire situation, and greatly increases requirements to operate. Risks to personnel may be extreme, especially if they have to use tunneling techniques to reach trapped occupants. For the effects of structural collapse when a plane crashes into a fire-resistive building, refer to the discussion of the Empire State Building fire in Chapter 6.

Effects of structural collapse after arrival are also quite apparent. Too often, one of these effects is that firefighters are buried in debris within the structure or injured by collapsing walls, heavy canopies, theater marquees, or fire escapes. The lives of other firefighters are then greatly endangered by the perilous measures taken to rescue the trapped personnel. The effects of other factors on structural collapse are less apparent but obviously of great concern to fire officers and especially to the commanding officer.

In assessing the effects of other factors on structural collapse, officers should carefully consider the type of construction involved in the fire. Nonfireproof or brick-joist construction is usually susceptible to collapse, for example, and has presented some of the most serious problems. Age of the structure intensifies structural defects. Duration of the fire—how long it has been in progress and how much water has been poured into the building—and location and extent of the fire present obvious problems. For instance, the fire may be located where it will quickly weaken structural members, and weakening is more likely since extensive fires generate much heat. Other important considerations are conditions on arrival, particularly where an explosive smoke or back-draft condition is present, or where an explosion has already occurred; presence of heavy machinery; and the nature of the burning or exposed material—is it explosive? absorbent? Still other points to consider are proper supervision (the stairs must not be overloaded); the span of floor between supporting members (wide spans are more susceptible to collapse); and whether supporting metal structural members are unprotected (they may fail rapidly if heated and then struck by cold water from streams, especially if they are made of cast iron). In vacant buildings, officers can anticipate that floor beams have been weakened by vandalism and, quite often, by previous fires.

Signs of imminent collapse are a rumbling sound that may accompany a wall disturbance or collapse, cracking of bulging of walls, water or smoke seeping through the walls, twisted or warped columns and beams, and floors sagging or pulling out from walls. It is folly, however, to wait for such signs to occur before changing from an interior to an exterior operation, especially when the fire building is unoccupied. When the major objective originally is confine, control, and extinguish, commanding officers should carefully consider *before* sending units into the fire building to achieve the final phase of extinguish. The responsibility to check for any signs of imminent collapse is then. If there are such signs, officers can proceed as suggested in Chapter 2.

In earlier suggesting an inquiry into the causes of casualties among fire personnel, it was noted that structural collapse was a major casualty factor. In many of the most disastrous situations the fire building was not even occupied. Certainly many casualties could be avoided by careful attention to the causes and signs of structural collapse, in conjunction with the specification that if there is no life hazard for occupants, personnel should never be jeopardized unnecessarily.

In an occupied building, the commanding officer is of necessity guided by the specification that if there is a life hazard for occupants, then risks to personnel, ranging from merely unusual to extreme, may be warranted. Hence, even if structural collapse seems possible and even probable, an attempt to enter the building for rescue purposes is in order. In later decisions, the commanding officer must consider the reports of rescue-work officers concerning the feasibility of continuing the effort or abandoning it because it has become uselessly dangerous—for example, when the fire floor has collapsed.

If on arrival the fire building has already collapsed and is so involved that usual rescue measures would only uselessly increase the casualty list, then there would definitely be greater limitations on risks to be taken. Even though rescue would continue to be the objective, confine, control, and extinguish tactics would have to precede efforts to search for victims of the collapse and fire.

Effects on Secondary Factors

In addition to what has already been said about decision making, supervision and communication are unusually important. Supervising officers should be alert to signs of impending structural collapse and should remove their units from the danger area without waiting for orders from commanding officers. However, subordinate officers should promptly communicate such developments to their commanding officers so that they can take appropriate action.

Effects on Rescue Work

Removing persons buried under several floors of debris is a difficult, dangerous, arduous undertaking, complicated at times by residual heat and smoke and the need for restraint in using water. In one case where an officer and three firefighters working on a roof were buried by a structural collapse, large numbers of personnel struggled to the top of about three floors of debris through which smoke and heat continued to rise. There they worked for about 10 hours, without masks, before the last man was extricated—and all the while knowing that they were endangered by the front wall, the only one still standing. In some cases victims must be reached by tunneling through the debris, a perilous technique even when used by personnel with special training.

TIME ELEMENTS

Time elements include the factors of times of origin, discovery, alarm, and response.

Time of Origin

Time of origin tells when a fire occurs: time of year, holiday time, or day or night.

The time of the year ordinarily reflects the usual seasonal tendencies relative to topography in woodland areas—humidity, rain, snow, dry spells. It even reflects business cycles with their individual as well as collective effects on fire incidence, occupancy contents, street conditions, time of response, and water supply. Effects of weather are discussed in Chapter 11.

Major holiday seasons maximize the hazards associated with churches and department stores. Extended weekends increase transportation problems and the related fire potential. A fire in a car on a congested highway can present a problem out of all proportion to the extent of the fire, assuming that only one car is involved. Procedures for such fires should be formulated before they occur. A booster-equipped apparatus responding against traffic will suffice in most instances.

Night Fires. At night, visibility is poor and it takes longer to evaluate factors that are pertinent in determining objectives and activities. At fires in residential occupancies, the objective is almost invariably rescue. The life hazard is maximized because such occupancies are more crowded at night and persons waking up to cries of "Fire!" are more prone to panic than those who confront a fire wide awake. In addition, fires are likely to be more extensive when discovered because occupants were asleep. This condition in turn worsens heat and smoke conditions, and therefore, exposure hazards.

At night fires in unoccupied buildings the selection of objectives and activities may be more difficult in borderline cases because of the darkness. It is likely that these fires, too, have been burning for some time before discovery, thereby worsening the effects of all related primary factors—such as location and extent of fire on and after arrival; heat, smoke, and visibility conditions; and exposure hazards. Consequently, night fires usually require more hose streams, apparatus, and personnel for control and extinguishment, and overhauling of both structure and contents is more extensive than at day fires.

Life hazard for personnel is increased by the dangers inherent in rescue work and overhauling at night fires. Statistics indicate that a surprisingly large number of injuries occur during overhauling at such fires. However, there are some advantages to night fires: Traffic is lighter and therefore less apt to interfere with response and placement of apparatus, with use of hydrants, and with the stretching and use of lines. In addition, pressure in public water mains is generally better because of the reduction in demand.

Daytime Fires. The fact that occupants are awake can result in quicker discovery, alarm, and response with favorable effects on the life hazard, location and extent of fire on and after arrival, heat and smoke conditions, and exposure hazards. In ordinary weather visibility is good, and fire officers can more quickly recognize and evaluate primary factors pertinent in selecting objectives and activities. On the other hand, traffic is heavier during the day. Street conditions can hamper response and placement of apparatus, use of hydrants (which may be blocked), and the stretching and use of lines. In addition, pressure in public water mains is generally lower because of the increase in demand.

Time of Discovery, Alarm, and Response

Delay in discovery or alarm obviously delays response of the fire department. This can in turn start an unfavorable chain reaction in relation to location and extent of fire on and after arrival, heat and smoke conditions, exposure hazards, structural collapse, life hazard for occupants and personnel. Early discovery or alarm is conducive to a favorable chain reaction of events. Fire officers responding to alarms are of course not likely to be aware of any delays and therefore do not evaluate these primary factors—since they cannot at the time recognize them as such.

Fires occurring when certified security personnel are present in certain occupancies (such as amusement parks, theaters, and industrial plants) can affect the promptness and efficiency with which auxiliary appliances in such occupancies are used.

The time of the day and the time a fire starts affect wind patterns—direction and velocity. Wind patterns are described in *Fire Research Abstracts and Reviews* in the following manner: "The atmosphere also has tidal waves similar to the tides of the sea. However, the period of the atmospheric tides is exactly twelve hours whereas ocean tides have a period of half a lunar day. The tide of the earth's atmosphere is almost certainly produced by the diurnal variations in the heating by the sun."[1] *Diurnal* means *daily*. Studies of diurnal wind variations have been conducted on location at Wildcat Canyon in California. It has been discovered that "a definite diurnal rhythm of air movement occurred, up-canyon during the day, down-canyon at night."[2]

AUXILIARY APPLIANCES

Auxiliary appliances include portable extinguishers of various types and sizes, fixed systems (standpipe, sprinkler, carbon dioxide, dry chemical, fog, foam, fog-foam, halogenated hydrocarbon), fans, and alarm systems. A new member of this group may be pressurization of stairways.

[1] D. L. Turcotte, review of *National Aerodynamics* by R. S. Scorer, *Fire Research Abstracts and Reviews*, vol. 2, no. 3, p. 153.

[2] G. R. Fahnestock, review of "Local Wind Patterns in Wildcat Canyon" by C. M. Countryman and D. Colson, *Fire Research Abstracts and Reviews*, vol. 2, no. 3, p. 158.

Portable extinguishers and fixed and alarm systems will not be described here since that task has already been excellently done in the *Fire Protection Handbook.*[3] The concern here is with providing information for evaluating the appropriate auxiliary appliances in selecting objectives and activities. Selection of extinguishing agents is discussed in Chapters 7 and 15.

Portable Extinguishers

These appliances are appropriate to extinguish or to check the spread of incipient fires. If there is a life hazard, the principle governing the use of hose lines in such cases applies: the contents of the extinguishers should be directed between the fire and the endangered occupants or their means of escape. Otherwise, the contents should be directed at the burning material. If extinguishers can be used successfully, the objective of the fire department on arrival would likely be extinguish, achievable by ventilating, checking for extension, overhauling, and possibly the use of a hose line. Much depends on the manner in which extinguishers are used.

Suggestions for Use. The use of extinguishers in industry or institutions should be governed by the following considerations—the intent of many of which also applies to fire personnel.

1. The type of extinguisher that can best cope with potential hazards should be selected. Advice from the neighboring firehouse should help the public in the selection. The fire service can also advise about the the number of extinguishers required. The number needed will vary with the presence or absence of sprinkler, standpipe, or other fixed systems, the nature of the occupancy, the construction, the area, the public fire service available, and the like. Of course, the classification of the extinguisher is an important matter.

2. The physical limitations of the people available to operate the extinguisher should be considered in its selection. In larger organizations operation should be assigned to specific individuals, with substitutes to cover in the event of sick or vacation leaves. Selected personnel should be instructed about suitability of extinguishers, method of operation, how to direct the stream, storage, maintenance, any associated hazards, and recharging processes. A good occasion for instruction is the periodic examination, when extinguishers are usually discharged. Personnel should know the limitations of extinguishers and should be instructed to notify the fire department as soon as possible if fire breaks out.

[3] National Fire Protection Association, *The Fire Protection Handbook,* 1976.

3. Extinguishers should be placed in a conspicuous and accessible location, and any change should be quickly made known. They should be hung at a reasonable height and should be easily removable. They should be close to potential hazards but not so close that it would be difficult to reach them in case of fire.

4. A sufficient quantity of charges or needed materials should be on hand to ensure multiple operations or to recharge the extinguisher at once, if necessary. Only the chemicals or ingredients stipulated by the manufacturer should be used.

5. Extinguishers should be examined regularly, and at least several times a year, to check for possible tampering, damage, or clogging of nozzles, and to make sure they are full. If there is doubt about their effectiveness, they should be checked by the manufacturer. Records should be kept of the inspections, showing results and including the name of the examiner or inspector. Hydrostatic tests should be conducted as required.

6. In some areas legislation prohibits the use of certain types of extinguishers because of the toxicity hazard. Inquiries definitely should be made on this point.

7. Signs legible for a distance of at least 25 ft [7.6 m] should indicate the classes of fire for which the extinguisher is suitable, and personnel should be instructed about the classes of fire. Classes of fire are discussed in Chapter 15.

Sprinkler Systems

Long-term records show that automatic sprinklers either extinguished a fire or held it in control more than 96 percent of the time in a wide variety of occupancies. Reasons for unsatisfactory performance were found to be mainly water to sprinklers shut off, only partial sprinkler protection, inadequate water supplies, faulty building construction, obstruction to distribution, hazards of occupancy, inadequate maintenance, antiquated systems, and slow operation. In addition, some costly mistakes could include premature shutting down, supplying the wrong inlet, delay in restoring the system to service, inopportune dismantling in buildings being demolished, and inadequately or tardily supplying systems.

At a large majority of fires, automatic sprinkler systems (particularly the water-flow-alarm type) have a favorable effect on discovery of fire, transmission of alarm, and response of the fire department. They also have a favorable effect on the location and extent of fire on and after arrival, heat conditions, exposure hazards, possibility of a smoke explosion or back draft, and hence on the life hazard for occupants and personnel. Smoke and visibility conditions, however, can present some minor problems. In such cases, the major objective is likely to be extinguish, achievable by ventilation, checking for extension, overhauling, and the use of a small hand line. Problems can be more troublesome if highly piled stock limits the effective range of the sprinkler discharge or if material burning on shelves or under workbenches cannot be reached by the sprinkler water.

Faulty construction can also allow the fire to involve other floors, thus negating the effectiveness of sprinkler systems which, in multistory buildings, are designed to extinguish fire on any one floor but not on several floors at once. Some occupancies involve dusts and fibers, resulting in flash fires and the opening of an excessive number of sprinkler heads. Others involve flammable liquids whose vapors can explode and damage the sprinkler system piping or cause a flash fire—which opens an excessive number of sprinkler heads. In such occupancies, the sprinkler system is less likely to be an asset and should be evaluated accordingly.

Systems that discharge other than water serve notice on fire officers that it may be dangerous or inadvisable to use hose streams because electrical equipment may be involved or water might dilute the foam cover and render it ineffective. In such situations it is well to consult private plant fire-security personnel. Presumably such personnel know about deenergizing equipment that is involved, the nature of occupancy contents, the operation of valves for transferring contents of tanks, exposure hazards, and supplying extinguishing systems.

Fans

The use of fans has been discussed in Chapter 6; refer also to Chapter 15.

Alarm Systems

Alarm systems are designed to detect a fire in its incipient stage, alert the occupants, hopefully initiate a calm, well conducted evacuation of the occupants, and in some cases activate an extinguishing system and notify the fire department directly or indirectly. Such systems can favorably affect the life hazard, time of alarm and response, location and extent of fire on and after arrival, and other related primary factors to such a degree that the major objective is likely to be extinguish, achievable by routine activities. Of course, other primary factors, such as construction and occupancy contents, can alter the situation considerably. Alarm devices in air-conditioning systems have already been discussed.

Pressurization of Stairways

Tests of this auxiliary appliance are described in an article in *WNYF* magazine entitled "Stairwell Pressurization" by Joseph W. Rooney. These tests provide data for future evaluation; some observations are offered below.

Test #1. An office on the seventh floor of a 22-story building was fire-loaded with combustible material to 6.3 psf [2.8 kg/m^2 (kilogram per square meter)]. The glass was removed from three windows to provide sufficient oxygen "for the generating of high temperatures and maximum fire pressures to challenge the stair pressurization." The remaining windows on the floor and auto-exposed window on the floor above were covered on the inside with sheet metal. The fire was started and the stairway pressurized.

The seventh-floor door was opened first. Then doors were opened on lower floors until a total of four stair doors were open. Pressurization failed when the fourth door was opened, and smoke entered the stairwell.[4]

Observation. It will be difficult to limit the number of doors opened by panic-stricken occupants subjected to severe smoke conditions on many floors. Such conditions prevailed on nineteen floors at a fire in a new high-rise building. If an attempt is made to prevent pressurization failure by increasing the pressure, occupants may be physically unable to open doors and this can worsen their mental state. Researchers are aware of this possibility and are taking steps to cope with it.

With only the seventh-floor door open, "a forced draft situation caused fire to leap out open windows and up the side of the building to the ninth floor." Smoke that entered the stairwell during the test was barely perceptible with three doors open, but when the fourth door was opened smoke was obvious in the stairwell.

Results of this part of the test are inconclusive since there is no mention of the direction and velocity of the wind. Ordinarily, if the wind were favorable the open windows would facilitate the advance of hose lines without the aid of stairway pressurization. If the wind were unfavorable, stairway pressurization would help to offset the disadvantages of working into the wind.

Test #2.

> A second fire test to research the development of carbon monoxide, carbon dioxide, and oxygen fluctuations within an air-conditioned office was staged on the 10th floor. This test was instrumented for temperatures, fire pressures, and gas analysis. Air conditioning machinery was not operated; all doors and windows were closed. This condition was arranged so as to

[4] Joseph W. Rooney, "Stairwell Pressurization," *WNYF*, vol. 33, no. 2, pp. 5–6.

force the fire gases to flow through the plenum. Wood desks, chairs, and paper in an interior office were ignited and observed through a 14″ × 14″ wire glass panel in a fire door. After three minutes of rapid fire development, the visible flame almost immediately and dramatically diminished to a complete blackout and the fire, having consumed the oxygen, had almost extinguished itself. The pressurized stair door was opened and the force of the pressure cleared the smoke from the corridor. Then the fire door, opening on the fire area was reopened cautiously, causing the flow of air to the fire which allowed a supply of oxygen to restore the fire to a free burning situation. The test was concluded when gas readings went off scale on all instruments.[5]

Observation. Supplying an enclosed, oxygen-depleted fire area with oxygen suggests the possibility of a back draft or smoke explosion even though neither occurred during the test. The possibility remains, however, and warrants further experimentation.

Test #3. Pressurization was not involved, and so will not be considered here.

Test #4. A test fire was set up in the southwest wing of the tenth floor. A 10,000 cfm exhaust fan on the roof exhausted smoke up the stairwell to simulate a smoke shaft. It took seven minutes for smoke to exhaust at the roof level. Light smoke did enter the elevator lobby and then the elevator shaft, but in an adjoining office firefighters detected only traces of smoke in the plenum area. After this test was concluded, the stairwell was pressurized to facilitate extinguishment. Rooney notes the immediate reaction to pressurization.

Simultaneously with pressurizing the stairwell, smoke, no longer venting up the stairwell, traveled through all voids. The firemen in the adjoining office observed the smoke being blown through the plenum toward them. Smoke started flowing from the supply diffuser in the elevator lobby of the 10th floor. Within six minutes the elevator call buttons on the 10th floor were energized and elevators were called to this floor. In the next seven minutes smoke was reported in the 1st floor lobby and throughout the building. Researchers were forced to leave their control station in the 8th floor elevator lobby.[6]

Observation. A smoke shaft may be helpful but, hopefully, a better means for ventilating the fire floor will be developed.

[5] Ibid., p. 6.
[6] Ibid., p. 7.

Extensive research on the subject of pressurization has been under way for some time in Canada. In a pamphlet titled *The Pressurized Building Method of Controlling Smoke in High-Rise Buildings,* G. T. Tamura and J. H. McGuire conclude that a method architects and building designers appear to favor is the total pressurization of a building with venting of the fire floor. The basic concept of building pressurization as a means of controlling smoke movement is to create a lower pressure in the fire floor than in adjacent spaces by increasing all building pressures above the outside pressure, except for the fire floor which is vented to the outside. Stairway pressurization is only one phase of this system.[7] Satisfactory ventilation of the fire floor to the outside presents some unusual and difficult problems, as noted by the researchers. Effectiveness of building (or stairway) pressurization has not yet been validated by adequate experimentation and actual application at fires.

REVIEW QUESTIONS

1. What are the effects of structural collapse on other primary factors when it is caused by a plane crash?

2. What are often the most disastrous effects of structural collapses that occur after fire operations get under way? Assume there is no life hazard for occupants.

3. What primary factors should fire officers, and especially the commanding officer, carefully assess in contemplating the possibility of a structural collapse?

4. What are the signs of an imminent structural collapse? Should the commanding officer be guided by signs indicating a *possible* structural collapse or an *imminent* structural collapse before withdrawing units from an unoccupied building? What is the responsibility of the commanding officer *before* sending units into an unoccupied fire building insofar as structural collapse is concerned?

5. Is it logical to maintain that structural collapse cannot be anticipated because it occurs without warning?

6. What are the effects of a possible and probable structural collapse on rescue efforts? By what should the commanding officer be guided in such cases if an effort is made to enter the building for rescue purposes? When would there definitely be greater limitations on risks to be taken? What tactics should be used in such cases?

[7] G. T. Tamura and J. H. McGuire, *The Pressurized Building Method of Controlling Smoke in High-Rise Buildings,* National Research Council of Canada, Ottawa, 1973, p. 1.

7. While overhauling an old, nonfireproof, commercial, unoccupied building, an officer notes that some floor beams have pulled away from the walls. In such a case, should the officer back all units out of the building forthwith and then communicate with the commanding officer about the action taken, or first communicate with the commanding officer about the condition found and then await orders about the action to be taken?

8. How does the primary factor time of origin convey information about or affect other primary factors?

9. What primary factors are likely to be unfavorably affected by night fires in unoccupied buildings? How are secondary factors affected as a result? What primary factors can be favorably affected at night fires?

10. What primary factors are likely to be favorably affected by daytime fires? How are secondary factors affected as a result? What primary factors can be unfavorably affected at daytime fires?

11. What are the overall effects of delayed discovery of fire and transmission of alarm? On arrival at fires, are officers necessarily aware of such delays?

12. What do auxiliary appliances include?

13. What are portable extinguishers for? How should they be used if there is a life hazard? Otherwise? How can the successful use of portable extinguishers affect the selection of objectives and activities? What are some other suggestions for using these extinguishers?

14. Why do sprinkler systems perform unsatisfactorily at a small percentage of fires? What costly mistakes can be made by the fire service in relation to sprinkler systems?

15. How does the water-flow-alarm automatic sprinkler system almost invariably affect other primary factors? Objectives and secondary factors?

16. Why should fire officers consult fire-security personnel when fixed systems are discharging special extinguishing agents on the fire?

17. What are some purposes of alarm systems? How can they affect other primary factors and the overall operation?

18. Discuss pressurization of stairways. Do you agree or disagree with the observations offered?

10

Heat, Flame, and Combustion

Scientific research continues to produce knowledge and advance new ideas about heat, flame, and combustion. The physics of heat and flame involves problems beyond the scope of this book, but some results of scientific research are of interest to the fire service, and some scientific concepts are of importance to firefighters.

HEAT

Heat can be considered mathematically and abstractly as *disordered molecular energy*. To understand this concept, we must appreciate that energy can exist in either potential or kinetic form. *Kinetic* energy is the energy of motion; it exists to some degree in all moving bodies. *Potential* energy, on the other hand, is the energy stored in a body that is under stress, compression, torsion, or pressure. It causes no motion of molecules and therefore is not directly responsible for heat that can be felt.

Potential energy may be converted into kinetic energy. For example, when a valve is opened, water under pressure begins to flow. When the uranium 235 atom fissions, potential energy is converted to kinetic.

Kinetic energy may be converted into heat. When the nucleus of a uranium 235 atom is struck by a neutron and fissions, the kinetic energy of the flying fragments is converted by collisions into random (disordered) motions of the electrons and other atoms in the surrounding material, that is to say, into heat. Conversion of potential energy into heat is the working principle of nuclear reactors.

Heat is produced by adding energy to disorder, as when one compresses the air in a bicycle pump by pumping vigorously. The air heats up, causing the pump to become hot to the touch. The molecules of air still move randomly, but they are under greater pressure (have more energy). In consequence of the work done (pumping), more energy has been pushed into the system, and the observed production of heat is simply the effect of adding energy to the preexisting disorder.

Quantitative Measures of Heat

Two measures of heat are of interest here: a measure of the quantity of energy, and a measure of the quantity of molecular disorder.

Energy is measured by a practical unit called the *calorie*, which is the amount of heat required to raise the temperature of 2 g of water 1°C under standard conditions. The ultimate unit by which all forms of energy are specified is the *erg*, which is defined as the energy of a mass of 1 g moving with a velocity of 1 centimeter per second. If one converts kinetic to heat energy, 1 calorie is found to equal 42 million ergs. Other important heat units are the large calorie (*Calorie*) and the British thermal unit (Btu). One Calorie is the amount of heat required to raise the temperature of 1,000 g of water 1°C. One Btu is the amount of heat required to raise the temperature of one pound of water 1°F. One Btu is equal to 252 calories.

To measure molecular disorder quantitatively, and in order that certain points relative to heat and of interest to the fire service may be considered, one must understand the concept of *entropy*.

Briefly, entropy may be defined as a number that indicates the variety of ways in which atoms may be arranged in a system. Roughly speaking, entropy measures the number of degrees of freedom possessed by a system. (The word *system* refers to the quantity of matter under consideration. Everything else is spoken of as the *surroundings*. A *closed system* has no interchange between it and the surroundings, whereas an *open system* does have an interchange. *Process* refers to any change that the system may undergo.) A high-entropy system is free to be in many different physical arrangements. An example of such a system is a liquid: its molecules may arrange themselves in a huge variety of ways with regard to each other. An example of a low-entropy system is a crystal lattice: its molecules are arranged in a highly ordered way, and it is possible to know precisely which molecule goes where (Fig. 10-1). For the purpose of this book, entropy simply refers to the amount of molecular disorder in a system.

Thermodynamics is the science of the relation between heat and mechanical energy, or the study of changes in which energy is involved, and it is based on two simple but well-known laws. The law of conservation of energy states that the total energy, including heat, of any closed

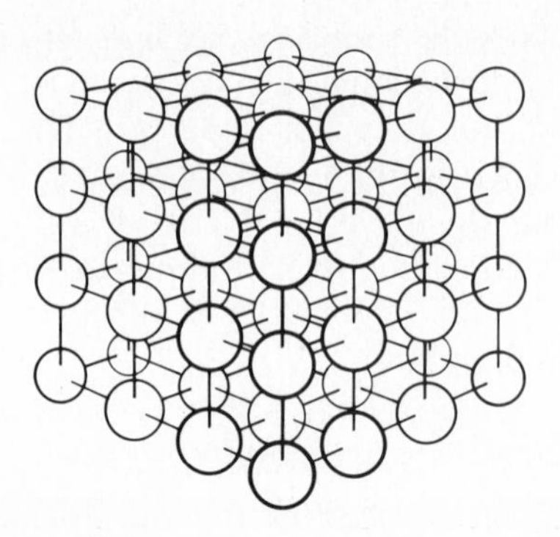

FIG. 10-1. Atomic pattern of matter in a crystalline solid.

system remains constant. The second law states a kind of "conservation of disorder": the total entropy of any closed system must remain constant or increase as time goes on; it can never decrease. The tendency is for entropy to increase rather than remain constant.

The effects of the second law of thermodynamics are that heat is transferred in such a way as to increase entropy, that is, to increase the number of ways in which molecules of a substance may arrange themselves, and that heat must flow from areas of higher to those of lower temperature. As temperature drops, the amounts of energy and disorder both decrease. But the energy always decreases more rapidly than the disorder, so that the amount of disorder per unit of energy grows larger as the temperature falls. A given amount of energy carries more disorder at lower temperatures. Hence heat tends to move from a higher to a lower temperature, and, in so doing, each unit of energy acquires greater disorder (or entropy).

Heat Transfer by Radiation

Radiation is energy in the form of electromagnetic waves, which are traveling disturbances in space and which include light, heat, radio, and cosmic rays. Electromagnetic waves are produced by acceleration of an electric charge and include an electric field at right angles to a magnetic field, both moving at the same velocity in a direction normal to the plane containing the two fields. They behave like other forms of energy, except that they can exist in empty space in the absence of matter. When electromagnetic waves are disordered, they convert to heat and are subject to the laws of thermodynamics.

The existence of heat radiation implies that no material body is ever completely isolated; the space around every object will contain radiation. If there is a temperature difference between the object and the radiation, energy will flow into or out from the object in order to equalize the temperature. Thus, by the process of radiation, energy is transferred from hotter to colder objects.

It is characteristic of radiation that the quantity of heat radiated from a given source varies as the fourth power of the absolute temperature, other conditions being equal. Thus, the quantity of heat radiated from a given source at 600°F [31.5°C] compared with the same source (under the same conditions) at 500°F [26°C] can be found by use of the equation

$$\frac{Q_2}{Q_1} = \frac{(600 + 459)^4}{(500 + 459)^4}$$

The rate at which a piece of matter can radiate away its energy is likewise related to temperature and varies as the fourth root of the energy for heat radiation. The practical effect of the fourth root is illustrated as follows:

If a piece of matter at 600°F can radiate away half its energy in one hour, the same piece of matter at 6000°F [3315.5°C] having about (6000 + 273) K divided by (600 + 273) K, or 7.18, times as much energy, would radiate away half its energy $(7.18)^4$ or about 2,660, times as fast. Hence the half-energy emission would occur in 3,600 seconds × 7.18 ÷ 2,660, or 9+ seconds.

The escape of heat via radiation places a practical limit on flame temperature, which in fire areas may range from 3000 to 5000°F [1650°C–2750°C]. However, effective temperatures, as measured by the melting of metals and other effects, have a practical upper limit of 3000°F, and they are usually less.

Heat may radiate downward from thermal columns bent over by the wind and traveling ahead of a conflagration. Such columns contain highly heated vapors in which actual burning is taking place. In some cases, radiation from such columns has caused ground fires well ahead of the conflagration. This explains why it is practically impossible to operate successfully ahead of or on the lee side of a conflagration.

The major portion of radiant heat passes through drops of water and through ordinary window glass (Fig. 10-2). The latter may intercept only about 15 percent of radiant heat energy. This must be kept in mind by firefighters when covering or protecting exposures. It may be more appropriate to wet down exposures rather than protect them by water curtains (Fig. 10-3). It has been proved that much less radiant heat will pass through window glass that is kept wet, by a fog stream, for example.

Radiation is largely reflected from bright, smooth surfaces. This fact makes aluminized firefighting garments effective against heat radiation. Such garments protect the wearer by reflection rather than by insulation.

Heat rays are similar in many ways to light rays. Both travel in straight

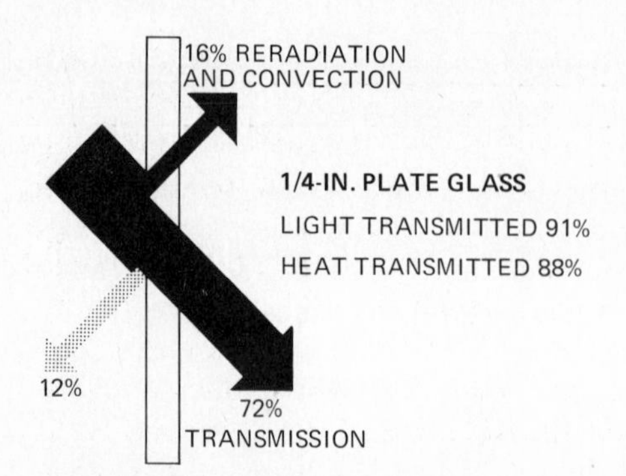

FIG. 10-2. Transmission of radiant heat through glass.

lines, at the same rate of speed, and spread out in all directions unless this pattern is disrupted by absorption or scattering, which may be the result of reflection and diffraction (or bending).

In most instances, the fire service is concerned with radiation emitted from *point sources,* that is, from fires of comparatively small extent, at which the intensity of radiation will vary inversely as the square of the distance from the source. Hence, at 60 ft [18.3 meter (m)] from the point source, the intensity of the radiation is one-quarter of that at 30 ft [9.1 m]. This variation does not apply, however, when the source of radiation is *linear*, as exemplified by grass fires with a long front and flames of limited height. Here, radiation varies inversely as the distance (not as the square of the distance): at 60 ft from the linear source, the radiation intensity is one-half of that at 30 ft. When the fire source is an *area,* such as a forest or a lumberyard, where the fire is extensive with a high flame front, radiation intensity decreases with distance, but not in accord with the

FIG. 10-3. The direction of the wind and the proximity of the exposure are the two factors considered in protecting this structure against radiant heat.

rules governing point and linear sources. At distances considered sufficiently large in comparison with the height and width of the flame front, the area source starts to resemble the point source in regard to the relation of intensity and distance. At intermediate distances, particularly near the area source, attenuation of radiation will be only slight.

The distance is the major consideration where structures are endangered by heat radiation. Consequently, proximity of exposures to the fire building in some cases determines the order of priority in covering exposures, despite the direction of the wind. The wind influences the situation, however, when it changes the direction of the convection currents so as to weaken or strengthen heat radiation intensity, by increasing or decreasing the distance the radiant energy must travel to the exposures. If hot surfaces are the source of radiation, the angles of emission from the source and of reception at the exposure affect intensity received. The intensity is greatest at the point opposite to the surface in a perpendicular direction as compared with other equidistant points. The fire service is more often concerned at fire operations with the effects of radiation emitted from convection currents than with effects determined by hot surfaces.

Much of the following treatment of heat radiation in nuclear explosions pertains to heat radiation in conventional fires.

Radiation in Nuclear Explosions. Fires caused by nuclear explosions present potentially stupendous problems to the fire service.

In a conventional (TNT) explosion, nearly all the energy is released as kinetic energy, which is immediately converted into blast and shock. In a nuclear explosion, however, about 85 percent of the energy released is kinetic, of which only about 50 percent is converted into blast and shock; the remaining 35 percent is converted into heat and light rays. Furthermore, very high temperatures are involved (several million degrees, compared with several thousand degrees in a TNT explosion).

After a nuclear explosion, a large quantity of heat is radiated in a very short time. Consequently, the intensity of radiation, that is, the rate at which it reaches a particular surface, is very high. The heat is absorbed very rapidly, with very little time for it to be conducted away internally.

Heat-radiation intensity is reduced by absorption and by scattering, as well as by distance.

Atoms and molecules in the air can absorb, and thus remove, certain rays, particularly ultraviolet rays. Oxygen and ozone are important in this connection. Normally, the protection of ozone in the air is quite small, but appreciable amounts are produced by the effect on atmospheric oxygen of gamma radiation from a nuclear explosion.

Scattering of radiation, that is, diversion of rays from their original paths, can be caused by molecules, of nitrogen or oxygen, for example, in the air or, more importantly, by particles in the atmosphere, such as dust, smoke, or fog. Radiation of all wavelengths can be scattered, with the energy attenuation varying with the wavelength. Visible light rays and the shorter infrared rays are more susceptible to attenuation by scattering, and by absorption, than are the longer infrared rays.

In the case of a nuclear explosion, a dense smoke screen between the point of the bomb burst and a given target can reduce the radiation received at the target to as little as one-tenth the amount that would otherwise be expected. Artificial white (chemical) smoke acts just like a fog in reducing radiation intensity. If the explosion occurs above a dense layer of cloud, smoke, or fog, an appreciable amount of radiation will be scattered upward from the top of the layer. This scattered radiation may be regarded as lost, as far as the ground is concerned. In addition, most of the radiation that penetrates the layer will be scattered; very little will reach a given point by direct transmission.

If the explosion occurs in moderately clear air beneath a layer of cloud or fog, some of the radiation that would normally proceed outward into space will be scattered back to earth. As a result, the radiation received on the ground will actually be greater than for the same atmospheric transmission conditions without a cloud or fog cover.

Objects may also be protected from radiation by shielding. Any solid opaque material, such as a wall, a hill, or a tree, between the bomb burst and an object will act as a shield against radiation. However, if atmospheric conditions are hazy, much of the radiation will be scattered; consequently, it will arrive at an object from all directions and not merely from the point of burst. Shielding protection is greatest if the shield completely surrounds an object.

Even after radiation intensity has been reduced by absorption and scattering and by shielding, some radiation will reach a given target. When radiation reaches an object, part is absorbed, part may be deflected, and the remainder will pass through and ultimately reach other objects. It is absorbed radiation that produces heat in the object and causes damage. Highly reflective and transparent materials absorb little radiation and hence suffer little damage.

If much radiation is absorbed, organic materials (such as wood) may ignite. Ignition depends largely on thickness and moisture content: under the same conditions, a thin piece of wood may flame and a thicker piece only char. Although the surface temperature in the thicker piece may temporarily exceed the ignition point, the heat is rapidly transferred away by conduction, convection, and radiation: the temperature drops, and combustion eventually ceases.

Heat Transfer by Conduction

Conduction is the process by which heat is transferred within a material from one particle to another or from one material to another in contact with it, without any visible motion. Heat is conducted because faster-moving atoms, electrons, or molecules in the hotter part of an object induce activity in adjacent atoms, electrons, or molecules, and thus heat flows from the hotter to the colder parts. The amount of heat transferred by conduction varies with the conductivity of the material and the area of the conducting path. Thermal conductivity is the ability of a material to conduct heat away internally. In a given material, conductivity increases with density. Wood, for example, has greater density, and hence greater conductivity, when wet. Wet wood is less apt to ignite than dry wood when exposed to radiant heat, because heat is transferred more rapidly away from the igniting source. Consequently, sustained-combustion temperatures are less likely to be reached. Thermal conductivity is expressed in Btu per hour per square foot of surface per degree Fahrenheit of temperature difference of the surface per inch of thickness; it may also be expressed in metric equivalents.

It is important for firefighters to know the conductivity of various materials in order to determine the possible spread of fire and to prevent rekindles (Fig. 10-4). For example, heat conducted by metal beams may cause fire, delayed or simultaneous, at unexpected distances from the source. When metal structural members or pipes have been heated, they should be checked at appropriate distances.

Silver is the best heat conductor of the ordinary metals and lead the poorest, only one-fourth as good as silver. Granite, limestone, ice, water, brick, glass, and plaster are of medium conductivity. Wood, asbestos paper, sawdust, paper, linen, silk, cotton, wool, and air are poor conductors. It should be noted that the transmission of heat cannot be completely stopped by any heat-insulating material.

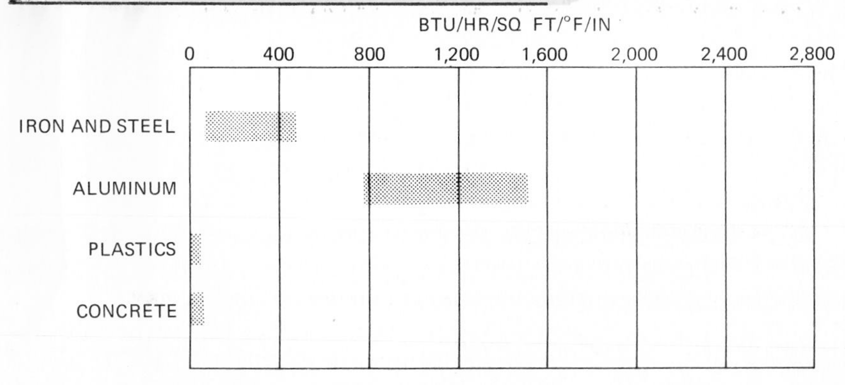

FIG. 10-4. Thermal conductivity range for various materials.

Heat Transfer by Convection

Convection is the process in which heat is transferred by a circulating medium in the gas or liquid state. Thermal columns (pillars of heated air and gaseous products of combustion) in conflagrations, or heated gases, air, and smoke circulating otherwise at fires are examples of convection in the gas state. The flow of molten metals or water heated by contact with the fire are examples of convection by a circulating liquid. Gaseous products of combustion combine with air, whose density is reduced by the heat. Such mixtures are invariably lighter and more buoyant than air alone. Their usual tendency is to rise but convection currents can be carried in other directions as well. The direction of the current is determined by pressure differences, which are influenced by elevation and temperature as well as by obstructions.

When not acted upon by a fan or other impeller, air or any fluid flows from a region of higher pressure to one of lower absolute pressure. (Absolute pressure equals the atmospheric pressure plus gage pressure, if the latter is positive, or minus gage pressure, if it is vacuum pressure.)

Atmospheric pressure at sea level is about 15 psi [103.5 kN/m^2 (kilonewtons/meter^2)]; this is also often expressed as 30 in of mercury, 34 ft of water, or 408 in of water. This last measurement is used in gages to indicate the difference between atmospheric and duct pressures. Atmospheric pressure decreases 1 in of water [250 Newton/meter^2 (N/m^2)] for each 69.2-ft [21 m] increase in height above sea level. Thus, 69.2 ft above sea level atmospheric pressure is 407 in [102.2 kN/m^2] of water, and 69.2 ft below sea level atmospheric pressure is 409 in [101.7 kN/m^2].

Temperature affects the density (mass of a substance per unit volume) of the air. The density of standard air at 70°F [21.1°C] and 29.92 in of mercury [101.7 kN/m^2] is 0.075 pcf [1.2 kilogram/meter^3 (kg/m^3)]. When the air is heated to 340°F [170°C], the density is 0.05 pcf [.8 kg/m^3]. A 70-ft [21.3 m] column of air and gases with a density of 0.05 pcf will exert a pressure of 0.05×70 or 3.5 psf [168 N/m^2] at the bottom inside the shaft (duct, incinerator, piped recess, or similar vertical channel), compared with a pressure of 0.075×70, or 5.2 psf [249.6 N/m^2] on the outside. The difference is 1.7 psf [81.4 N/m^2] or 0.0118 psi, which may seem insignificant until it is realized that a pressure differential of 0.01 psi [69 N/m^2] is sufficient to cause standard air to flow at a velocity of 2,100 ft [10 meter per sec (m/s)] per minute.

Based on scientific facts as well as fire-service observations and analyses, it is apparent that problems at fires start with the conduction of heat, which can then be transferred by radiation and convection. These problems are solved only when conduction of heat no longer presents such a possibility, regardless of whether liquids, solids, or metals are involved. Until that stage is reached, however, activities must be carried out to cope with hazards created by radiation and convection. Ultimately the temperature of the burning material must be reduced by the proper extinguishing method so that conduction of heat no longer poses a threat from radiation and convection.

Exterior exposure hazards are best protected against radiation by wetting down the exposed surfaces, with water fog if feasible. This increases the density of wooden surfaces, thus increasing thermal conductivity, increasing the amount of heat absorbed in vaporizing the water, and delaying ignition. Even if the exposed surface is not wood, the suggested technique is advisable, since most radiant heat passes through drops of water limiting the effectiveness of water curtains. If the hazard from radiation is not severe, it may be preferable to alternately wet down the exposure's vulnerable surfaces and direct the stream at the fire. In ventilating the fire building, much heat can escape by radiation and convection and care must therefore be taken that the heat escapes as harmlessly as possible so that exposure hazards are not created or intensified.

If the rise of convection currents in a shaft is checked by some obstruction, and if the stoppage is complete and sufficiently prolonged, a positive pressure will build up and will be greatest immediately beneath the stoppage. Unless the pressure is relieved by ventilation or by the fire burning through the obstruction, heated gases and smoke will flow toward areas of lower pressure (Fig. 10-5). This flow generally results in a horizontal extension of fire (mushrooming) or, in some instances, in limited downward spread through vertical channels in which pressures are lower. Since the positive pressure decreases at lower levels, downward spread is limited. Model cases can be seen at fires involving clogged incinerators,

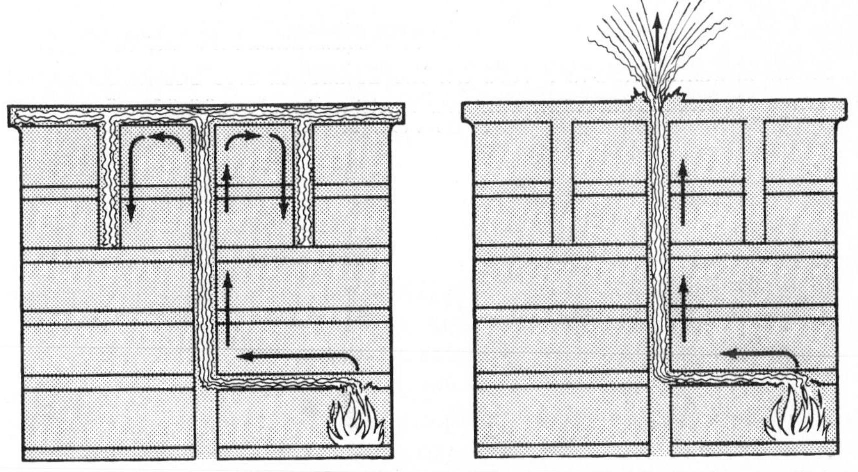

FIG. 10-5. Unless the pressure created by convection currents is alleviated by ventilation (right), heated gases and smoke will flow toward areas of lower pressure (left).

where smoke conditions are worst just below the obstruction and create less of a problem on lower floors, as indicated in Chapter 1.

On the basis of expectations indicated by these scientific principles, together with the fire service's own observations and analyses, principles for ventilating have been formulated—for example, to localize and thereby prevent horizontal spread of fire. Other ventilating principles to control the flow of convection currents are discussed in Chapter 18.

Wooden walls and partitions in frame dwellings fall easy prey to hot convection currents. In many dwellings the roof has a considerable overhang at the eaves. Roof overhangs, as well as those of small fixed awnings and over shutter and door openings, tend to pocket heated air currents and materially further the spread of fire by convection currents.

The spread of fire by convection can be modified by the location of the fire within a building. Fires in basements, cellars, and subcellars are usually slow-burning and smoky, due to oxygen deficiency, and tend to produce gases of comparatively low buoyancy and to retard heat transfer by convection. If the fire is located near drafty vertical channels, however, it can spread rapidly to the upper parts of the building. In any event, spread of the fire by convection would be retarded until the gases were sufficiently heated and buoyant for the required circulation.

Certain materials, such as oils, greases, fats, rubber, wax, tar, and some plastics, often produce large volumes of smoke that are largely unburned vapors, as indicated in Chapter 7. The heat of such smoke is low, as is its buoyancy. Therefore, regardless of where such fires are located, heat transfer by convection is retarded. If, on the other hand, the burning material is susceptible to rapid ignition and combustion, and sufficient oxygen is available, heat is produced quickly. This, in turn, increases the rate of combustion; it is well known that the rate of chemical change involved in the burning of ordinary combustible material approximately doubles for every 18°F [−7°C] rise in temperature. Thus, in such fires, highly heated and buoyant gases develop and circulate rapidly.

Effects of wind and weather on convection currents are discussed in Chapter 11.

Convection may abet the spread of fire, but it may also assist the fire department during the ventilation process as well as in the actual extinguishment of the fire. During ventilation, heated gases and smoke escape by convection, making it possible for firefighters to conduct interior operations more efficiently. Water supplied to the fire absorbs heat and thereby becomes a medium of helpful convection. Water in the form of fog is the most helpful, but it is possible that very hot convection currents will be created if the fog vaporizes to steam. Operations should be conducted so that the heat moves away from, and not toward, advancing fire personnel. Controlled ventilation (opening up the structure or using mechanical means to control the direction of the convection currents) is an urgent matter in such situations.

Convection and Smoke

Convection can occur without smoke. For example, heat is transferred in an air-conditioning system by convection. At fires, however, smoke is usually present. Smoke has been defined as the volatilized product of incomplete combustion of an organic compound, such as coal or wood, charged with fine particles of carbon or soot. Owing to the heat at fires, many chemical elements combine and others are liberated. The resulting product is a smoke or gas of varying mixture and undetermined toxicity. When many materials burn, free carbon is given off in the form of soot, or as minute particles that produce the blackness of smoke. Smoke behaves much like a gas; both are subject to some of the same natural laws.

Visual observation of smoke and convection currents may enable the commanding fire officer to deduce the following: (1) the kind of material burning; for example, white, yellow, or black smoke could indicate the involvement of phosphorus, gunpowder, or petroleum and petroleum products; (2) suspicion of arson; for example, if the smoke is caused by fast-burning materials strange to the occupancy; (3) location of the fire; (4) the extent of the fire, which at times can be gaged by the area of the thermal column rising above the fire; (5) the temperature developing; rapidly rising convection currents may indicate that high temperatures are developing and are quickly reducing the density of the air causing such an ascent; (6) whether the pressure within a structure is positive, as revealed by jets of smoke issuing from loose-fitting doors or windows, or negative, as indicated by the withdrawal of smoke toward the interior when a door or window is opened; (7) the severity of the exposure hazard; (8) whether ventilation will be difficult; for example, if smoke is rising very slowly; (9) when hose streams are hitting the fire, generating white smoke; (10) the direction and velocity of the wind.

It may be noted that item (6) provides information about the possibility of a back draft, or smoke explosion. In effect, the foregoing observations partly describe the effects of the primary factor of smoke conditions on other primary factors and on secondary factors.

HEAT AND SMOKE CONDITIONS

Before explaining the relationships of the primary factor of heat conditions, it should be noted that the terms *heat* and *temperature* are not synonomous. However, in the explanations offered about heat the implication is that temperatures are high enough to produce the results described. The Btu's released and transferred by conduction, radiation, and convection determine the temperatures.

Effects of Heat Conditions on Other Primary Factors

Heat can create a life hazard and has often driven trapped people to leap to their deaths. Heat makes rescue work both difficult and dangerous.

Heat reduces the density of the air and gases and makes them buoyant, resulting in rising thermal columns whose cross-section areas and rate of ascent may indicate the location of the fire and, to some degree, its extent on and after arrival.

Heat transferred by conduction, radiation, and convection can worsen interior and exterior exposure hazards, extend time required for control, contribute to structural collapse, make street-front positions inadvisable, and necessitate more than the usual requirements to operate.

Heat affects the inception of fire and therefore has a bearing on the time of origin. Fire grows and sustains itself by the heat it produces: heat raises the temperature, and the temperature, in turn, increases the reaction rate until heat is produced as fast as or faster than it is lost to the surroundings.

Specific temperatures can activate certain fixed systems and transmit an alarm of fire. In this context, heat can hasten time of discovery, alarm, and response.

Heat, because it raises the temperature and consequently the reaction rate, can affect the production of smoke and, thereby, visibility.

Heat can provide an ingredient that makes accumulated gases explosive with an inflow of air, resulting in back drafts or smoke explosions sometimes accompanied by structural collapse.

Heat produced can indicate other than a class A fire and the need for special extinguishing agents. For example, temperatures of over 5000°F [1650°C] are developed during combustion of magnesium, which has a melting point of about 1200°F [650°C].

Heat can have a bearing on the major aspects of occupancies by jeopardizing the human element and damaging the contents.

Heat produces the well-known updraft and an inflow of air to the seat of the fire, thereby affecting the direction and velocity of the wind.

Effects on Secondary Factors

The effects of heat conditions depend in large measure on the degree to which the fire building can be efficiently ventilated. These effects are more pronounced at fires in windowless high-rise buildings because of masonry materials that can absorb and retain much heat, especially when ventilation and other activities have to be attempted from the interior. Where roof ventilation to alleviate heat conditions is inadvisable because it would unnecessarily jeopardize personnel, the major objective would more likely be confine, control, and extinguish rather than extinguish.

Refer also to earlier comments in this chapter about coping with problems caused by radiation and convection of heat.

Effects of Smoke Conditions on Other Primary Factors

Smoke and gases may panic and very often overcome occupants, impair visibility, and make search and removal efforts arduous and perilous. Early smoke detection can save lives and give the fire alarm more quickly.

"Where there is smoke, there is fire" is generally true, but not necessarily in the sense that smoke pinpoints the location of the main body of the fire. Smoke can travel a considerable distance through vertical and/or horizontal structural channels before it becomes visible. Smoke first seen issuing from a roof cornice may be caused by a cellar fire. However, as stated before, the cross-section area and rate of rise of columns of smoke may indicate the location of the fire and, to some degree, its extent.

In other respects, the effects of smoke conditions are quite similar to those of heat transferred by convection. However, in view of the number and variety of gases that currently can be encountered at fires, it is suggested that each fire officer be provided with an authoritative source of information about the toxicity, explosive ranges, and ignition temperatures of hazardous gases. An explosimeter for taking readings is a necessary tool.

Effects of Smoke Conditions on Secondary Factors

Smoke conditions in the street can be so dense that it is difficult to see and properly evaluate pertinent primary factors, or even to locate hydrants. Within the fire building, much depends on ventilation. If it is ineffective, visibility will be poor, hampering all activities.

The manner in which toxic gases affect ventilation and other activities has been described in Chapter 7. The effects of weather conditions on smoke will be discussed in Chapter 11.

Recognizing and evaluating heat and smoke conditions is an ongoing process, starting on arrival and ending when there is no longer a life hazard for occupants or personnel, or possibility of further damage to property. To make proper determinations, officers must know why, how, where, and when heat and smoke conditions improve or deteriorate as a result of pertinent primary factors as well as of activities (secondary factors) undertaken. Among other things, such knowledge enables officers to anticipate the hazards created by heat and smoke conditions and thereby avoid or minimize casualties among personnel, as suggested in Chapter 4.

FLAME

In a recent article in *Scientific American,* flame was described as a subtle process wherein molecules are rearranged and give up radiant energy. Since then, much scientific effort has been spent to provide new information relative to flame and heat.

Progress has been slow, however. Flame is extremely complicated and is only partly understood. The high temperatures and short time periods involved in combustion of a given molecule of fuel make direct experimental studies very difficult.

When wood is sufficiently heated, it first undergoes thermal decomposition (destructive distillation), then combustible gases or vapors are evolved, which burn as the familiar flames. After these volatile products are driven off, the combustible residue is essentially carbon (charcoal). On further heating, the charcoal surface reacts with the oxygen of the air to produce considerable heat (glow) but usually little flame. During the first stages of a fire (evolution and burning of gaseous decomposition products), flames generally travel rapidly over the surface of wood or similar combustible solid materials. Glowing, flameless combustion of the charcoal embers follows.

In his article titled "Fire and Fire Protection," Howard W. Emmons, Professor of Engineering at Harvard University, makes the following statement about steps leading to ignition and burning of materials.

> As the temperature of a material rises, its atoms and molecules acquire increased kinetic energy. When the atoms of a solid vibrate too violently, the chemical bonds between them are broken and smaller molecules are evolved. This process, whereby a complex solid is thermally decomposed into simpler solids or liquids and ultimately into gases, is pyrolysis. Pyrolysis of a material is the essential first step in the ignition and burning of the materials from which buildings are made.[1]

The luminous flames commonly associated with fires are sometimes described as "diffusion flames." This is because the combustion that produces them occurs in the zone where the unburned vapors or decomposition products intermix with, or diffuse into, the surrounding air as they evolve. Because the rate of mixing, among other things, influences the characteristics of diffusion flames, they differ in many respects from flames that propagate in gases, vapors, or dusts premixed with air before ignition.

[1] Howard W. Emmons, "Fire and Fire Protection," *Scientific American*, vol. 231, 1975, pp. 21–7.

COMBUSTION

Combustion can be defined as a heat-producing process that may take place at almost any rate and at almost any temperature. It may be as slow and mild as the rusting of iron (not all textbooks agree on this point) or as violent as the explosion of hydrogen with oxygen, which involves temperatures of 3000°C. Combustion does not necessarily involve oxygen, but the fire service is mainly concerned with combustion that does require oxygen. Some metals, such as magnesium, may burn in nitrogen, and certain substances, such as hydrazine (N_2H_4), hydrogen peroxide (H_2O_2), and ozone (O_3), can burn in the absence of any medium except themselves; that is, at sufficiently high temperatures they decompose and give off heat without combining with another substance. Ozone, in burning, gives only a single product: two molecules of ozone yield three molecules of oxygen plus heat. Other examples of combustion without oxygen are the hydrogen-fluorine torch and the burning of hydrogen in an atmosphere of chlorine.

When combustible organic material, such as wood or flammable liquids, burns freely in an atmosphere containing plenty of oxygen, the volatile products of combustion are mainly carbon dioxide and water vapor. In an atmosphere that is deficient in oxygen, appreciable quantities of carbon monoxide, smoke, and other products of incomplete combustion may also be present.

The chemical composition of common woods (dry basis) is essentially as shown in Table 10-1. Figure 10-6 shows the stages of burning of a piece of wood.

Many common flammable liquids consist of carbon and hydrogen chemically combined in various proportions (hydrocarbons) or of carbon, hydrogen, and oxygen in combination (alcohols, aldehydes, ketones, organic acids, and ethers).

During combustion, hydrogen and carbon react with oxygen in the air to produce carbon dioxide and water vapor. The chemically combined oxygen in these materials is not available for the usual combustion reactions, but it is available in a substance such as cellulose nitrate, which

TABLE 10-1

	PERCENT BY WEIGHT
Carbon	48.0–51.0
Hydrogen	5.9–6.3
Oxygen	39.0–45.0
Nitrogen	0.1–1.0
Ash	0.2–3.4

is capable of undergoing exothermic decomposition. (*Exothermic reactions* are accompanied by the liberation of heat, as opposed to *endothermic reactions,* which are accompanied by the absorption of heat.)

What is fire? It is any combustion intense enough to emit heat and light. It may be quietly burning flame or a climactic explosion. Fire grows and sustains itself in the reacting medium by the heat it produces: heat raises the temperature, and the temperature, in turn, increases the reaction rate until heat is produced as fast or faster than it is lost to the surroundings. However, heat is not always the sole, nor even the principal, agent that initiates flames and explosions. Fire may be started by a chem-

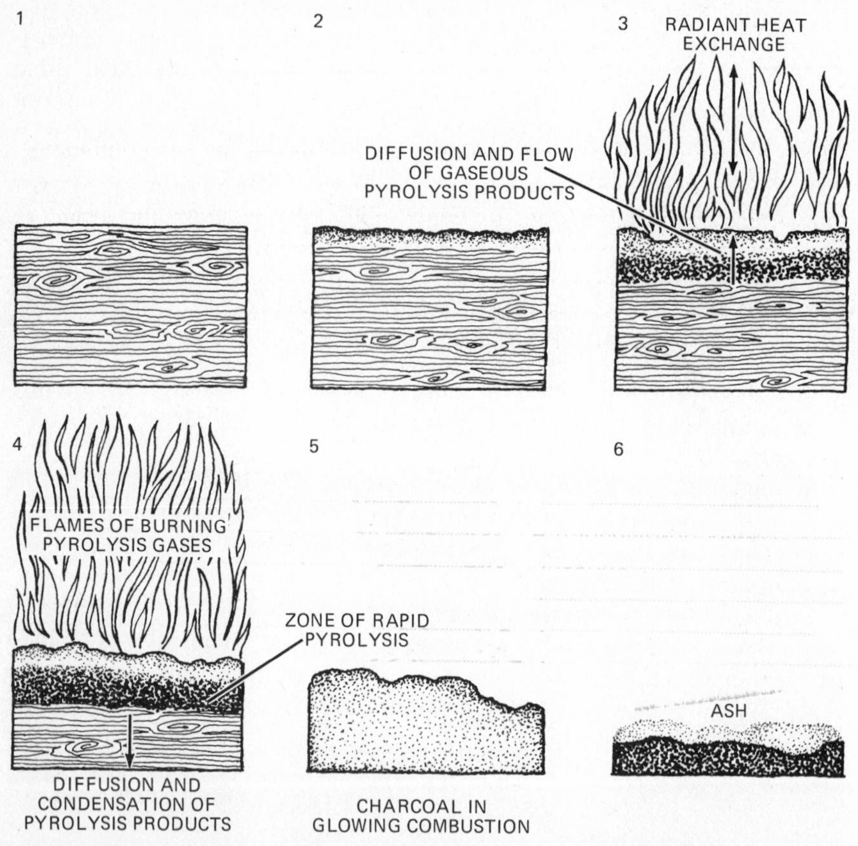

FIG. 10-6. Stages of Burning. A piece of wood (1). The wood is heated and the surface blisters (2). Active pyrolysis begins and moves slowly through the wood, with a radiant heat exchange between the wood, flames, and atmosphere (3). Some of the pyrolysis products (gases) escape through cracks in the char layer to mix with air and burn (4). After the pyrolysis products are driven off, the residue (charcoal) continues to glow (5). A small amount of minerals remains as ash (6).

ical process known as *branching of reactions chains*. The study of branched reactions is obviously of interest and of use to the fire service, but it is only briefly mentioned here. Branched reactions are similar to the nuclear chain reaction of the atomic bomb. (See Fig. 10-7.)

The Combustion Wave

A combustion wave is a propagating zone of intense burning. In gas-phase combustion, a spark starts the chemical reaction in a small zone of the fuel mixture; as this flames up, heat flows into the adjoining layer of unburned gas, and so on, like a wave. The combustion wave is divided into three zones: (1) the preheat zone, (2) the reaction zone, and (3) the completed-reaction zone (Fig. 10-8).

In the preheat zone, the temperature of the fuel and air is raised by the excess enthalpy (excess heat content) of combustion. The spark starts the chemical reaction, and the temperature rises.

In the reaction zone, the mixture is further heated by the combustion of the fuel. When a hot enough part of the wave reaches it, the gas breaks into rapid chemical reaction, or flame. This burning generates heat, at a rate that first rises and then declines as the fuel is used up.

In completed-reaction zone, the fuel is completely burned. The heat still being generated passes forward to the advancing wave front as much heat as it absorbed before it began to liberate heat. In this way a com-

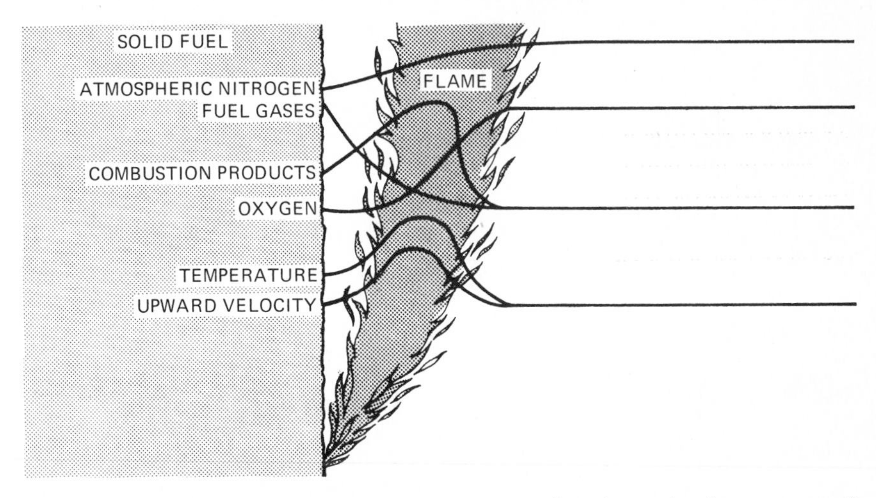

FIG. 10-7. Processes of combustion. In a vertical fuel surface the combustion processes all occur in a thin layer near the surface. The curves indicate by their height where the various processes are most intense. Heat from the flames is conducted to the surface, producing fuel vapors that diffuse away from the surface. They meet oxygen that is diffusing inward and react to sustain the flame. Buoyancy of hot gases moves flame and combustion products upward and maintains a fresh supply of air near the surface.

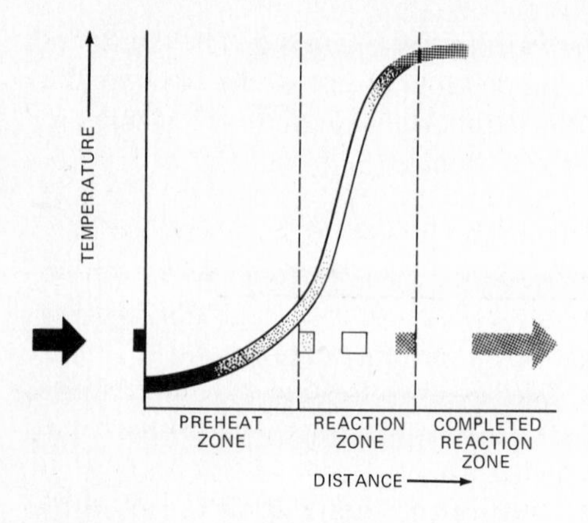

FIG. 10-8. The combustion wave is divided into three zones. In the preheat zone the temperature of the fuel and air is raised by the excess enthalpy of combustion. In the reaction zone the mixture is further heated by the combustion of the fuel. In the completed-reaction zone the fuel is completely burned. The small rectangles suggest changes in volume with burning.

bustion wave continuously "borrows" and "repays" heat from its excess enthalpy.

Combustion-wave thickness and temperature gradient vary greatly, depending on the fuel: in a mixture of hydrogen and fluorine, the temperature rises about 4500°C in the space of about one-thousandth of an inch.

The velocity at which the wave propagates is called the *burning velocity*. Burning velocities range from a few inches per second for weak hydrocarbon-air mixtures to several hundred times this value for mixtures such as hydrogen and fluorine. The difference in burning velocities is important to firefighters because in simple combustion, involving gas-phase combustion alone, extinguishment can be readily effected by interrupting flame propagation in a zone for as long a time as it normally takes a flame to propagate through the zone.

If an explosive mixture flows continuously from an orifice, a combustion wave can, under suitable conditions, propagate against the stream at a rate that matches the flow velocity of the stream. This is seen in the stationary flame of the kitchen gas range. In the kitchen range, a jet of fuel gas, after entraining air, flows under pressure from numerous orifices in the burner head. The flame is stabilized by the controlled flow distribution and the quenching effects.

The combustion wave anchors itself at an equilibrium position whose distance from the burner rim depends on the rate of gas flow. When the velocity of the gas flow is reduced to less than the burning velocity, the flame theoretically flashes back into the piping. In large oil and gas installations, tanks are provided with *flame arresters* to guard against such a possibility. When the velocity of gas flow exceeds the burning velocity everywhere in the stream, the flame blows off. It is well known that gas- and oil-well fires, again involving gas-phase combustion alone (assuming no exterior extension), can be extinguished by the explosion of TNT as near to the center of the fire as possible. The explosion interrupts flame propagation, and it seems reasonable to speculate that the shock and concomitant wind increase the velocity of the flowing gas over the burning velocity of the gas and the flame blows off.

Combustion waves lose heat to solid bodies with which they come in contact. A solid, therefore, quenches burning in a gas for some distance from it. If the diameter of a duct is made small enough, an explosive mixture cannot burn in it. The critical quenching diameter depends upon the composition of the fuel mixture, the pressure of the mixture, the temperature, and the shape of the duct. A mixture of hydrocarbons and air at very low pressure will not burn in a duct as much as several inches in diameter, but a mixture of oxygen with hydrogen or acetylene can propagate a flame in a fine tube as small as one-thousandth of an inch in bore.

The minimum spark energy required for ignition depends upon the composition of the explosive mixture, the pressure, and the temperature. Certain weak mixtures may require as much as a calorie, whereas hydrogen and oxygen in proper ratios can be touched off by less than one-millionth of a calorie—a spark far less energetic than the static electricity a human being generates by walking on a carpet on a dry day. It is possible for a low-energy spark to pass through an explosive mixture without igniting it, even when the temperature in the path of the spark is of the order of several thousand degrees. This can occur if the potential combustion wave is not given a sufficient initial boost of heat by the ignition source; as a result, the temperature drops so much that the wave dies.

REVIEW QUESTIONS

1. For practical purposes, radiation at fires can be defined as the transfer of heat energy in the form of electromagnetic waves. How does the escape of radiation at fires affect flame temperatures? How can heat radiating from convection or thermal columns at conflagrations affect fire operations? What should be considered in protecting window glass exposed to radiation? Why is proximity of exposure such an important factor insofar as radiation is concerned? What influence can the primary factors of wind direction and velocity have on radiation intensity received at exposures?

2. Why are the hazards from heat radiation so much greater in a nuclear explosion than in a conventional (TNT) explosion?

3. For practical purposes, conduction at fires can be defined as the process by which heat is transferred within a material from one particle to another or from one material to another by direct contact. With

what does the amount of heat transferred by conduction vary? What is thermal conductivity? How does it increase in a given material? What is a major problem at fires, so far as conduction of heat is concerned?

4. What must ultimately be done to eliminate the dangers caused by the transfer of heat at fires?

5. Why does a knowledge of conduction of heat help protect exterior exposure against the radiation of heat?

6. Convection can be defined as the transfer of heat by a circulating medium in the gas or liquid state. Give some examples of convection in the gas state, and in the liquid state. What is the usual tendency of convection currents in the gas state? Why?

7. How can obstructions affect the direction taken by convection currents traveling through vertical shafts at fires? How does the basic scientific principle involved help the fire service formulate principles for ventilating?

8. How is the transfer of heat by convection affected by primary factors? By secondary factors?

9. What can officers deduce from visual observations of smoke and convection currents at fires?

10. How can the primary factor of heat conditions affect other primary factors at fires? Secondary factors?

11. How can the primary factor of smoke conditions affect other primary factors at fires? Secondary factors?

12. Define flame. Why are experimental studies of flame so difficult? How does flame evolve when wood is heated sufficiently?

13. Define combustion. What kind of combustion is the fire service mainly concerned with?

14. What is fire? Discuss the process involved. What is pyrolysis?

15. What is a combustion wave? What is meant by its burning velocity? What is the significance of the burning velocity to firefighters?

11

Visibility and Weather

VISIBILITY

Visibility here refers to the range within which an officer with normal sight can recognize the primary factors and properly supervise the pertinent secondary factors. It is deemed more practical to explain how poor visibility rather than good, visibility affects other factors.

Effects on Other Primary Factors

Visibility that is impaired by smoke, snow, rain, sleet, fog, or time of the fire's origin may delay discovery of the fire, transmission of alarm, and response of units, and thereby cause the favorable or unfavorable chain of events described in Chap. 9.

Effects on Secondary Factors

Impaired visibility makes it more difficult to recognize and properly evaluate pertinent primary factors, thereby hampering decision making and increasing the possibility of error. Poor visibility is a serious handicap in searching for trapped occupants, in determining the order in which exterior exposures should be covered, or in carrying out any fire activity. In addition, supervision becomes a more critical matter because of the increased dangers for personnel.

Effects of Blackouts

Prolonged power failures covering wide areas uniquely affects the life hazard, street conditions, response of units, occupancies (especially the human element), auxiliary appliances, and requirements to operate.

A serious life hazard can develop because the public often becomes unusually apprehensive. In some occupancies people become panic-stricken, particularly when they are trapped in stalled elevators or subway trains, in theaters or, in hospitals or institutions, are plunged into darkness, and so forth. Confused and congested vehicular and pedestrian traffic delay responding fire apparatus, compounding the problems to be faced on arrival.

Many trouble bells in alarm systems will be activated, increasing alarm incidence to the point that only skeleton assignments can respond initially. Occupancies such as hospitals may require emergency lighting or standby pumpers to supply fixed systems if no auxiliary power sources are available. The lighting situation could be particularly acute in operating rooms.

More fire companies may be needed, necessitating the use of spare apparatus and recalling off-duty personnel. Some fire companies can be located strategically out of quarters, thus exerting a calming influence on the public by being highly visible as well as obviating the need to open fire-house doors manually and facilitating emergency response.

The following aspects of communication should be considered in planning for such emergencies:

1. Provide for automatic operation of auxiliary gasoline, propane, diesel, or other power sources to ensure uninterrupted alarm service.

2. Restrict the use of communication devices to avoid unnecessarily clogging contact with the dispatcher. For example, delegate more authority to line or staff officers who can handle many emergencies without contacting the dispatcher. Such officers and units under their command can keep in touch by portable communicating devices.

3. Provide vocal alarm systems in company quarters. There may be telephone delays.

4. Provide for contact with the state fire network so that official information can be received and acted upon quickly.

It is obvious that the disadvantages of poor visibility will be maximal at fire operations during a blackout.

WEATHER

Snow

Falling snow can reduce visibility, delay discovery of fire and transmission of alarm, and worsen street conditions, which slows response and therefore tends to make fires more extensive on arrival. It can also hinder the use of hydrants and hose lines, placement of apparatus and ladders, and the movement of personnel, thus hampering efforts to rescue occupants and control the fire. Such handicaps magnify exposure hazards, prolong fire operations, and increase requirements to operate. Most of the foregoing observations also apply to conditions following a snowstorm.

In some cases snow has prevented apparatus from getting within effective range and has contributed in large measure to heavy involvement and collapse of the fire building. It has also intensified exterior exposure hazards. Slippery footing, of course, is a constant menace.

Effects on Secondary Factors. In addition to delaying the use of hydrants, stretching lines, and placing apparatus, snow can impede efforts to ventilate. For example, a snow-covered roof is difficult to open up and thus increases chances of a back draft or smoke explosion. Decisions to initiate interior operations should be considered accordingly. On the other hand, snow melting on the roof may indicate hot spots, where openings should be made. Supervisors must be constantly aware of the unique hazards caused by snow.

Low Temperature

Reference is to temperature below the freezing point of water.

Effects on Other Primary Factors. Low temperatures may cause frostbite; frozen hydrants; ice-covered hose lines, apparatus, and equipment; and slippery footing. It may freeze water, thereby expanding it and bursting sprinkler piping systems without warning until a thaw occurs and causes transmission of a water-flow alarm.

Low temperatures can worsen the plight of occupants fleeing from their homes in flimsy night clothing. It also generally means that windows will be closed, which may contribute to a delay in discovery, alarm, and response. Closed windows may temporarily reduce the exterior hazard but can also create conditions in the fire building conducive to the development of a back draft or smoke explosion.

Low temperatures increase fire incidence because more heating devices are used. Consequently, the possibility of having to fight simultaneous fires is greater. Likewise, the possibility of structural collapse increases because less water runs off.

Effects on Secondary Factors. Extremely low temperatures retard the initial development of fire, but once a fire has started they impair the efficiency of the operation in general in that they necessitate such things as heavy, encumbering clothing, which slows actions.

In addition, when not in use, nozzles on hose lines must be kept cracked to avoid freezing. Hose may be coated with ice, causing it to stiffen and reducing maneuverability. Care must be taken to avoid damaging the hose when it is placed on apparatus.

Frozen hydrants and the formation of ice on apparatus and equipment may necessitate the use of thawing devices during the extinguishing as well as the finishing, or "taking up," phase. Snow accentuates the disadvantages of low temperatures.

High Temperature

Reference is to temperatures in the eighties or ninties.

Effects on Other Primary Factors. A major effect of high temperatures is that they make personnel prone to heat exhaustion. Fire officers can take steps to avoid or minimize this hazard by relieving personnel at shorter intervals, and transmitting additional alarms, if necessary.

In addition, windows are usually open during the warm season, thus supplying oxygen that will accelerate combustion, worsen the life hazard for all concerned, the extent of fire on and after arrival, heat and smoke conditions, and, in general, other primary factors.

Open windows also have favorable effects: Discovery, alarm, and response are likely to be quicker, and a back draft or smoke explosion is unlikely to occur.

Street conditions are better and daylight is longer in the summer months, which affects time of response, visibility, and activities accordingly.

Prolonged hot dry spells may strain water supplies in surburban areas and cities because of the watering of lawns and the use of hydrants for street showers.

Humidity

Relative humidity is the ratio of the quantity of moisture (water vapor) actually present in the air to the greatest amount possible at the same temperature. Water vapor is part of the air: the higher the humidity, the lighter the air per unit volume. Humidity, therefore, has a controlling influence on wind and air-mass movement. The average prevailing humidity in any section determines the moisture content of wood and other combustible materials in roofed structures. The average air-dry seasoned lumber used in building construction in the United States has a moisture content of 12 percent. The moisture content of wood in closed, artificially heated buildings is less. Before wood can burn, this moisture must be driven off as steam, but since this requires relatively little heat it has a comparatively minor influence on the rate of fire propagation.

The humidity of the air, however, has a distinct bearing on the development and propagation of fire. High humidity combined with inversion condition at the inception of a fire retards the establishment of a draft or pillar of heated convection currents. Inversion refers to stratification of layers of air having different temperatures. The line between the two air strata is referred to as the line between the floor of the upper stratum and the ceiling of the lower stratum. This line forms a barrier to convection currents until such time as temperatures develop to cause a break in the ceiling (a lapse).

Effects on Other Factors. Fires under high humidity and inversion condition are often characterized by dense smoke and poor visibility. High humidity, and a consequent high moisture content in material exposed to fire, make it more difficult for a vigorous fire to become established but do not slow its spread once it is well started.

Rain greatly reduces the probability of fire spreading from building to building because it wets exposed sides of structures and provides some protection against radiant heat. Rain droplets cool convection currents and help to extinguish flying sparks and embers that may land on roofs, but they do not intercept the major portion of radiated heat. Steaming sections of a roof during a fire might indicate the location of hot spots, where openings should be made.

Wind

Wind velocity is an important factor in the initial stage of fire; 2 mph [1.2 kilometers (km)] is considered optimum at this stage. Under favorable conditions, it is possible that fires of moderate proportions can be controlled by determined defensive measures if the wind velocity is no more than 15 mph [9.3 km], although wind also affects the use of hose streams (Chap. 6). As velocity rises from 15 to 30 mph [9.3 km to 18.6 km], the rate of fire propagation from building to building increases enormously; at 30 mph even a relatively minor blaze, involving only two or three dwellings, may constitute a serious threat to all the downwind area. The exposure hazard to the lee of the fire is more severe, and it becomes increasingly difficult to set up a successful defensive line there. The spark-and-ember hazard is magnified, necessitating protecting lines well in advance of the fire.

Ground winds of more than 30 mph are conducive to the development of a *conflagration,* which is characterized by pillaring of heated air and products of combustion, rising several hundred feet [meters] in the air or more, depending upon wind velocity. The higher the velocity, the more the pillar slants from the vertical in the direction of the wind. In a conflagration, velocity on the lee side is materially reduced and frequently reversed in direction by the forces that produce the pillar of heated gases. This phenomenon may sometimes be noted at large brush fires and can possibly be utilized in making a firebreak.

Fire Storms. A fire storm may develop in the absence of ground winds sufficiently strong to promote a conflagration (Fig. 11-1). Data from wartime experiences indicate that an area less than 1 square mile [0.6 km] is probably incapable of sustaining a fire storm (Fig. 11-2). In addition, building density (the total ground area of buildings divided by the total area of the zone) must be greater than 20 percent.

A fire storm is basically a wind storm. It may produce rain if the rising columns of hot smoke meet a stratum of cold air, causing the moisture in the air to condense around motes of soot and fall in large black raindrops. To the fire service, however, the fire storm is comparable to a conflagration in size but is definitely different in other ways. It results from the merging of numerous smaller fires into one massive inferno and is much more likely to be a wartime than a peacetime phenomenon. Minimum area and building density are essential, and absence of a strong ground wind is necessary. The thermal column (convection currents) rises almost vertically (that of a conflagration is bent over by the prevailing wind), and the rising column creates a powerful centripetal force that draws air along the ground at velocities that may exceed 100 mph [62.1 km], toward the expansive low-pressure area at its base. The true fire storm should not extend beyond its perimeter because of the centripetal pattern of the air currents created. High temperatures prevail, and combustible building material and plant life are consumed, with only brick and similarly resistant walls and charred trees remaining.

FIG. 11-1. A conflagration.

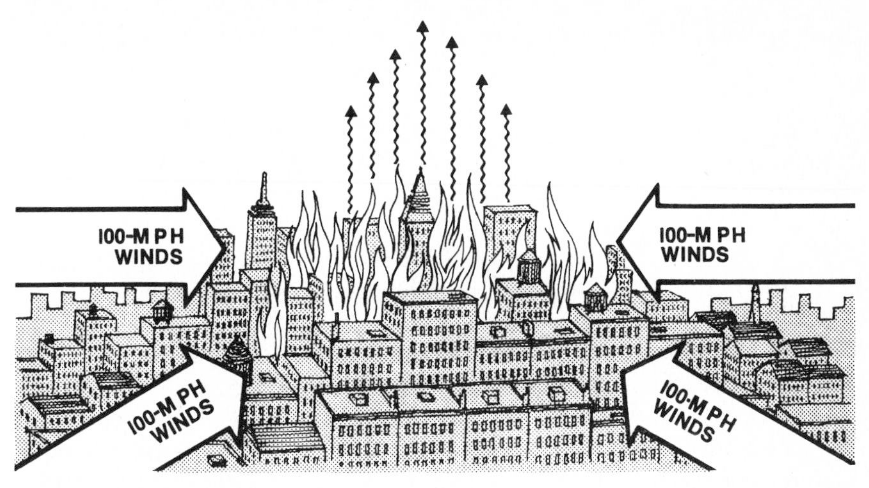

FIG. 11-2. A fire storm.

Underground shelters offer little or no protection, as judged by the Hamburg fire storm of July, 1943. Many persons, even though shielded from the searing heat, were found dead of oxygen deficiency. Conventional firefighting tactics were completely ineffective in coping with such a fire—which is certainly no reflection on the fire service or the civilians in England and Germany who performed so well at fires created by less extensive incendiary attacks.

Nuclear attacks add new dimensions to the problems of the fire service: higher temperatures, greater and more widespread devastation by fire, and initial and residual radiation. In Hiroshima, the effects of an atomic bomb—comparatively small by present standards—wiped out most of the fire equipment and killed or injured 80 percent of the firefighters on duty. Those that remained were ineffective. At Nagasaki, the damage to the fire department was relatively minor but their efforts were equally ineffective because of the failure of the water supply. Elaborate plans to cope with an incendiary attack were completely disorganized.

Prospects in the event of a nuclear attack are gruesome. Scientists are already discussing the feasibility of developing a system which could be ready by 1980 and would make possible the delivery of warheads over intercontinental ranges (about 1,500 miles or 2,400 km) with an average error of 40 ft [12 m]! Because 50 or 100 targets could be hit simultaneously, the fire service could not begin to cope effectively with the potential holocaust if it used only conventional techniques.

In the United States, less than 1 percent of the more than 20,000 fire departments have reasonably satisfactory staff services for personnel, training, fire prevention, maintenance, and other functions, in case of such an event, and in the remaining departments, such services have been classified as ranging from rudimentary to nonexistent. These were the conclusions drawn by qualified analysts, as indicated in *A Study of Fire Problems* published by the National Academy of Sciences-National Research Council in 1961. It is possible that Civil Defense, by means of aerial infrared mapping, may be able to advise about affected areas—but who is going to put such fires out—and how?

The fire service can—and in all likelihood will—be effective if the fires are reduced to manageable proportions. This may be possible if scientists can develop an effective antidote for the fires created by nuclear weapons. The existence of such an antidote is suggested by tests conducted by the Russians in the Lake Balkash area in September of 1953, as reported by the British Intelligence Digest. During such tests, jet-fighter-bombers flew over a test area at about 2,000 ft [600 m], seeming to eject a light spray. The result was an intense, killing cold that lasted for half an hour. All vegetation was killed, and trees became so brittle that they burst from the ground and froze solid. Obviously, such a spray has a powerful cooling and extinguishing effect, and it is hoped that planes circling the perimeter of a fire storm could effectively utilize the winds created by such a fire. The development of such a form of extinguishment would necessitate the organization of a special air force unit and coordination between Civil Defense and the fire service. The air force unit could operate at strategically located bases, at safe distances from likely targets.

Effects of Wind on Other Primary Factors. Wind can affect the time it takes for a fire to start, its rate of growth, and thus the extent of fire on and after arrival, heat and smoke conditions, and hazards to occupants, personnel, and exposures. If the wind is blowing toward the rear of the fire building, it can affect the location of the fire and delay discovery, alarm, and response. This is more likely in an unoccupied building, and during the night.

Wind direction and velocity can be decisive factors in maximizing the exterior exposure hazard, at times to the proportions of a conflagration. These factors in turn affect requirements to operate and duration of the operation. Thermal columns driven and bent over by the wind may carry sparks and embers downwind for miles and radiate heat downward from the convection currents, thereby starting other fires.

Wind, in the form of an inflow of air, can provide the oxygen required to ignite hot, accumulated combustible gases in a fire area, causing a back draft or smoke explosion.

Wind entering through windows, particularly where curtain boards (draft stops) are not provided, can dissipate heat and reduce the effectiveness of sprinklers.

Effects on Secondary Factors. Wind can affect the priority for covering exposures, ventilating, using streams, and placing aerials.

High winds may necessitate the use of solid rather than fog streams. However, the same winds may make fog preferable when personnel are operating from the windward side of the fire.

REVIEW QUESTIONS

1. How can extensive, prolonged power failures and resulting blackouts affect primary factors?

2. What aspects of communication should be considered in planning for emergencies caused by blackouts?

3. How can snow affect other primary factors? Secondary factors?

4. How can cold weather affect other primary factors? Secondary factors?

5. How can humidity affect other primary factors? Secondary factors?

6. How can wind affect other primary factors? (The manner in which it affects secondary factors is explained in the appropriate sections of the text.)

12

Requirements to Operate

The term *requirements to operate* pertains to the water, apparatus, equipment, personnel, and special extinguishing agents required and available for an effective fire operation. All these items form a balance. If the water supply required is available, for example, the need for additional personnel, apparatus, and so forth is ordinarily decreased. If the water supply required is not available, the other needs are ordinarily increased. The same "rule" holds true for each of the other primary factors comprising requirements to operate.

Individually and collectively, requirements to operate have such an obviously favorable effect if available, and unfavorable effect if unavailable, that explanations of their relationships with other primary factors appear redundant. It is not always easy, however, to correctly determine *how much* water, apparatus, equipment, personnel, and special extinguishing agent will be required at a fire.

On arrival, officers estimate requirements to operate according to the primary factors pertinent in selecting objectives and activities. Thereafter, the commanding officer must carefully check the effects of activities on each other (such as ventilation on the advance of hose lines) and on the pertinent primary factors. If personal observation or communication with subordinate officers indicates that the action plan (activities) is not measuring up to expectations, the initial estimate may have to be revised. The matter of availability will be considered later in this chapter.

Incorrect evaluations of primary factors, notably extent of fire on and after arrival, lead to incorrect estimates of requirements to operate. As a result, the commanding officer may be late in calling for help, may prematurely declare fires under control, and send units back to quarters.

Some officers prefer to call one or two additional units at a time instead of ordering a second or greater alarm. This practice may have some merit when simultaneous fires are in progress in a community, assuming officers sending such calls are highly skilled in evaluating the factors that are pertinent in determining requirements. However, when calls for individual units have to be *followed* by greater alarms, there can be a harmful delay in the arrival of needed help. Until control of the emergency is clearly established and while the issue is still in doubt, it is good practice to order a greater alarm rather than to call for individual units. A specific quantitative analysis of requirements is not yet available for complex fire operations, and apparatus can always be contacted by radio if not wanted.

Some officers are reluctant to call for help because they feel it reflects unfavorably on their ability. Others may be uncertain about the validity of their decisions and therefore apprehensive of criticism. Still others "don't want the high brass breathing down my neck at greater alarms." These faulty attitudes result in procrastinating and end in the delay of the transmission of greater alarms. Positive attitudes should be taught and recognized in training programs. Such training programs can be helpful if they stress the policies that help should be called even when the need is probable rather than actual and that one can give the fire a better fight with fresh personnel than with exhausted personnel.

Calculating requirements is one thing but availability is something else. Fundamentally, availability depends on the ability and willingness of a community alone or by mutual-aid programs to provide adequate fire protection.

Officers in different communities have no choice but to evaluate pertinent primary factors and select objectives and activities on the basis of the available requirements to operate. Thus officers in smaller communities, because of limited firefighting facilities (requirements available), more often have to choose confine, control, and extinguish as the major objective and operate from the exterior of the fire building, whereas at similar fires in larger communities, because of greater firefighting facilities, officers can logically choose extinguish as the major objective and initiate an interior operation—assuming in each case that there is no life hazard for occupants. The choice depends on the specification that personnel are not to be jeopardized unnecessarily. It should be remembered, however, that the format of the action plan does not change from fire to fire or community to community. Only the primary factors change.

WATER

It is important to know how to use available requirements.

To use available water most effectively and thereby limit the need for other requirements, officers should know the benefits and limitations of fog and additives, when and how to employ master streams, and the principles governing the use of hose streams, apart from the effects of ventilation, selecting hydrants, mechanics of stretching hose lines and so forth.

Fog

To exert any practical extinguishing effect, the droplets of a water spray must have sufficient carry (trajectory) to travel from the source of the spray to the heated area surrounding the fire, and the droplets arriving there must arrive with sufficient kinetic energy (velocity) to overcome the turbulent hot gases moving upward and away from the fire. Under comparable conditions of pressure and temperature, the density of water vapor is only 0.62 that of air. In some cases, the updraft of heated gases is less of a problem. For example, the inflow of gases at the base of the fire can be utilized to carry water vapor and water droplets to the seat of the combustion.

In order that the droplets may have the necessary reach or carry, and to overcome the usually prevailing updraft from the fire, it has been calculated from the extensive and useful research on the characteristics of water droplets that the optimum diameter for extinguishment is of the order of 350 microns (a micron is a very small unit of length, equal to 0.00003937 in or 0.001 mm). In some experiments, droplets smaller than 350 microns achieved extinguishment. Various factors, such as ventilation, confinement of the fire, kind of material burning, gravitation, prevailing drafts, distance to be traveled by the droplets, and method of application (vertically downward or horizontal), may affect the optimum diameter.

The establishment of a means of describing or characterizing water sprays used for extinguishing fires is an important research development. More scientifically manufactured fog and spray nozzles, designed to deliver water droplets of optimum diameter appropriate for various fires, may result.

Advantages. Fog can be used effectively with master-stream appliances, with wetting agents, and also with foam. It has greater and quicker absorption of heat per gallon than plain water. Water has its maximum cooling and extinguishing effect when applied as a cold fog and evaporated into steam. One pound [.5 kg] of water applied at 60°F [15.5°C] and evaporated into steam at 212°F [100°C] has a cooling effect of 152 Btu (212 − 60) + 970.3 Btu [100°C − 15.5°C + 970.3 Btu], the specific heat plus the latent heat. *Specific heat* is the heat (thermal) capacity of a substance and is the number of Btu required to raise the temperature of one pound of a substance 1°F, or the number of calories required to raise 1 g 1°C. Water is a valuable extinguishing agent mainly because its specific heat is higher than that of other substances. *Latent heat* is the quantity of heat absorbed or given off by a substance in passing between the liquid and gaseous or the solid and liquid states, and it is measured in British thermal units or calories per unit weight. The latent heat of water (normal atmospheric pressure) at the freezing point (32°F) [0°C] is 143.4 Btu per pound; at the boiling point (212°F) [100°C], it is 970.3 Btu per pound. It is substantially greater than that of most other common substances.

Theoretically, fog applied at 100°F [37.8°C] would more quickly reach the temperature at which the benefits of its latent heat are realized than would fog applied at 50°F [10°C]. Quantitatively, however, 50 Btu less per pound of water used would be absorbed by the fog applied at the higher temperature.

Fog causes less water damage to property and the contents of fire buildings. This has a favorable effect on the public, as the salvage problem is simplified and business can be resumed and homes reoccupied more quickly. In addition, fires are extinguished more quickly.

Further advantages of fog are as follows: it reduces smoke toxicity; one man can handle nozzles since fog exerts considerably less pressure reaction at the nozzles than solid streams with comparable pressures; there is less drain on the public water supply; directing the fog toward the exterior through selected openings will more effectively ventilate partially enclosed areas; there is less chance of destroying arson evidence with fog; it can be used effectively on some class B fires, exerting both a smothering and a cooling effect; it increases the effectiveness of apparatus with booster equipment, stretching the amount of water in tanks; it can be used successfully in fixed systems; it protects the firefighter who is advancing or operating lines in hot spots.

In conjunction with the Navy all-service nozzles, fog can be used very effectively with applicators of various sizes. For example, it can be used to combat fires in ceiling spaces. Openings can be made by regular 6-ft [1.8 m] or larger hooks for the insertion and use of applicators. The length of the hook to be used varies with the height of the ceiling. With this method, ceilings do not have to be pulled at an initial stage. Openings must be made at strategically selected spots, and the application of fog must be controlled as to time and volume to prevent the ceiling from being blown down by the expanding steam and to allow the expanding steam to work effectively on the fire in the ceiling area. Ceiling openings should follow a systematic pattern in order to cover the area as advantageously as possible. An additional and important advantage of this operation is that the fog, as it vaporizes to steam, follows the fire through hidden channels, exerting a powerful smothering and cooling effect and thereby minimizing the possibility of vertical and horizontal extension. Solid streams can be held in readiness if deemed advisable and, as a matter of fact, are available simply by removing the applicators. Pulling ceilings and then hitting the exposed fire with solid streams is a slower and less effective method, and the possibility of vertical and horizontal fire extension is greater.

Applicators can also be used through openings in walls. The openings need be only large enough to insert the head and pipe extension (preferably 45° rather than 90° angle) of the applicator and to permit sufficient freedom for effective use; they should be placed to ensure the fog reaching areas in which vaporization will occur. Such openings can be readily made in walls or partitions by rescue companies employing concrete breakers or similar equipment. Applicators may then be inserted from a room or structure adjoining the fire area and fog applied to otherwise inaccessible spots. Care must be exercised to warn personnel operating in the interior of the fire structure to withdraw in order to avoid injury from the expanding steam. This operation is particularly appropriate where one or more floors (lower) are involved in such a way that roof ventilation is of little or no assistance, and an interior attack is made unusually difficult by heat, smoke, and restricted access.

This technique is recommended as a supplement to an interior attack already under way; it is not offered as a substitute. It differs from the technique of exclusive use of fog from the exterior, in that the fire has been located, the building has been vented, thus minimizing the possibility of back-draft explosion, and the openings for fog injection have been located with definite knowledge that the fog will be applied where it can be effective and not be nullified by interfering walls and partitions.

The use of applicators as recommended (through subdividing walls) may very well prove to be a worthwhile alternative to the continuous use of master streams at exterior operations. If the fog can be applied effectively, the need for heavy streams may be reduced more quickly.

Fog from applicators can also be used with good results on fires involving cocklofts, voids, unventable rooms, ducts, and chimneys. Low-velocity fog from applicators is recommended for fires in occupancies that contain radioactive materials. Fog applied from applicators has successfully extinguished fires in the lining of chimneys serving large oil-burning systems in tall buildings. The fog can be injected through the lowest-level access opening located between the heating unit and the chimney, which is ordinarily intended for inspection and maintenance. This method is much more effective and far less hazardous than trying to force water down through the top of a chimney, some of which rise 20 ft [6.1 m] above the roof and are located uncomfortably close to the edge. Trying to operate a nozzle from a ladder, particularly on a cold and windy night, is very dangerous, and in this situation it should be avoided. For this technique to be effective, however, there must be sufficient heat to vaporize and expand the injected fog. Such heat will rise and induce an upward draft. In some cases, it may be advisable to keep the heating unit working to ensure adequate heat and a favorable draft.

Fog applied by applicators centered over a tank is recommended for tank fires whose entire surface can be covered by fog. The size of the tank that can be successfully handled in this way is necessarily limited.

Fog can be applied from applicators to cover firefighters operating a solid stream through openings such as fire doors. In some cases, it will enable men to advance the solid stream so that they command a sweep of 180° through the door rather than the 40° or 50° possible without the help of protective fog.

Fog, upon being vaporized to steam, exerts a powerful smothering effect, which appears to be the primary factor in extinguishing fires in some flammable liquids such as gasoline, kerosene, and ethyl alcohol, as indicated by comparatively recent research. This smothering effect is also a factor in extinguishing or greatly retarding flaming combustion in wood fires. The smothering effect supplements the cooling effect realized when fog vaporizes to steam.

Fog can be a boon to the fire service and the public in small communities where initial response to an alarm consists of one pumper equipped with a water tank and limited personnel, assuming 1½-in [37 mm] lines are preconnected. Lines can be stretched and charged quickly. One person can handle a line, if necessary, allowing others to ventilate. A 500-gal [1.9 m^3] water tank can supply two fog streams discharging about 60 gpm [.22 m^3] for at least four minutes—enough to effectively handle four out of five structural fires in any community. A built-in wet-water system would make the fog even more effective. If ventilation is not feasible because the apparatus does not carry ladders, or for some other such reason, and the structure is unoccupied, consideration definitely should be given to the use of indirect attack. The number and size of openings made should be limited until the fog can exert its effect. In such communities, a shuttle tank truck for replenishing the water tank supply is a good investment.

Disadvantages. It has been proven many times that personnel are uselessly endangered and injured when they try to advance fog lines into seriously involved unventilated fire areas; the steam created pushes back through the means of entry. Unless ventilation can be effected to prevent such an occurrence, another technique must be used.

Fog cannot be aimed as well as a solid stream. Whereas the latter can throw 75 percent of the water within a 10-in [250 mm] circle (or 90 percent within a 15-in [375 mm] circle) when it reaches the seat of the fire, much of the water from a fog stream will not reach the seat of the fire if turbulent currents have to be overcome. The paths of the light particles of the fog stream are readily changed by such currents or adverse winds, both of which render the fog streams less effective. Fog is also diverted from its path by the tendency of the light particles to rise through vertical channels in which there is an updraft of any kind, a tendency that makes fog less effective for flooding floors (a technique sometimes used at stubborn cellar fires).

Under ordinary conditions, fog lines do not have as good a vertical or horizontal range. This could be a disadvantage in extinguishment or in covering exposures. The difference in range exists because much more of the surface of the water in a fog line is exposed to friction, as compared with a solid stream of the same volume. Fog streams do not have the impact of solid streams and are therefore less effective in venting windows from the exterior of the structure, or in turning over material in the extinguishing or overhauling phases at fires involving demolition debris.

The use of fog at very high pressure (over 400 psi [2.76 meganewton/meter^2 (MN/m^2)]) could possibly be effective at demolition debris fires but it also has disadvantages. High-pressure fog coming from the windward side at grass or bush fires may drive hot convection currents and burning embers beyond the extinguishing range of the fog and spread the fire. Care must also be taken when high-pressure fog is used at structural fires because convection currents can be driven rapidly through the structure, possibly endangering fire personnel as well as the structure.

A serious disadvantage is that some firefighters do not realize that fog is no more a panacea than solid streams are. It is important to know how to use each kind of stream to maximum advantage. Officers who favor one technique for all fires have closed their minds to the possible alternatives. Fire officers should be familiar will all recognized techniques and should learn to select and apply the most appropriate one in each situation.

Placement and Use of Fog Lines. In covering life hazard, extinguishing fires, and protecting exposures, the principles that govern solid streams are applied, bearing in mind the limitations and peculiarities of fog. For example, where life hazard is present, the first line must of course be placed to operate between the fire and the endangered occupants or between the fire and the means of escape. However, the possible effects on the occupants of the fog as it vaporizes to steam must be kept in mind. Accordingly, where there is a life hazard, the use of fog in indirect attack is inadvisable. However, Lloyd Layman's theory[1] of indirect attack is one of the foremost contributions to the fire service in the past three decades because it resulted in a greater appreciation and use of fog, despite differences of opinion about some aspects of the theory.

WATER WITH ADDITIVES

Much experimentation has been carried out in recent years to improve the effectiveness of water as an extinguishant by providing additives that reduce its surface tension, increase its viscosity and reflectivity, reduce its friction loss in hose, and so on. All can better the effectiveness of water in one way or another and merit consideration accordingly.

Wetting Agents

Wetting agents reduce the surface tension of hard or soft water from approximately 70 to 25 dynes (units of cohesion). Either salt or fresh water can be used. The solution may be applied premixed from tanks on apparatus, or it can be injected directly into a hose line by means of a proportioner, which injects the amount of wetting agent properly proportional to the flow of water in the line.

Advantages. The advantages of using a wetting agent with water include the following:

1. It has greater penetrating qualities and less runoff, and consequently less water is required.

2. It can be used with fog. There is quicker conversion of fog to vapor, resulting in quicker absorption of heat, and consequently quicker extinguishment. In this form, it is particularly effective on some class B fires involving high-flash-point products.

3. It is effective on smoldering and hidden fires, as in baled cotton, paper, and rags; fires in sawdust or where charring might ordinarily repel water penetration; brush, grass, and duff fires. It is also effective in apartment fires in residential buildings, where it is particularly appropriate for fires in upholstered furniture (it has been estimated that one-fifth to one-third the usual amount of water will suffice when a wetting-agent solution is used on such fires); and fires involving natural

[1] Lloyd Layman, *Attacking and Extinguishing Interior Fires,* National Fire Protection Association, Boston, 1952.

and foam rubber. In general, it is effective against the same kinds of fires as water.

4. It has a definitely favorable effect on the overhauling phase of an operation, on preventing rekindles after the department has left the scene, on protecting exposures, and on reducing life hazard for fire personnel. Less structural overhauling is needed, since the solution can penetrate charred structural members such as wooden floor beams. This reduces the extent to which a structure has to be opened up for examination. Overhauling of contents is also reduced since baled cotton and paper, and upholstered furniture do not have to be opened up to the same extent necessary when only ordinary water is used.

The likelihood of a fire being rekindled after the fire department leaves the scene is lessened when a wetting-agent solution is used because of its great penetrating powers and more permeating extinguishing effect. Thus, it is more likely to extinguish hidden fires that might escape the notice of a fire officer.

It also gives better protection than ordinary water for exposures endangered by radiant heat, because the exposed combustible parts absorb more of the solution than they do of water alone. Consequently, more radiant heat is needed to dry out and ignite surfaces that have been saturated by a wetting-agent solution. Exposure protection is also improved by the use of insulating air-water foam which contains wetting agents. This combination can be used as an insulating and reflecting thermal barrier for surfaces exposed to radiant heat and direct flame.

Personnel benefit because a wetting-agent solution absorbs heat and extinguishes fires more quickly, thereby lessening the development of toxic gases.

5. It also means that, because fires can be extinguished more rapidly and thoroughly, fewer units are required at the fire. In addition, the fire companies that respond can be returned more quickly. This means better overall fire protection for the community.

Disadvantages. The disadvantages of using a wetting agent with water include the following:

1. It is sometimes corrosive. However, wetting agents listed by the Underwriters Laboratories are no more corrosive to brass, bronze, and copper than ordinary water. In some cases, the detergent action of concentrates accelerates corrosion.

2. It may increase the electrical conductivity of the stream. If electrical equipment comes into contact with wet water it must be flushed clean before it is restored to service.

3. It should not be used with mechanical or chemical foam because the results are harmful. (The wetting agent breaks down the foam.) Wetting agents can, however, be used with air-water foam (see section on wetting-agent foams.)

4. Its lower surface tension tends to increase the breaking up of a stream and therefore can adversely affect the range and pattern of fog streams as well as the range of solid streams. However, efforts are being made to lessen such possibilities.

5. It should not be used in soda-acid and antifreeze extinguishers, which depend on a balanced chemical reaction for operation.

6. It can reduce the effectiveness of canvas-type salvage or protective covers, which are susceptible to its penetrating qualities and may leak.

7. Its use in extinguishing fires involving flammable liquids and greases is limited to materials insoluble in water.

8. It should not be used on fires at which water will react with the materials involved.

Conclusions. Obviously, the advantages far outweigh the disadvantages. Such solutions make a little water go a long way and therefore are strongly recommended for areas where water supplies are limited. (See above.) They are recommended for use in fabric-industry occupancies, since wetting agents are very appropriate for fires in fabrics. They penetrate to extinguish hidden fire but do not injure the fabrics any more than plain water does, and they minimize overhauling of the contents.

Additives Increasing Viscosity of Water

Tests have shown that additives can increase the viscosity of water and thereby its extinguishing power by retarding runoff. Experience in laboratory and field fire tests have shown that viscous water "knocks down" a class A fire much more rapidly than ordinary water does. In one case, a series of suppression and retardant tests were conducted to determine whether chemical additives could be used to help control forest fires with ground equipment. A report on this effort from the U.S. Department of Agriculture, Forest Service, Division of Forest Fire Research states:

> Viscous water reduced suppression time under many conditions and was outstanding in keeping fires from rekindling. The residual film of algae-thickened water seemed to be particularly effective in extinguishing usually difficult-to-extinguish fires in fuels such as baled hay and sawdust. Although there were operational difficulties such as spoilage and slight cor-

rosion of metal parts, most problems can be solved. The dry powder that makes the water thick can be mixed on the fireline in one to five minutes using the jet-type mixer which is easily installed on the truck.

The chemical additives referred to have been in full production for the past decade and quite likely have been improved upon during that time. Initially, some difficulties were experienced with the spray distribution pattern. For example, a nozzle which gave a full spray with ordinary water gave a hollow cone-spray pattern as viscosity was increased. Some nozzle modification was therefore indicated to obtain fully effective use of viscous water. Overall, it has been established that viscous water absorbs at least as much heat as plain water, provides a coating that makes rekindles less likely, and increases rather than decreases horizontal and vertical range from straight-stream nozzles. However, viscous water does not penetrate burning fuel as well as ordinary water, and it increases friction loss.

Opacifiers

Experiments have been conducted to improve the opacity (opaqueness) of water so that it can better protect an exposed surface by intercepting and reflecting more radiant heat. The results of these experiments have not yet demonstrated the practical use of opacifiers, but any agent with such a potential warrants extended consideration.

Rapid Water

In some departments, pumpers have built-in systems for injecting an additive into water flowing through $1\frac{3}{4}$-in [44 mm] hose. The solution, referred to as *rapid water,* reduces friction loss and increases flow equivalent to that in a $2\frac{1}{2}$-in [63 mm] line. Rapid water does not necessarily stretch the water supply but it does reduce the horsepower required to supply lines.

Nozzle Pump Operator

Some beneficial technological changes have been made in water delivery systems through the use of constant-flow nozzles, rapid water, and automated pumpers. Now a firefighter at the nozzle can adjust an electronic nozzle pump operator (a radio transmitter built into the nozzle) which in turn automatically signals the remotely located pumper, changing the water flow as desired. This system is still being evaluated. Potentially, it will enable the nozzle operator to be in direct control of the water supplied to each line by a pumper, thereby improving the capability, as well as the safety, of the firefighter.

Master Streams

A master stream is a heavy-caliber stream with pressure, discharge, and range adequate for fires too large to be fought with hand lines. Ordinary master streams require nozzle pressures of 60 to 120 psi [414 to 828 kN/m^3] and discharge approximately 350 to 1200 gpm [1.3 to 4.5 m^3]. Fireboats and special apparatus, such as the superpumper, can provide much higher pressures and discharge, as well as greater horizontal and vertical range. Master streams are frequently required to combat the large fires that cause such great loss in life and property. (Only a small portion of 1 percent of the total fires cause over 60 percent of the property destruction in this country.) To use the water from master streams to maximum advantage, it is essential to employ them properly and *in time,* bearing in mind such features as reach, penetration, sweep or mobility, and control. Most large fires that are lost by the fire department are caused by the failure to use large streams early enough.

Reach. Reach should be sufficient to exert effective extinguishing action on the fire, and/or protective action on exposures. Reach, both horizontal and vertical, increases with nozzle pressure. Both horizontal and vertical reach also increase with nozzle diameter, assuming the same nozzle pressure. The rate of increase is greater at higher pressures.

Fundamentally, a nozzle is a device for transferring pressure energy into velocity—which is also the means by which the reach of a stream is altered. This transfer is more efficient when appliances have looped barrels to countercheck thrust forces, and stream shapeners in the discharge pipe, thus minimizing the effects of water turbulence. The layout of hose supplying the water and the inner surfaces of nozzles and discharge pipes can also have a bearing on turbulence. Theoretically, the greater the velocity at the nozzle, the greater the reach from the nozzle. But there is a maximum effective velocity beyond which the stream tends to break up into fine particles, dissolving in midair. In effect, this means there is such a thing as maximum effective nozzle pressure where the reach of streams is a concern.

Effect of Wind. Wind has an effect on solid-stream reach. A faint tail breeze will increase horizontal reach by about 10 percent in some cases, and a moderate tail breeze will lower vertical reach by 10 to 15 percent. A head breeze will raise the vertical and shorten the horizontal reach, and a strong wind will cut the stream off sharply. If it increases in velocity, it will destroy the stream at the tip by turning it into spray, which is driven to the rear of the nozzle. A strong tail wind will carry the water forward but break it up into rain and spray. In adverse winds of 10 mph

[16 km] or more, the effective stream range may be reduced by as much as 40 percent.

Wind also has a strong effect on resistance. Tail winds reduce resistance; head winds increase it. In both cases, the stream wastage is increased.

Obviously, the effect of wind on fog streams is more pronounced than it is on solid streams—so much so that strong head winds could render fog streams practically useless. However, tail winds may increase the effectiveness of fog streams even more than they do that of solid streams—except where streams are being directed toward the base of rapidly rising convection currents.

When the water leaves the nozzle, gravity pulls it toward the earth. The stream then takes on a parabolic curved form which increases the surface exposed to friction (more so with fog than solid streams) and thereby decreases the reach. A stream operated from higher ground (fire escape or roof) has greater reach than the same stream operated from a lower level—something to consider in selecting vantage points for operating lines. The horizontal effective reach is at its maximum at an angle of 30 to 34°. Above that angle, it decreases.

Penetration. The water of a fire stream must get close enough to the burning or exposed material to absorb heat at an effective rate. Obviously, other conditions being equal, the stream with the greatest striking power will achieve the greatest penetration. Penetration will be adversely affected by piled-up stock, partitions, walls, blocked windows, and street conditions that prevent favorable placement of apparatus. Stream penetration is also affected by the direction and velocity of the wind, the angularity of the stream, and whether or not the fire area is ventilated. Fog streams have much less penetration than solid streams if the fire area is not ventilated or if convection currents are rising rapidly in the structure.

Sweep or Mobility. This feature has been definitely improved in the past decade. In addition to master streams provided by fireboats, ladder pipes, and deckpipes, the fire service can now use master streams from tower ladder apparatus, squrt apparatus, superpumper satellites, and comparatively new devices that are portable or can be used from fixed positions. All feature mobility.

Streams from fireboats can be more mobile because fireboats can maneuver better since slips are wider at many new piers. In addition, newer fireboats are designed for better maneuverability.

Metal aerials with their ladder pipes made it possible to place a master stream directly in a window and sweep the floor or fire area. However, rear-mount ladder units are now more common and do not have a bed-ladder pipe previously affixed to the underside of the bed ladder, 32 ft [9.7 m] from ground level. The sweep of the ladder pipe at the tip of an aerial for frontal or side coverage depends in large measure upon the elevation of the nozzle, distance from the surface to be covered, mechanism for directing the stream, wind conditions, and so forth. Overextension at angles below 60° from horizontal can be guarded against by placing the center line of the truck not more than 35 ft [10.6 m] from the fire building. If overextension and frontal coverage do not present problems, the center line of the truck can be placed nearer the building.

Deckpipes permanently mounted on the decks of pumpers have long been a standby for supplying master streams. Their mobility can be enhanced by hose layouts that permit limited maneuverability of the pumper. However, deckpipes are being phased out of some departments in favor of Multiversal or Stang-type appliances which can either be operated from a fixed position on a pumper or are portable, and hence more mobile. Another alternative is to carry a portable deluge set in two pieces in a compartment.

Snorkels or tower ladders have master stream appliances permanently mounted on platforms. This type of apparatus is being used increasingly, which would seem to indicate that some authorities feel it is superior to the aerial ladder in several ways, including mobility in operating heavy-caliber streams.

The squrt apparatus has snorkel-type features, including an upper and lower boom. The possibility of inserting the upper boom into the fire building through a window contributes to better mobility and penetration.

Control. Stream direction and pattern on ladder pipes are governed by manual, hydraulic, or electric controls.

Manual controls can be manipulated either from the ground or at the nozzle location on the ladder tip. From the ground, control for elevation of the stream is via rope guidelines and direction of the stream right or left is by rotation of the turntable. A firefighter at the top of the ladder at the nozzle location can regulate the pattern of the stream and, in an emergency, swing it right or left manually 15°. The nozzle operator can observe whether the stream is being used effectively and give directions presumably limited by the smoke conditions.

Hydraulic controls of the nozzle are operated from the turntable. Two simple switches control stream elevation and stream pattern, from full fog to solid stream. Such ladder pipes cannot be thrust over backward, as has happened with rope-controlled nozzles. All operations can also be controlled manually.

By means of pushbuttons at the electric control panel on the turntable pedestal, the operator can control the angle of operation and pattern of stream from fog to solid stream. Hydraulic and electric controls eliminate the need for an operator at the nozzle location. This is now a much safer operation.

The elevation and rotation of portable monitors, also known as deluge sets or portable turrets, and by various trade names, are controlled by hand, by a worm and gear, or by combinations of both. Such devices also have a safety notch or lock in the elevating mechanism to maintain the nozzle at the minimum angle necessary to keep the appliance stationary. If the elevation is less than the safe angle designated, the nozzle thrust may cause the entire device to move backward and whip out of control. In newer models, safety stop locks within vertical and horizontal friction locks operate automatically at predetermined positions. If necessary, some appliances can be operated down to 35° below horizontal by releasing the pin located atop each lock. Portable monitors are refinements of the deluge sets but can also be used as fixed deluge sets.

Remote control and direction innovation can be applied with equal success to portable turrets and deluge guns. With slight modifications, control levers can be made portable, thus permitting a deluge gun to be operated by remote control from a safe area.

Conclusions. Where water supply is limited, consideration should be given to the ways in which it can be stretched by the use of fog or additives. Where water supply is ample and master streams are needed, they should be used quickly and with proper direction. Prefire plans should serve to develop procedures for supplying master-stream appliances under various conditions, bearing in mind the performance limits of available pumping equipment. These comments also apply to heavy-caliber cellar or subcellar pipes.

FOAM

Foam is applied primarily to extinguish fires in flammable liquids by blanketing the liquid surface, sealing off the escape of vapors, insulating the liquid from the heat of the fire, and cooling the surface. The fire service is familiar with two types of foam: chemical and mechanical.

Chemical and Mechanical Foams

Chemical foam is produced by the reaction in water of a mixture of two chemicals, one of which may be sodium bicarbonate containing a foaming agent and stabilizer. Carbon dioxide produced by the reaction is contained by the bubbles comprising the foam blanket. Mechanical foam is produced by mixing and agitating air with water that contains foam-making ingredients. Whether the bubbles contain inert gas or air is not significant for extinguishment. Ordinary foam, whether chemical or mechanical, is effective on hydrocarbons that are liquid at ordinary temperatures and pressures, but it cannot extinguish fire in liquefied compressed gases. Special foams are usually required for alcohols and ketones and for the more volatile esters, as their solvent action tends to break down ordinary foam.

Foam must have lower density than the flammable liquid it is used on, so that it will float on the surface. It must also have low enough viscosity to spread or flow readily across liquid surfaces and around obstructions, and it must not break down rapidly under fire exposure or when in contact with the flammable liquid. In addition, it must be long lasting. Chemical foam is affected by the proportions of water and foam-making chemicals, by water and air temperatures, by the size and length of foam-delivery lines, and, in some cases, by water-flow pressure. Mechanical foam is affected by the type of foam compound (hydrolized protein base or synthetic-foam fluid), by the proportions of water, foam fluid, and air, by thoroughness of mixing, and, to a lesser extent, by water and air temperatures. Both hard and salt water can be used to make mechanical foam but may adversely affect chemical foam.

There are three main kinds of foam-liquid concentrates: low expansion, high expansion (Fig. 12-1), and alcohol resistant. It is not essential to know the chemical ingredients of each, but one should know that the types should never be mixed—as well, of course, as when, where, and how to use each kind.

The low-expansion protein-base type is more widely used for the protection of tank farms, aircraft-crash fires, and hydrocarbon fires in general. This type of concentrate produces foams with expansion rates up to 10:1. High-expansion concentrates are more appropriate in areas where water supplies are limited. They are particularly effective for spill fires where rapid coverage is desired. They have a high rate of expansion (16:1 to 20:1) and a high rate of flow and coverage because the foam is more fluid.

Today, the term "high-expansion foam" pertains to a relatively new agent with expansion ratios as high as 1,000:1. In some departments it is referred to as high-X foam. It has already proven to be quite effective at below-ground fires at which the fire department was having difficulties in ventilating and advancing hose lines. A smothering as well as a cooling effect is achieved by high-X foam. If structural members are obstructing the flow, it may be advisable to reposition the generator and apply the foam at another opening. However, generators should not be placed where heated air or products of combustion and pyrolysis might be used in the foam making; the results would be unsatisfactory.

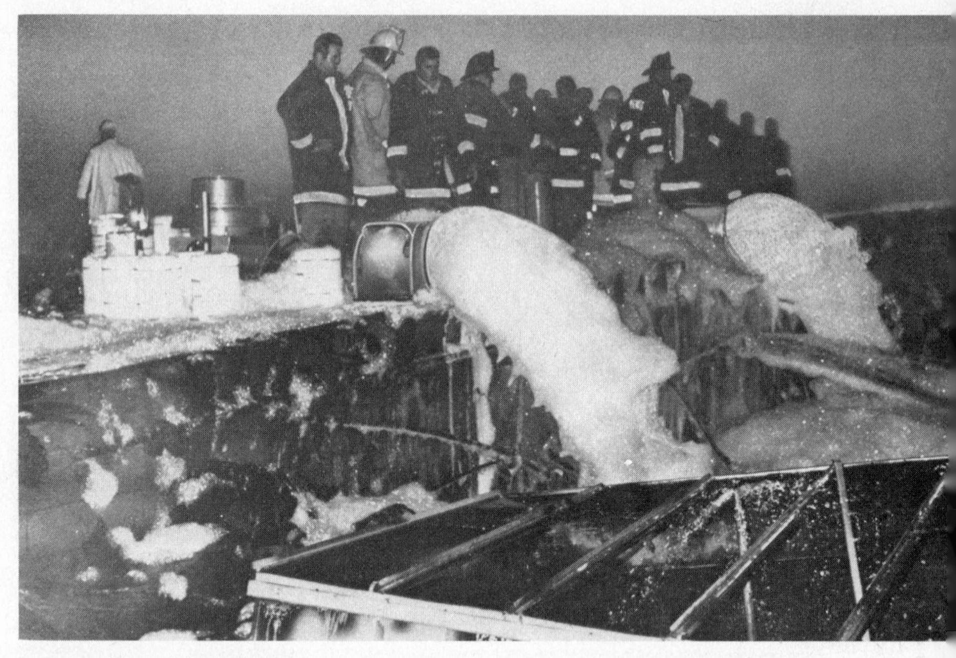

FIG. 12-1. High-expansion foam.

The quantity of foam required for extinguishment varies widely. For fires in small indoor tanks of flammable liquids, a few inches may suffice; in large outdoor tanks, several feet of foam may be necessary. The amount of foam needed will be affected by (1) the required rate and time of application, and (2) the quality of the foam and the effectiveness with which it can be applied. At oil-tank fires, effectiveness may depend on the time it takes to reduce the heat wave so that foam can be applied without causing a slopover.

Foam delivery rates at aircraft-crash fires are likely to be limited only by the capacities of the apparatus and equipment on hand and the supplies available. Techniques at these fires may involve the combined use of mechanical or aqueous film-forming foam with either foam-compatible dry chemical or carbon dioxide.

Mechanical-foam playpipes or turrets can deliver a straight foam stream, a fog foam stream, or both. Some can be used to discharge water spray. Their effective range varies from 20 to 200 ft [6 to 60 m], depending on use, pressure, capacity, and design. The capacity varies from 15 to 1,500 gpm [.057 to 5.7 m^3] water rate.

Advantages of Mechanical Foam. The advantages of mechanical over chemical foam include the following:

1. The equipment required is less expensive and simpler to operate and can be placed in operation more quickly. It does not require the constant attention and the personnel necessitated by the use of generators. A chemical-foam hose line requires, in addition to the ordinary equipment and apparatus, a generator with hopper attached and an adequate supply of foam powder. With $2\frac{1}{2}$-in [63.5 mm] hose, the minimum size for chemical foam, an open $1\frac{3}{4}$-in [64.4 mm] nozzle is used. When the generator gage registers 100 lb [4.8 kN/m^3] of flowing pressure, the foam powder is poured into the hopper. The discharge line from the generator to the nozzle should be limited to 100 ft [30 m] for best results, and the powder must be stirred sufficiently to ensure proper settling.

2. The mechanical foam can be switched quickly to a plain water stream merely by removing the pick-up tube from the can of foam liquid.

3. It flows readily, flows around obstructions, does not obstruct itself, and achieves quick coverage, and therefore quick extinguishment.

4. Soft water, hard water, and air and water temperatures have less adverse effect on mechanical foam.

5. Methods of producing mechanical foam at fires are more numerous and more flexible than those for chemical foam. It can be applied (*a*) premixed; (*b*) at the end of the hose line with the pick-up tube and aspirating nozzle through various arrangements of hose, adapters, and so forth; (*c*) at any part of the hose line with a suction proportioner; and (*d*) at the pumps with a suction proportioner.

6. Mechanical foam can be used with small-sized hose, whereas chemical foam requires larger sizes.

7. In addition to normal application, it can be applied as regular foam below the surface.

8. Vigorous agitation of an oil surface has less adverse effect on mechanical foam.

9. The foam blanket does not harden with time and therefore is comparatively easy and inexpensive to remove.

10. It can be used to extinguish secondary fires resulting from the use of tricresyl phosphate or boric acid in triethylene glycol as extinguishers on magnesium fires.

11. It requires less space for storage and transportation. A 5-gal [.019 m^3] can of liquid will produce about 1,650 gal [6.2 m^3] of foam. An equivalent-sized can of chemical powder will ordinarily produce less than one-third as much foam.

Mechanical foam obviously has advantages in storage, transportation, and application (simplicity, speed, variety, and personnel needs) over chemical foam. However, some experienced fire officers feel that, in some cases, the chemical-foam blanket is a superior extinguishing agent. The final word has not been said on this point and cannot be until adequate research and experimentation has been done.

Wetting-Agent Foams

Wetting-agent foams are made from wetting-agent solutions and air. They break down into their original liquid state at temperatures below the normal boiling point of water, and in this respect differ from the mechanical and chemical foams. If a wetting agent of the synthetic detergent type is used, the opaque cellular structure formed can intercept and reflect radiant heat, and thereby provides effective protection for exposed surfaces of exposures. A cooling effect is achieved when the flowing foam carries off the absorbed heat.

The effectiveness of wetting-agent foams as a blanket on class B fires is limited because of the comparatively quick breakdown when heat is absorbed. However, the resulting liquid retains the penetrating qualities of the wetting agents and this aids in creating a cooling action.

Surfactant Foams

Surfactant means a *surface-active* material. The surfactant foam referred to as "light water" is a fluorochemical material, and is described as a fluorinated surfactant. Combined with potassium- bicarbonate-based dry chemical, it extinguishes flame and provides an effective sealing and smothering effect at the surface of flammable liquid fires. The surfactant used alone is comparable to protein mechanical foam. It produces an aqueous film-forming foam when mixed with air either in a foam pump or at an aspirating type nozzle.

APPARATUS AND EQUIPMENT

Apparatus and equipment can be more effective if it is selected in accordance with the potential fire problem in the community as indicated by such primary factors as life hazard, possible location and extent of fire on and after arrival, heat and smoke conditions, exposure hazards, construction, occupancy, height, area, auxiliary appliances, weather conditions, topography, street conditions, time of response, and, naturally, the water, personnel, and special extinguishing agents available. This observation is substantiated in effect by an article entitled "Fire Department Stations—Planning the Location," which appeared in the *AIA Bulletin No. 176*. The article states, "Many points should be considered when choosing the location so that the company or companies to be housed in the new station will provide good coverage of the area to be protected and quick response to alarms of fire or other emergency calls."

Good coverage includes, among other things, properly selected apparatus and equipment. In effect, the bulletin explains that the points to be considered in selecting locations of fire stations include life hazard, occupancy, exposure hazard, street conditions, time of response, construction, water supply, apparatus and equipment required, and topography.

Apparatus and equipment are only as reliable as their care and maintenance. Large and small communities arrange for the care and maintenance problem in different ways, as indicated in the *NFPA Handbook*. At least one reserve pumping engine should be provided for each eight pumpers in service or major fraction thereof, and one reserve ladder truck for each five in service or major fraction thereof. The *AIA Bulletin* No. 175 contains pertinent information about the care and maintenance of apparatus and equipment.

Apparatus and equipment are obviously more effective if they arrive at their destinations on schedule. Consideration should therefore be given to the selection and training of chauffeurs and to routes taken to the fire. Thought should be given to one-way streets and avenues to minimize responding against traffic. In some cases, particularly at greater-alarm fires, radio messages can indicate the approach route.

Placement of apparatus is a matter of prime importance. One poorly placed apparatus can seriously impair the usefulness of others (Fig. 12-2). Misplacement is a particularly severe handicap at the start of an operation and should be guarded against when much of the assignment is approaching the fire from the same direction.

In hooking up to hydrants, fire officers should be sure that the end of the pumper will not obstruct or prevent the passage or the proper placement and effective use of other apparatus. This is a crucial concern when the other apparatus is needed to raise ladders or tower-ladder platforms for rescue purposes. Again, officers of engine companies should have reasonable reservations about the practice of dropping off the required hose in front of the fire building and proceeding to the hydrant beyond the fire. In one such case, the hydrant beyond the fire was defective and what might have been controlled by the first-alarm assignment became a second-alarm fire. Passing by an available hydrant 50 or even 100 ft

FIG. 12-2. The bane of a commanding officer is poorly placed apparatus. Note the pointless position of the aerial and the unused hydrant.

[15.2 m—30.5 m] ahead of the fire could be a poor decision. It is a good practice to err on the side of safety and performance whenever a life hazard is possible, especially if temperatures are below freezing, or during or after a snowstorm, or if the neighborhood is such that a blocked hydrant is not uncommon. Hose lines can be readily protected by hose bridges against incoming apparatus.

Officers of ladder and tower-ladder units should also avoid blocking available hydrants, especially one that can be used by the first-due engine company. Much time and difficulty can be encountered in releasing the tormentors or ground jacks to reposition apparatus so equipped. Where feasible, street conditions can be improved for placing apparatus by keeping chiefs', squad, and rescue vehicles out of the fire street, or at least away from the front of the fire building. The commanding officer can use the radio of a favorably located apparatus for communication with headquarters.

Effectiveness can be measured by the success with which apparatus and their equipment are used in carrying out the functions of the unit to which the apparatus is assigned (Fig. 12-3). Officers must know how best to use the apparatus and equipment, and must be aware of all specified limitations. Limitations in this instance pertain to such things as attainable heights by aerials and platforms of tower-ladder trucks, allowable weights on aerials and platforms, pumping capacities, and usable time for breathing apparatus. Officers must know what equipment is carried, as well as how, when, and where it is to be used. This is quite a responsibility even for regularly assigned officers, since modern apparatus have multiple uses and a wide assortment of tools and equipment. Difficulties may be great for officers detailed or transferred to units having unfamiliar apparatus.

If the hydrant pressure is adequate, modern pumpers are able to utilize water from supply lines and can be stationed nearer the fire. This will allow hose lines to the fire to be shorter, stretched more quickly, and require fewer firefighters. In some departments, smaller pumpers supply larger pumpers for this reason.

Pumpers, particularly the first to arrive, should stretch two lines instead of one when, for example, the use of a hydrant beyond the fire is logical. This can save time at a crucial stage.

Even where hydrants are adequate and available, the capabilities of $1\frac{1}{2}$-in [37 mm] preconnected hose, wet-water fog, and booster-tank water could be used to advantage much more often than they are. Fire authorities in some communities, however, are reluctant to sanction only hydrant hookups at building fires. In other communities, booster water is successfully used more often. As a precaution, two lines are stretched from a hydrant ahead of the fire and supplied by the next arriving pumper. Then, heavier lines can be used by the pumper at the fire, if necessary,

FIG. 12-3. The ability to use modern equipment to maximum advantage characterizes the skillful officer.

or the $1\frac{1}{2}$-in [37 mm] lines can continue to be used as developments dictate. Officers trained to recognize and properly evaluate major primary factors can be expected to make quick and correct decisions about initially using a line from a hydrant or from a booster tank. If there is any doubt, the hydrant is the choice.

In their manuals, individual departments prescribe the procedures—or evolutions, as they are sometimes called—for stretching hose lines into buildings, or via ladders or fire escapes, or to the roof via the outside of the building, or into fixed systems, or into master stream appliances, or to relay water. In addition, associated nozzle pressures, hose size, connections, and other equipment are specified. These procedures or evolutions also stipulate how additives such as foam powder or liquid, wetting agents, and rapid water are to be mixed with water. Directions are given about making knots and the use of roof ropes and ladders. The effectiveness with which apparatus and equipment are used depends in large measure on the skills that are taught, developed, and maintained, in carrying out these official procedures and evolutions.

Tools

Electric-powered. Power for this equipment is obtained from built-in generators on apparatus, from portable generators carried on apparatus, or from public or private utility sources. Where power is taken from a public source, such as a lamppost on a highway, a specially arranged hookup may be required. The generators used may have a capacity of 3,000 watts. This can supply concrete breakers; chain, circular, and bayonet-type hack saws, which are particularly advantageous at aircraft and transportation fires; circular cutters, which assist in ventilation and make openings through which cellar pipes and distributors may be operated, and augers, which help drainage.

Hydraulic-powered. This group includes hydraulic jacks, some of which power a ram and special attachments for pulling, spreading, and clamping, with capacities up to 20 tons. In specially made rescue kits, the capacity may be 50 tons. Heavy-duty jacks have capacities of 100 tons. A recent addition to this group is a portable spreading and pulling tool that can be operated by one person, and that operates with a maximum output of 10,000 lb of force at the tip of the jaws. This tool could be advantageous where people have to be extricated from plane, railroad, and car accidents.

Pneumatic Air Hammer Set. This tool provides a small, hand-held air chisel, which has a rapid hammer action and is driven by the compressed air from a cylinder for the breathing apparatus. The quick-action can open the roof of a car like a can-opener.

Protective Clothing

Especially designed suits enable firefighters to walk through flame and are invaluable for rescue work at airplane crash fires. Wearers cannot always carry victims out immediately, but they can advance closer to the fire with an extinguishing agent and apply it with maximum effectiveness, thereby increasing the chances of survival for trapped passengers. Some aluminized garments offer amazing protection against radiated heat, protecting wearers by reflection rather than insulation. Such garments enable wearers to approach and shut or open valves in an area subject to a high level of radiated heat. This could well be the turning point in a class B fire, where numerous tanks with highly flammable contents might be exposed. Such clothing is referred to as *approach, proximity*, or *fire-entry suits*.

Reflective stripes are used on fire apparatus and on helmets and turnout coats of fire personnel to make apparatus stand out more noticeably at night, thereby minimizing the number and severity of accidents, and to enable fire personnel to maintain better contact at night operations and more quickly to find personnel not accounted for.

Plexiglass eyepieces are very effective protection for the eyes of firefighters. Many apparatus are provided with safety straps for the officer and chauffeur. A hose strap has also been developed, which enables the user to cope more safely and effectively with sudden and unexpected surges of pressure at the nozzle.

Masks

Despite the increase in the availability of improved masks, the number of injuries to personnel caused by smoke and exhaustion is still considerable. Masks in themselves do not prevent burns and exhaustion. However, injuries can be reduced if firefighters are trained in the use, maintenance, and limitations of masks—and if officers are taught to anticipate the effects of other primary factors on heat and smoke conditions. Because of the large number of unusual gases that may be met in modern occupancies, supervising officers are advised to carry standard pamphlets issued by nationally recognized authorities describing such gases and their associated hazards.

It is important that masks be maintained and available for immediate and safe use. Since fire personnel have greater confidence when they know all about their equipment, they should be taught how the masks function and how to make an effective inspection of them (Fig. 12-4). The masks should be inspected daily, and records kept of the inspections.

FIG. 12-4. If personnel are properly trained in the use of equipment, they will perform more skillfully at fire operations.

This minimizes the chance of defective masks being used at fires. When not in use, masks must be enclosed to protect them against weather, water, vandalism, and so forth.

Fire personnel need to be taught in practical drills how it feels to pull hose or swing an axe while wearing masks. They should also be instructed in how to take a mask off in smoke when visibility is zero (if such is ever absolutely necessary) and how to put it on in the dark (not in smoke). At fires some personnel are inclined to wait too long before putting masks on, and then they may put them on in a smoke-affected area. This emphasizes the need for continued training and supervision.

If personnel are injured while wearing masks, and the injury seems to be connected in some way with the mask, the mask should be carefully checked to determine the cause, if any, and thereby avoid a repetition of such injuries. In some fire departments, the mask is immediately impounded and thoroughly examined for possible failure. Some departments have apparatus equipped to charge cylinders expended at fires. Resuscitators and pnealators are also helpful to fire personnel overcome by exhaustion and smoke inhalation.

The fire officers in command should specify how, when, and where masks are to be used just as they specify how other equipment, apparatus, and personnel are to be used. In congested areas of larger cities, prefire planning is limited by the fact that second-due units may arrive at the fire before first-due units. It could be ordered that all units except the engine company starting the first hose line should report in with masks. But even this order should be qualified, because if there is no life hazard personnel on the first line should don masks at structural fires. If the first line is stretched without masks because there is a life hazard, and if smoke conditions warrant, an incoming unit with masks should be assigned to take over the first line. If smoke conditions are severe, personnel with masks are much more likely to attain the objective of the first line: to protect occupants pending their rescue. The members of the first unit can then get their masks and be used as conditions warrant. They can take over their original line, or assist with another one, or, if feasible, relieve others from further duty.

Hose

Much effort has been expended to develop the ideal hose. There have been definite improvements in weight, flexibility, compactness, drying capacity, and reduced turning, warping, and rising of hose under pressure.

Leading hose manufacturers have developed a lightweight product with synthetic fiber. This has many advantages: hose can be stretched faster; lines can be stretched with fewer people; extinguishment is quicker, with less water and fire damage; less energy and time are required to replace hose on apparatus and to hang it in drying towers; units will be ready for service more quickly.

More flexible hose is easier to handle when being stretched or placed on apparatus. Improvements in weaving have minimized the tendency to kink, seen in some cases when synthetics originally came into use. This improvement, in turn, reduces the possibility of bursts in the line.

Hose that is more compact can be stored and carried in greater quantity in the same amount of space taken up by bulkier, more rigid hose; hence more hose will be available from each apparatus, more lines can be stretched from the pumper near the fire, rolled-up lengths of hose are prepared easily and can be carried more easily, and the control and extinguishment of the fire is facilitated.

Some newer types of hose are treated to practically eliminate mildew. They are also readily cleaned in automatic hose washers and dried in electric dryers. These advantages tend to prolong the useful life of hose. Reduced turning, warping, and rising of hose under pressure ensures greater dependability during use. Time and experience may show that heavy and light hose can be used to advantage in combination, the former outside, where rougher usage may be encountered, and the latter inside the fire structure, where maneuverability may be a prime factor.

Photographic Equipment

Some fire departments have photography units, which can be used to advantage at fires and elsewhere. If a unit is strategically located, it can be on hand early in the operation. The resulting pictures will be more timely and practical for the purposes of improving public relations, training firefighters, and providing legal evidence. The public might understand better the problems confronting the fire department in the crucial early stages of an operation if pictures taken at that time were available.

For training purposes, the motion-picture camera can reproduce a fire operation practically from start to finish. A motion-picture film is an ideal means of appraising the performance of units at fires. It can also be shown in training schools to demonstrate how tools should be used, how procedures should be followed, how apparatus should be placed, and how companies should report in at fire operations. Photography can be helpful where arson is suspected, as investigation and prosecution will be facilitated if pictures can establish guilt. Photography can also provide testimony in cases of accidents to apparatus and members, and in cases of meritorious acts.

Probably there is no more effective way to depict safe and unsafe acts and conditions than by photography. This can be utilized at fire operations and in fire-prevention activities. Much could be accomplished if brief pictures were shown on television and in motion-picture theaters of the proper way to transmit a fire alarm, the possibly disastrous results of a false alarm, the many ways in which human carelessness magnifies the fire hazard, and so forth. Motion pictures are an effective medium for educating school children about fire drills and fire prevention.

Photographic reproduction can also be used in fire-service administration: personnel records and inventories of real estate, apparatus, and equipment can be kept on microfilm.

The Peeping Tom camera is a military development that might be of use in coping with fires after a bombing attack, since this camera has a lens that can cut through 20 miles [32.1 km] of haze. It is mounted on a tripod and can be carried and operated by two people.

The Photoscan system makes it possible to transmit photographs from aircraft to ground stations with no loss of detail. This is a sizable advance in aerial reconnaissance and could help the fire service where large areas are involved, such as at brush fires or wartime fires. Pictures could be relayed hundreds of miles in seconds, and units in communities joined together in a mutual-aid program could be quickly alerted and advised on a situation that required cooperation. The system weighs about 70 lb [31.7 kg]. It is best suited for conventional photographs, but it can also be used to transmit filmed data recorded by infrared, radar, or sonar equipment.

Modern Aerial Trucks

The transition from wooden to metal aerial ladders is fairly complete.

Advantages. The length of some modern aerial trucks is only slightly more than that of a conventional pumper; other types are a little over 50 ft [15.2 m] long. Smaller quarters will suffice for some, with reduced maintenance costs. Other units may be housed in quarters where metal aerials replaced the wooden type. This facilitates the consolidation of units, which is generally desirable. The shorter truck gets out of quarters more easily and quickly, is more maneuverable, has fewer accidents, causes fewer injuries, deaths, and lawsuits, arrives at the fire more often, and causes less obstruction at fires. The danger from overhanging ladders, which caused many accidents, has been eliminated.

Guard rails to the top of the aerial enable fire personnel to carry out rescue assignments with greater safety to themselves as well as to those being rescued and also to stretch and operate lines from aerials with greater safety and efficiency. Rubber-covered rungs minimize slipping. There can be no accidental kickoff. The greater strength of metal aerial ladders has less possibility of collapse when they are overcrowded, as sometimes happens at greater-alarm fires. The metal aerial will not exceed the perpendicular, and its stability in high winds is a favorable feature. Automatic devices prevent overextension at low elevations, and a strain gage indicates when the ladder is being overloaded. Automatic plumbing devices correct road crown or grade. Automatic devices also stop the ladder if it hits an obstruction while being raised, lowered, or extended.

Metal aerials can be operated by one firefighter, a considerable advantage, particularly if four or six firefighters would otherwise be needed to raise a long portable ladder. Because of the shorter bed ladder and power propulsion, metal aerials have an advantage when overhead obstructions are encountered, such as wires and tree limbs. They have better propulsion to ventilate by breaking windows.

Aerial ladder trucks have closed cabs so that it is no longer necessary for members to ride to and from fires on runningboards paralleling the ladders. Personnel are now better protected against inclement weather, injuries from accidents with other vehicles, and missiles hurled by individuals whose motives are difficult to understand.

Disadvantages. The width at the rear aerial-truck mudguard and the rear step increases the likelihood of accidents. This disadvantage is aggravated by the fact that the visibility of the tiller operator is impaired by the width of the truck at the top rear area. Tiller operators sometimes use dangerous means (including removal of safety strap) to elevate their position and improve visibility. Some tiller seats are too springy, and as a result the operator is bounced around too much, even on paved streets. Other seats are too small, particularly in the winter when heavy clothing is worn. In some cases the control pedestal on the platform impairs the visibility of the chauffeur backing up the apparatus. Some of these comments do not apply where tiller operators are not required.

Motor failure renders the metal aerial ladder much less useful for fire operations. At 375°F [180°C] duraluminum metal loses about 50 percent of its tensile strength. Between 400 and 600°F [200°–315°C], aluminum alloys lose about 50 percent of their ultimate strength (commonly expressed in psi [N/m^2], this is the maximum stress that can be developed in a material as determined by a cross section of the original specimen). Their strength decreases continuously as the temperature continues to rise. Metal aerials are subject to galvanic corrosion from rainwater, water received at fires, or condensation containing chlorides or salts. Even water passing over one metal and dripping or running onto another can cause this harmful corrosion. Aluminum is more vulnerable in this connection than any other common metal except magnesium and zinc.

Tormentors are metal supports extended from the side of the truck below the aerial ladder platform, and adjusted in position to carry part of the extra weight imposed at that point when the aerial is raised. They are awkward on some types of trucks and take too much time to put in place. Some designs provide for large hydraulic ground jacks.

Jackknifing is advocated for some types of trucks. This cannot always be done because of street conditions and the presence of other apparatus, and hence the advantages are lost. Where the truck can be jackknifed, it can readily interfere with the passage or placement of other apparatus.

The high guard rail on some aerials makes it difficult and even dangerous to step from the ladder to a roof when the tip of the aerial is extended too much. This is particularly noticeable when a firefighter is carrying a tool and wearing a mask. To remedy this difficulty, it is advisable to raise the aerial only a foot or so above the roof so that personnel can continue straight over the top rung rather than have to climb over the guard rail. If a fly ladder with such a guard rail is placed at an oblique angle into the ordinary residential window, it is extremely difficult for firefighters to enter from the ladder or remove a person through the window onto the ladder.

The foot space on the rungs of some aerials is limited, particularly on the comparatively narrow top fly ladders of long aerial ladders. In addition, other features of the raising mechanism interfere with footing. A firefighter must be in very good physical condition to ascend these long aerials without stopping on the way, assuming boots are worn. If, in addition, a tool must be hand carried, the effectiveness of a firefighter would be questionable after climbing a fully extended 150-ft [45 m] aerial. (Axes should be carried in an axe belt.) Perhaps there is a maximum effective length for aerials just as there is a maximum effective nozzle pressure beyond which streams become ineffective.

Metal aerials as well as portable metal ladders must be positioned, used, and supervised with unusual care when there is any possibility of contact with overhead wiring in order to avoid making a ground which will injure or kill fire personnel.

Snorkel or Tower Ladder

This is a very significant addition to fire-service apparatus. It features well-controlled placement of a platform or basket by means of a boom that may reach 50 to 90 ft [15.2 to 27.4 m] depending on the model. The capacity of the platform in any boom position is about 1,000 lb or 450 kg for the smaller sizes, and diminishes as the height and radius of the extension equipment increases. A built-in 3½-in [87 mm] water supply to the platform assures prompt application of water, and a ground-to-platform communications system facilitates coordination.

Another feature is maneuverability: it can rotate 360° from ground level to full extent. In addition, it can be moved around the fire building within a few minutes because it is equipped with quick-acting hydraulic jacks. The snorkel, or tower ladder as it is called in some departments, may not have the horizontal reach of a given aerial and it may be somewhat slower in matters of placement, but it has many advantages.

Fog or solid streams are readily available from the basket (see above). A dual-control feature enables either the operator in the basket or the one at the base of the boom to swing the boom back and forth or up and down. When platform pipes are no longer needed, rolled-up lengths can be carried to upper levels via the basket and stretched from the outlet ordinarily used for the turret pipe. In this way, less hose and fewer personnel are required for overhauling and taking up.

The snorkel can be used for quick ventilation, entry, search, and removal of occupants. It is especially advantageous for the removal of occupants who are unable to help themselves, for firefighters engaged in rescue work, or for a building collapse, when conditions make it possible to survey the debris from above in attempting to locate trapped occupants.

The snorkel can facilitate ventilation, since firefighters are able to ventilate a roof of moderate height without getting out of the basket, as by breaking a skylight. Water spray nozzles operating below the basket provide additional protection. This could be an invaluable advantage where roof ventilation by other means is extremely dangerous. Metal shutters may be opened more easily and safely from snorkel platforms than from aerials.

Miscellaneous

Modern apparatus is characterized by versatility, speed, economical use of personnel, increased safety, and greater capacity. Included in this group are rescue-company trucks, quads (apparatus with a pump, water tank, hose, and ground ladders), quints (apparatus that often carries a power-operated aerial ladder in addition to what is carried by a quad), and the modern fireboat. There is also apparatus especially designed to carry and operate master-stream appliances; mask-service apparatus to replenish the air cylinder supply during operations; welfare-service apparatus, such as ambulances, oxygen-therapy unit, and coffee wagons; photography units (see above, Photographic Equipment); and supervising engineer service units, which carry personnel assigned to check water supply on mains, selection of hydrants, the manner in which pumps are being used, and so on.

Helicopters can transport portable pumps, hose, water for firefighting purposes, communication lines, and helitankers. Helitankers are water tanks fitted with small pump, lightweight hose, and adjustable nozzles. They can be used for fighting brush fires.

Breakers for fighting brush fires also carry water tanks and pumps. In addition, they have 10-ton winches in front with 200 ft [60 m] of $\frac{5}{8}$-in [15 mm] cable, and they have a prowlike arrangement in front to push over fairly sizable trees and break the way.

Some important improvements have been made in tools for forcible entry and overhauling. A simple nail puller adapted to facilitate removal of lock cylinders has proved effective and less damaging on most types of doors. There have also been improvements in hooks designed to pull ceilings. Special mention should be made of explosimeters and eductors. Explosimeters make it possible to determine whether a gas is within or near its explosive limits. Without such a device, officers investigating "an odor of gas" are at a loss. The eductor, while not exactly a recent development, is an important one. It simplifies the dewatering process at levels and in areas otherwise beyond the capacity of apparatus pumps because of high lift or inaccessibility.

Glass removal kits are now being carried on some apparatus for use where glass has to be removed during overhauling or possibly ventilating at fires in high-rise buildings (Fig. 12-5). These kits consist of pressure-sensitive tape, glass cutters, and suction cups—the type used by professional glaziers. Other equipment designed for coping with fires in high-rise buildings includes the concrete core cutter (Fig. 12-6) and the hi-rise tool (Fig. 12-7). The concrete core cutter can make a smooth hole—8 in [200 mm] in diameter and 12 in [300 mm] deep—without adjustment or interruption of operation through any concrete, brick, cement block of pouring, including any steel or iron reinforcing rods, and it can do all this significantly faster than a pneumatic ram or other equipment. With adjustments, holes much deeper can be made. This core cutter can cut equally well in three positions: overhead through the ceiling, laterally through walls at a 90° angle, and downward through the floor slab. The hi-rise tool is capable of ventilating a fire floor from the exterior and instantaneously applying a strong stream through the window ventilated. This can be an invaluable tool in coping with high-level fires in high-rise buildings.

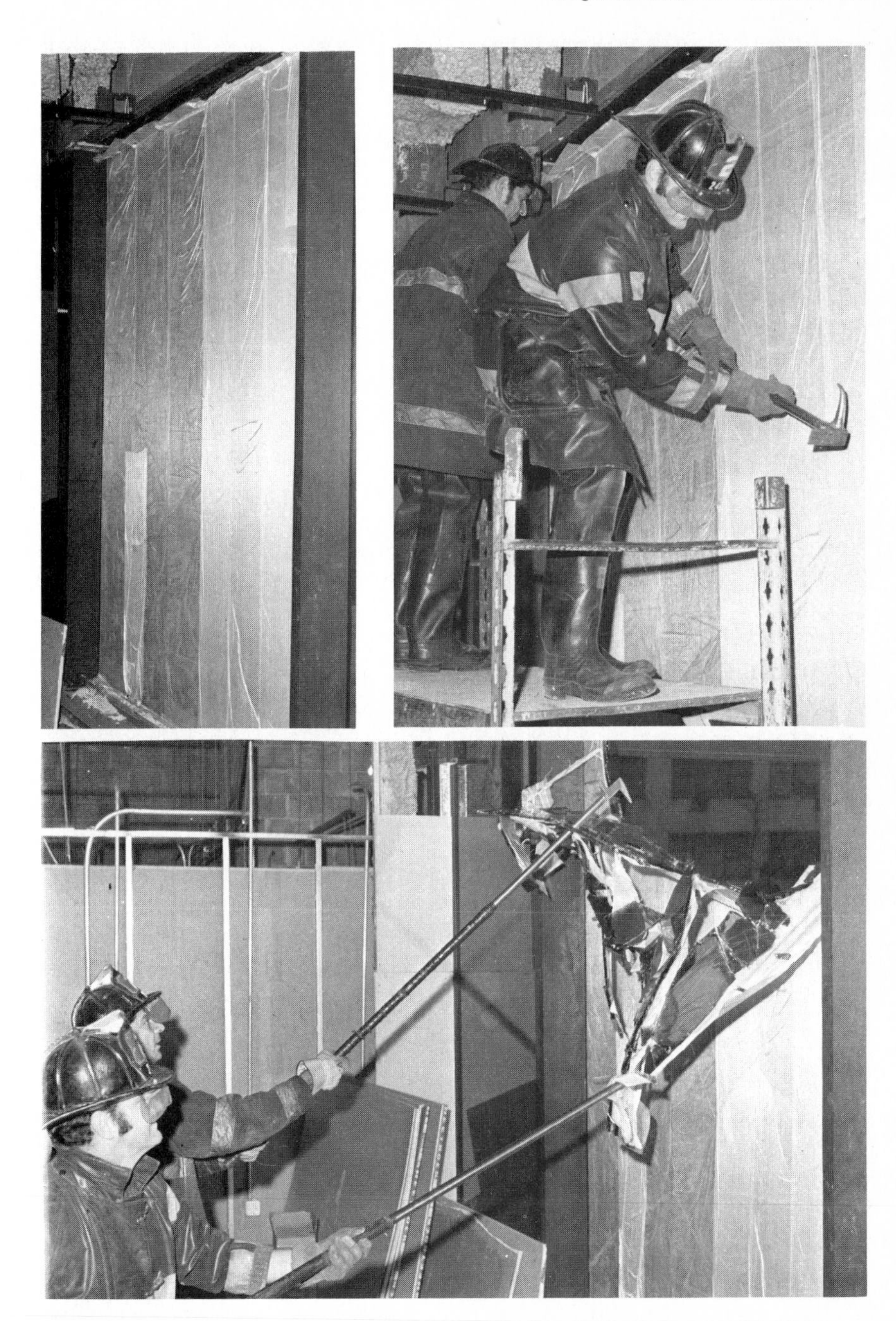

FIG. 12-5. Pressure-sensitive tape. The tape is applied over the entire window (left). Rescue company personnel carefully crack the glass with a Halligan tool (right). Using a Halligan hook, the officers safely peel the tape and glass into the building, without endangering anyone on the street below (bottom).

FIG. 12-6. Concrete core cutter. As operator applies steady pressure to feed control, the cutter begins boring into the concrete floor (left). Cutter is retracted and 7-in. thick concrete core removed (right).

A new and much better mask (referred to by NASA as the Fireman's Breathing System, or FBS) is now on the horizon. Eventually, too, it is expected that all members of ladder and rescue companies will be equipped with their own personal ropes and suspension systems for holding safety harness in turnout coats.

PERSONNEL

People are the most important part of any organization. As demonstrated so vividly in the lunar expeditions, the most sophisticated apparatus and equipment alone do not ensure success. In unforeseen emergencies, success depends on the training of the personnel. The same thought applies to the overall fire situation as clearly indicated in the report *America Burning:* "Those who bear responsibilities know that the key to their performance, and the performance of those under them, lies in training."[2]

[2] National Commission on Fire Prevention and Control, *America Burning*, p. 41.

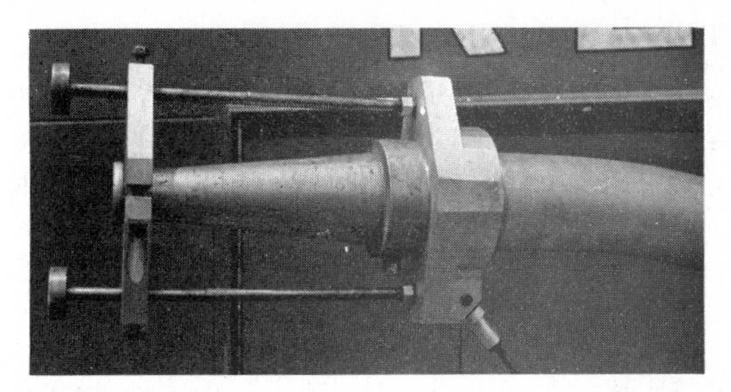

FIG. 12-7. High-rise tool. The tool can be operated from the floor above or below (left). Steel rod window breakers project from both sides of the nozzle (top right). A jack is positioned in the window; this enables personnel to control vertical and horizontal movement of the tool (bottom right).

In the preface of this text it was pointed out that the need for standardization in training can be met by providing officers with the prerequisites for using an action plan that is practical for officers of all ranks at all types of fires. Essentially, these prerequisites should be based on an acceptable science and art of firefighting, or fire suppression. Such training would ensure better use of personnel and, concomitantly, water, apparatus, and equipment.

If fire losses are to be reduced satisfactorily, the role of fire officers *in the fields* of firefighting and fire prevention should be adequately stressed in training programs. This fact should be kept in mind in formulating training programs if curriculums are to be truly relevant. Some curriculums scarcely touch on the role of fire officers in the field. Academic subjects may be necessary for degrees, but they may hardly help the officers where they need help most: in the field.

REVIEW QUESTIONS

1. On arrival at a fire, officers estimate requirements to operate according to the primary factors that are pertinent in selecting objectives and activities. How does the commanding officer proceed thereafter, insofar as requirements to operate are concerned? Why are the primary factors of extent of fire on and after arrival so important in this situation?

2. Discuss the practice of calling one or two units to fires after a full first-alarm assignment has been put to work. What policies would you recommend?

3. What does the availability of requirements to operate ordinarily depend upon? How does the matter of availability affect decisions of officers in various communities? Does the matter of availability affect the format of the action plan? Why?

4. How can the available water supply be used most effectively?

5. What are the advantages of water applied as fog? Disadvantages? What principles govern the placement and use of fog lines?

6. What are the advantages of using a wetting agent with water? Disadvantages?

7. What must be kept in mind in order to use water from master or heavy-caliber streams to maximum advantage? Discuss each of the items to be considered.

8. How can apparatus and equipment be used more effectively at fires? Discuss the matters involved.

9. How can prefire planning affect the use of water, apparatus, and equipment at fires?

10. How can injuries be reduced by masks? Explain.

11. Who should specify how masks are to be used at fires? Why is it

difficult in prefire planning to specify how masks are to be used? Discuss ways for coping with this problem.

12. Compare aerial ladder trucks and snorkel or tower-ladder trucks.

13. Where does the key to performance lie? How can training in the fire service be standardized? Why should relevancy be considered in fire training courses as it is in others?

14. How do conventional foams extinguish fires in flammable liquids? What are some limitations of ordinary foam?

15. Where is low-expansion protein-base foam widely used? Where are high-expansion concentrates more appropriate? Where has high-expansion foam, referred to as high-X foam, been found effective? What is an important restriction in positioning high-X generators?

16. What are the advantages of mechanical or air foam over chemical foam?

17. Why are wetting-agent foams effective in protecting exposures? What are the effects when this foam is applied to the burning surface of a fuel?

18. What is surfactant foam and what are its possibilities?

19. What affects the quantity of foam required for extinguishment and the rate of application?

13

Explosions, Topography, Exposure Hazards, Duration of Operation, and Street Conditions

SMOKE EXPLOSIONS

Smoke explosions or back drafts at fires are essentially caused by the ignition and rapid combustion of a mixture of flammable gas, vapor, mist, or dust and air, and under certain conditions. They can occur before or after arrival of the fire department. Smoke explosions or back drafts can occur before arrival if heat breaks windows, abetting an inflow of air to an unventilated fire area in which active combustion has ceased because of an oxygen depletion. The inflow of air replenishes the oxygen supply and can accelerate combustion of the accumulated smoke or gases with explosive effects (Fig. 13-1). This can also happen after arrival if injudicious forcible entry supplies air to otherwise unventilated and susceptible fire areas.

Effects on Other Primary Factors

Smoke explosions or back drafts can cause structural collapse. If they occur before arrival they can adversely affect the life hazard, location and extent of fire on and after arrival, occupancy, auxiliary appliances, smoke and heat conditions, exposure hazards, requirements to operate, duration of operation, life hazard for personnel, and street conditions, especially if there is a frontal collapse. If smoke explosions and back

FIG. 13-1. Heated gases rise but do not ignite because the oxygen available will not sustain active combustion (1). When the door is opened, the heated gases begin to escape (2). The inrushing air (oxygen) mixes with the heated gases and draws them back toward the fire (3). A back-draft explosion results (4).

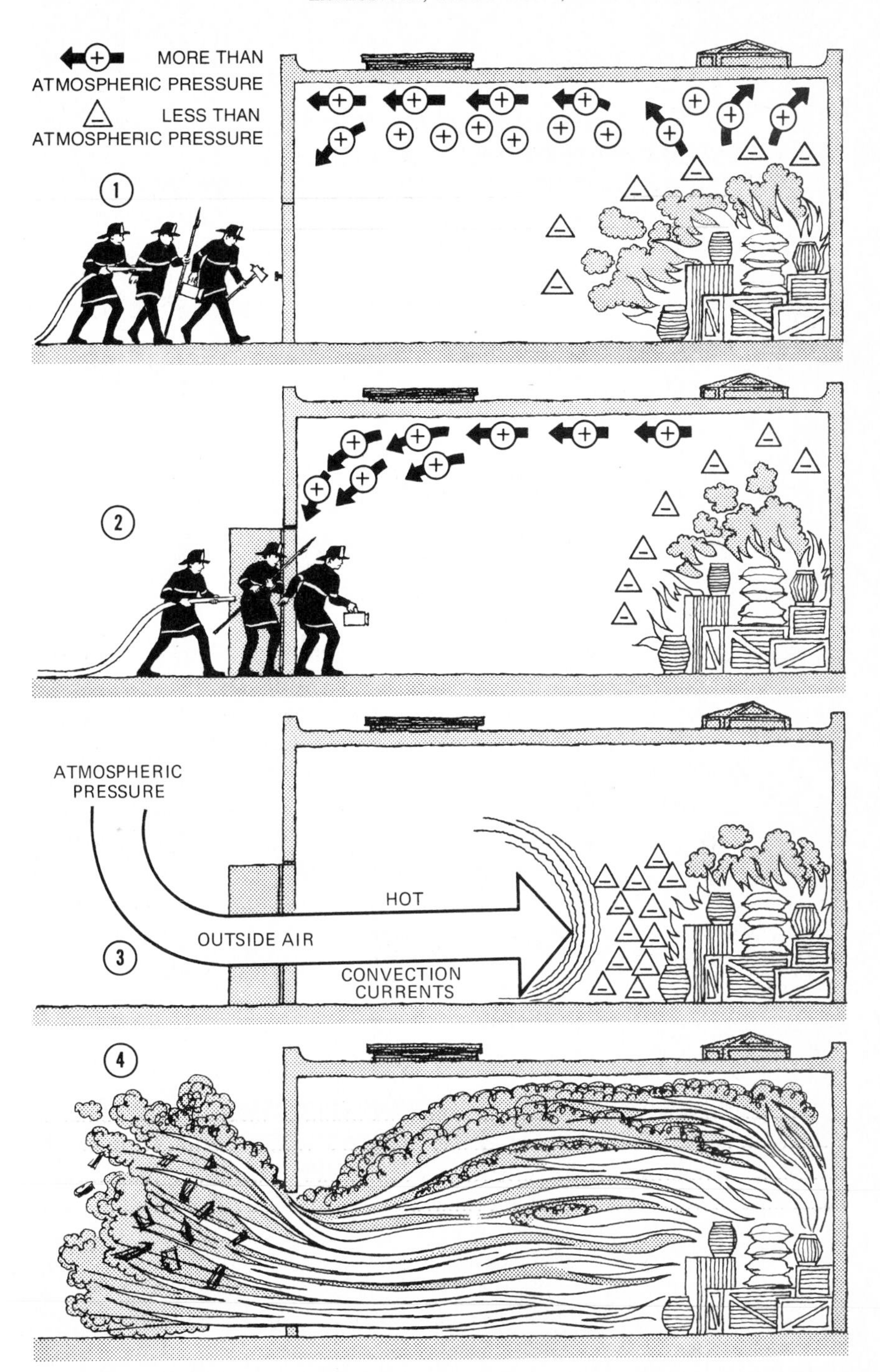
MORE THAN
ATMOSPHERIC PRESSURE
LESS THAN
ATMOSPHERIC PRESSURE
1
2
ATMOSPHERIC
PRESSURE
HOT
OUTSIDE AIR
3
CONVECTION
CURRENTS
4

drafts occur after arrival, the foregoing effects are intensified, especially for fire personnel who may be in the fire building or within range of collapsing walls. In some cases, the first explosion throws flammable dusts into suspension causing additional explosions.

Any enclosed and inadequately ventilated fire area should be considered susceptible to a smoke explosion or back draft. If the fire building is unoccupied, such areas should not be entered until they are ventilated in such a way—from above, side, front, or rear—as to prevent or minimize the effects of a smoke explosion or back draft. Failure to comply with this cardinal principle can result in disaster.

In relation to the matter of explosions and back drafts, the *AIA Bulletin No. 225* stated, about three decades ago:

> Under some conditions, ventilating a fire in a cellar or basement may result in a serious explosion; as an instance, while firemen were fighting a subcellar fire from a stairway, an attempt was made by other firemen to ventilate the area by opening the sidewalk trap door over the ash hoist elevator shaft. This allowed an inrush of air to the seat of the fire with a resulting explosion and back draft up the stairway, putting all the men on the hose line in the hospital.

AIA Bulletin No. 98, a paper of somewhat similar vintage and intent, stated:

> For night fires and for all fires which give evidence that the building is well charged with smoke, it is well for the commanding officer to investigate before opening up the building for fire fighting. If the window panes are hot or the smoke which rises from the building rises rapidly, the officer can be assured that the only thing necessary to turn the interior of the building into an inferno is to provide more air. When this condition exists, extreme care is necessary in opening up the building and ample hose lines must be available for fire fighting. The opening of a door or window on the floor on which the fire is burning or on the floor below, will probably result in a back draft explosion.

In view of these and other authoritative sources that could be cited, it seems reasonable to maintain that smoke explosions and back drafts are predictable and that the casualties they can cause are avoidable.

If officers err in opening a door to an unventilated fire, they should at least grasp the significance of the initially outgoing smoke reversing its flow back toward the fire, as this phenomenon can indicate the potential smoke explosion or back draft. In such a case, the officers should promptly close the door and back their subordinates out of the danger zone. However, the phenomenon described is not always discernible and therefore should not be regarded as *the* criterion for the advent of a smoke explosion or back draft. Nor does a smoke explosion or back draft necessarily occur immediately after a door is opened. The interval between the opening up of a structure and the occurrence can be influenced by (1) the type of gases in the fire area and their temperature, pressure, and content; (2) the volumetric areas involved and their location in the building; (3) the type, size, and location of the opening made by the fire department; and (4) the direction and velocity of the wind. Fire research scientists could help in this matter.

Explosions can also happen for other reasons—for example, because of a hazardous process associated with an occupancy. In such cases, explosions can affect other primary factors by pinpointing the location of the fire as well as the time of its origin, generally hastening the time of discovery, alarm, and response. They may cause gas leaks, collapse the structure, and render sprinkler systems useless. They can also worsen smoke, heat, visibility, and street conditions in the affected area. Much depends on the type of construction involved. Explosions caused by escaping gas in suicide attempts can administer a destructive blow to the structure but may start only minor fires in fire-resistive construction. This was demonstrated more strikingly by an explosion in a fire-resistive building in New York City on April 22, 1974. The force of the blast blew out a 50-ft [15.2 m] section of the western facade of the building from the first floor to the roof (25 stories high), caused heavy structural damage in another building, and broke windows three blocks away—but the fire itself was comparatively minor. On the other hand, explosions caused by leaking public gas mains have resulted in the ignition of nonfireproof structures over a wide area.

Rescue is the major objective at an explosion and fire when people are injured and endangered. The first line or lines will be operated as usual (between the fire and the endangered occupants). Ordinarily, the purpose of these lines is to protect the occupants pending their rescue. At the explosion and fire in the high-rise building described, the protective lines controlled and extinguished the fire long before rescue operations were completed. Other activities, such as searching for occupants and overhauling, were complicated by the explosion and structural factors. Elevators became unserviceable. Stairways became virtually impassable due to huge masses of rubble. The uncertain stability of the structure increased dangers for personnel as darkness came on. The building wreckage, and piles of debris often consisting of twisted steel structural members and broken concrete, made a search of elevators and floors unusually difficult and precarious. Heavy smoke conditions added to the problems. The dangers from escaping gas and exposed electrical wiring had to be handled promptly. Portable and aerial ladders and tower-ladder platforms were the main means of removal. In the 175-ft [52 m] exposure where most of the injuries occurred, evacuation was completed without delay because conditions there were comparatively favorable for removal.

Problems in overhauling were magnified by broken glass. Virtually all the windows in buildings in the immediate area were damaged. Some were blown out completely, while others were shattered, leaving jagged shards clinging precariously to the frame. While such a hazard was being removed, a "keep-out" danger zone was established in the street below. The magnitude of this problem can be gaged by the fact it engaged a number of units for a considerable time. One of the units employed a glass removal kit. Perhaps more units in high-rise areas will carry such equipment in the future.

BOMB EXPLOSIONS

Sporadically, certain groups explode bombs as a means of "sending their message." Warnings of the impending explosion may or may not be given. Commanding fire officers responding to the designated target must assume that the warning is authentic and conduct operations accordingly. If the suspected building is occupied, it should be vacated forthwith. Fire personnel can assist in the evacuation but should not participate in searching for a bomb since they have neither the protective equipment nor the special training required for such a task. At such alarms, pumpers should hook up to hydrants and preparations should be made to use hose lines as required. A full first-alarm assignment is recommended for bomb warnings. In such instances, commanding fire officers should not allow the return of occupants until assured by the commanding police officer that there is no bomb on the premises.

TOPOGRAPHY

Topography pertains to the physical features of a region or community and includes hills, valleys, rivers, streams, lakes, and nature of the terrain (grass, brush, woodland).

The nature of the terrain combined with prolonged hot dry spells tend to cause hot, smoky fires that frequently involve large areas. Often, such fires severely endanger residents on the lee side. Heavy smoke obscures visibility. Response may be delayed by one or more of the topographical features. The initial response in some rural areas may be completely inadequate to successfully handle these situations which quite frequently require considerable personnel, water, apparatus, and equipment. Special extinguishing agents, preferably thickened water applied from planes, are desirable but are not always available. The effectiveness of hose lines may be limited by the lack of roads and the inaccessibility of the fire. As a result, these fires turn into long, drawn-out affairs, with considerable hardships and danger for personnel.

Effects on Other Primary Factors

Topographical features under unfavorable weather conditions can adversely affect the location and extent of the fire on and after arrival, smoke, heat, visibility, and wind conditions, and the life hazard for all concerned. The nature of the terrain can provide a practically inexhaustible supply of fuel, necessitating unusual requirements to operate effectively, especially personnel. Operations are likely to be of long duration.

Effects on Secondary Factors

Effects on using hose lines and selecting and applying extinguishing agents may be as described above. Effects on determining location and extent of fire, supervision, communication, and coordination at woodland fires are discussed in Chapter 22.

Topographical features are in some cases advantageous. For example, a body of water can provide an effective firebreak as well as supplement the water supply. Again, response may be better in communities without hills or not divided by rivers.

In hilly communities, it is advisable to select aerial trucks with self-balancing turntables. When operating at fires on steeply graded streets, it may also be advisable to position aerial trucks or tower ladders on the high side of the fire to ensure maximum reach.

EXPOSURE HAZARDS

Interior Exposure Hazards

The possible spread of the fire in the originally involved building is called the interior exposure hazard. The relationships of this factor are therefore similar in most ways to the relationships of the primary factors of extent of fire on and after arrival. In other words, factors that alter the extent of fire on and after arrival will unfavorably or favorably affect the interior exposure hazard. In addition, the extent of fire on and after arrival and the interior exposure hazard are affected in the same way by other factors. Thus, explanations about the relationships of the interior exposure hazard would entail unnecessary repetition. Hence, relationships will only be briefly indicated.

The interior exposure hazard, depending on the degree of severity, can favorably or unfavorably affect life hazard, extent of fire after arrival, occupancy (human element and contents), structural collapse, heat and smoke conditions, wind direction and velocity, requirements to operate, duration of operation, smoke explosion or back draft, and exterior exposure hazard. The life hazard embraces occupants and personnel.

The interior exposure hazard can also be affected for better or worse by the life hazard, location and extent of fire on and after arrival, construction, height, area, occupancy, structural collapse, time factors, presence or absence of auxiliary appliances, heat and smoke conditions, weather and wind conditions, requirements available to operate, back draft or smoke explosion, and life hazard for personnel (warranted risks).

Exterior Exposure Hazards

Exterior exposure hazards pertain to buildings or occupancies that may be endangered by the original fire. Theoretically, every building is a potential exposure hazard or the cause thereof, and therefore many relationships with other primary factors are legal in nature. For example, because a fire in one structure may endanger its neighbor, many communities adopt building and fire-prevention codes, and zoning or other laws govern construction, height, area, allowable occupancies, distance between buildings, and auxiliary appliances that may be required.

Occupied exposures may create a life hazard or intensify the one already present in the original fire building, thereby increasing the risks that may have to be taken by personnel in rescue work. The existence of a life hazard can have detrimental effects on many other primary factors, as indicated in Chapter 4. If there is no life hazard, involved exposures can adversely affect such primary factors as location and extent of fire on and after arrival, structural collapse, heat and smoke conditions, wind direction and velocity, requirements to operate, duration of operation, smoke explosion or back draft, and simultaneous fires.

Interior and exterior exposure hazards are affected by the same primary factors, except that the exterior exposure hazard can also be affected by the proximity factor.

The relationships of extent of fire after arrival and exterior exposure hazard are quite similar but are regarded as separate and distinct factors because the former relates to an involved area whereas the latter relates to an endangered but not yet involved area.

Covering Interior Exposure Hazards

After assessing the effects of other primary factors on these hazards, commanding officers may have to force entry, ventilate, check for possible extension of the fire, and overhaul, as suggested in Chapter 18. They may have to use hose lines as suggested in Chapter 17, assuming the major objective is extinguish. If the fire building is heavily involved, the major objective may have to be confine, control, and extinguish and efforts to cover the interior exposure hazards may only be feasible by an exterior operation. In some such cases, the floor or floors above the fire may have to be flooded to check vertical spread of fire to achieve the first phase, confine. Then other steps would be taken to control and extinguish.

Covering Exterior Exposure Hazards

After assessing the effects of other primary factors on these hazards and assuming there is no life hazard, the commanding officer establishes the order of priority for coverage. The assumed major objective is confine, control, and extinguish. Initially, all available lines may have to be used for defensive purposes by supplying sprinkler and standpipe systems and using heavy-caliber streams to wet down the exposures, bearing in mind that most radiant heat passes through drops of water and through ordinary window glass that is not kept wet. (See Chapter 10, Heat Transfer by Radiation.) Other activities in exposures may include forcing entry, ventilating, putting out incipient fires, closing windows, and moving exposed stock. After the fire is confined, lines that were used for defensive purposes can supplement other lines being used offensively to achieve control and finally extinguishment.

DURATION OF FIRE OPERATION

Fire operations of long duration do not necessarily connote extensive property damage or great danger to civilians or personnel. For example, a fire in a garbage dump may go on for days and require only a watch line which can be handled by a few firefighters in comparative safety. On the other hand, a forest fire raging out of control for days is often characterized by serious property loss and danger for civilians and personnel. (See Chapter 22.)

If structures are involved, operations of long duration may be attributable to difficulty in (1) locating the fire; (2) ventilating so as to alleviate smoke, heat, and visibility conditions; and (3) effectively reaching the fire with the proper extinguishing agent, because the location of the fire makes it less accessible to attack. In addition, many primary factors can be pertinent.

Effects of Other Primary Factors

A life hazard can delay efforts to control the fire and can therefore let it get out of control. Location of the fire can make it difficult to locate, ventilate, and attack with hose lines, thereby worsening heat, smoke, and visibility conditions, as indicated above. As a result, the fire structure may collapse, intensifying exposure hazards. Adverse wind and weather conditions, occupancy contents, height, area, and construction can also be pertinent factors in prolonging a fire operation for reasons given in previous chapters. At fires in high-risers, it may be recalled that operations are prolonged by, among other things, the fact that firefighters could work for only short intervals because so much heat was being retained by the structural parts.

Effects on Other Primary Factors

For one reason or another, fire operations of long duration are generally difficult to cope with from the beginning. They prolong the exposure of occupants and personnel to smoke and heat. They are often inaccessible for venting, and thus smoke conditions impair visibility and pose the threat of a back draft or smoke explosion. They may feature heavy involvement and structural collapse, especially if the structure is old and the contents are water-absorbent. They may maximize the exposure hazard and cause other fires if sparks and embers created by structural collapse are carried by the wind. They indicate that auxiliary appliances such as sprinklers are absent or ineffective, and usually necessitate more than the usual requirements to operate.

If such fires develop a thermal column and pressure gradient, they can affect the direction and velocity of the wind, as indicated earlier in this text.

STREET CONDITIONS

Streets that are one-way, congested by vehicular traffic, or snow- or ice-covered, tend to delay response of the fire department. Hence, under such conditions, fires are likely to be more extensive than usual on arrival, intensifying the existing hazards.

At the fire scene, snow-covered streets and/or parked cars may interfere with locating gas shutoffs and the placement and use of apparatus at hydrants and elsewhere. Ice-covered streets can slow down the movements of personnel. Canopies, overhead wires, and tree-lined streets can handicap efforts to use portable and aerial ladders and tower ladders. Some modern housing projects with extensive terraces practically preclude favorable placement of aerials and tower ladders. The width of streets naturally has a bearing on proximity of exposures and therefore on the exterior exposure hazard. Marginal streets along the waterfront are sometimes used for the storage of material that really belongs on piers, thus making it difficult to connect pumpers to draft or to locate hydrants. Piers, dead-end streets, and buildings facing on only one street, restrict avenues of attack. Steeply graded streets can affect the placement of apparatus (see above, Effects of Topography).

Sidewalk conditions can be pertinent for various reasons. In some cases, the building line is 30 ft [9.1 m] from the curb so that if cars are parked there, aerial ladders or tower ladders would have to be placed about 40 ft [12 m] away, reducing their effective reach. This situation is more pronounced where the building lines of high-risers are 60 ft [18.3 m] or more from the curb. Sidewalk inlets to fixed systems, subway gratings, deadlights, and covers over inlets to below-ground areas may also be pertinent in deciding on activities. Inlets may have to be supplied promptly; subway gratings may have to be covered to minimize water damage and hazards in the subway system; sidewalk deadlights and covers may play a part in ventilating and carrying out other activities at below-ground fires. (See Chapter 6, Brick-Joist Construction.)

Modern Developments Affecting Street Conditions

One-side-of-the-street parking is a favorable development, but illegal double parking continues to be a handicap. The danger of glass falling from involved high-rise buildings has added a new and sizable dimension to the problems of the fire service (see above, Explosions). Some cities convert a main thoroughfare into a mall with sidewalk cafeterias and extensive garden trimmings. This creates considerable problems for the fire and rescue efforts.

REVIEW QUESTIONS

1. How can a back draft or smoke explosion affect other primary factors before arrival of the fire department? After arrival?

2. When is a fire area susceptible to a back draft or smoke explosion? What precautions should be taken in such cases, especially if the building is unoccupied? Remember, if the building is occupied greater risks may be warranted, but if possible and feasible the *same* precautions should be taken.

3. What may influence the interval of time between opening up a structure and a back draft or smoke explosion?

4. How can explosions from other causes affect primary factors? Secondary factors, including decision making? Refer to the case described in the text on page 183.

5. How should a commanding officer act when responding to a report of an impending bomb explosion in an occupied building?

6. How can the nature of the terrain in conjunction with prolonged hot dry spells affect other primary factors? How can topographical features under unfavorable weather conditions affect other primary factors? How can topography affect secondary factors?

7. How is the *interior* exposure hazard ordinarily covered when the major objective is extinguish? When the major objective is confine, control, and extinguish?

8. How is the *exterior* exposure hazard usually handled?

9. What secondary factors are usually adversely affected when fire operations in unoccupied buildings or structures are of long duration, even though the extent of the fire on arrival is limited?

10. What primary factors can contribute to prolonging fire operations? Why?

11. How can fire operations of long duration affect other primary factors?

12. How can unfavorable street conditions affect other primary factors? Secondary factors? What modern developments can affect street conditions and, thereby, fire operations?

14

Simultaneous Fires

Simultaneous fires, even of one-alarm proportions, can delay response of required apparatus, equipment, and personnel, and may overtax the water supply. Simultaneous fires of greater-alarm proportions can obviously make the situation even worse. On some occasions during World War II, simultaneous fires coalesced into fire storms. Delays are caused by the need to call on units that are farther away. In addition, only a minimum assignment may be available to respond since they may also have been called upon for other fires. Effects will be felt not only in areas where simultaneous fires are in progress, but also in the areas previously protected by the relocating companies.

Simultaneous alarms present similar problems, since units must respond. In one large city, only a skeleton assignment responds during certain hours in which the number of alarms increases. Sometimes further help can be called quickly on arrival at the fire or even approaching it, but the arrival of a full assignment will nevertheless be delayed. On the other hand, if the alarm is false (which it often is), the assignment remaining in quarters is available for other alarms.

The large number of calls, whether for false alarms, "junk" fires, or whatever, along with the matter of availability, does limit the options of the fire department. However, if a loss of life can be attributed to a delay, some serious and difficult questions must be raised about the responsibility of the community to make adequate fire protection available. This responsibility is hardly lessened by the sometimes-quoted statistics, however carefully they are compiled, because the conclusions derived are not necessarily in accord with some of the most important features of the "Grading Schedule for Municipal Fire Protection" (1973 edition), a booklet promulgated by the Insurance Service Office. These conclusions can be faulty because they are not sufficiently relevant to the overall problem confronting the fire service. Then, tainted perhaps by economic considerations, they can lead to unwisely reducing fire personnel in an era characterized by undiminished high fire losses in life and property.

Effects on Other Primary Factors

Simultaneous fires can result in a response that is both delayed and inadequate. This can adversely affect many primary factors, such as the life hazard, location and extent of fire on and after arrival, heat and smoke conditions, exposure hazard, structural collapse, requirements to operate, and duration of operation.

Simultaneous fires and civil disorders complicate matters considerably, particularly in applying the action plan. In such dual situations the action plan should relate to the community as a whole rather than to individual fires. In carrying out the first and second steps in the action plan (to recognize and properly evaluate the primary factors that are pertinent in selecting objectives and activities), fire administrators must be realistic. The foremost factor is life hazard for personnel, for reasons that have become all too obvious. Administrators can anticipate that limited assignments will have to cope with extensive fires on arrival. In addition, personnel may have to contend with physical assault, cut hose lines, or other interference. As a result, the fire service has no option but to modify its policies and procedures if it is to protect property, assuming there is no life hazard. For example, such major objectives as extinguish and confine, control, and extinguish may have to be modified to confine and control. Activities required for extinguishment may not be feasible during civil disorders.

Confine and control becomes the logical objective at each fire, and of the overall action plan for the community, because at such times the fire service must rely on hit and run tactics. This situation is indicated in the report titled "Fire Fighting during Civil Disorders," promulgated by the International Association of Fire Chiefs, which states: "Task forces should attempt to knock down and black out fires as quickly as possible with heavy streams. Small fires should be attacked with preconnected lines to maintain mobility. When fire is blacked out, pick up and get out of the area as quickly as possible. Do not overhaul or even think of salvage." These directions are realistic but do not, in effect, change the format of the action plan. If there is a life hazard, the major objective must always be rescue.

REVIEW QUESTIONS

1. How does the primary factor of simultaneous fires affect the primary factors included under requirements available to operate?

2. How advisable is it to use skeleton assignments during periods when the number of alarms is highest?

3. What are the overall effects of simultaneous fires on other primary factors?

4. What is the foremost limiting or strategic factor to be considered in carrying out the first two steps in the action plan at fires connected with civil disorders? Why is confine and control the logical major objective at such fires? (Rescue must of course be the major objective if there is a life hazard for occupants.)

15

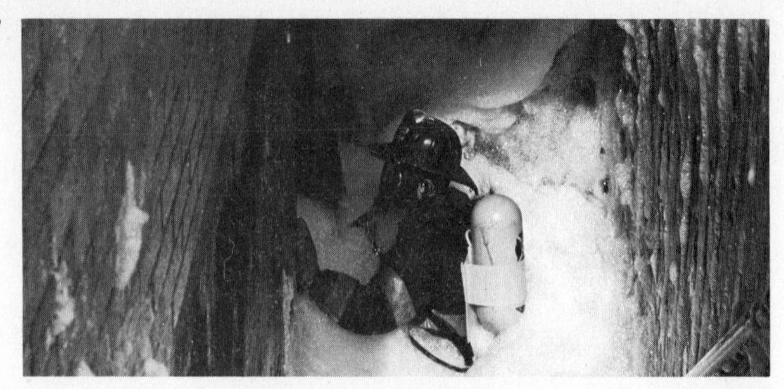

Classes of Fires and Methods of Extinguishment

CLASSES OF FIRES

Fires are divided into four *classes*—A, B, C, and D. Each class denotes a kind of material involved and exposed at a fire.

Class A fires involve ordinary combustible material such as wood, paper, cloth, rubber, and many plastics.

Class B fires involve flammable liquids, gases, greases, and similar materials. They are further classed as contained, uncontained, or a combination of both.

Class C fires involve energized electrical equipment. If the equipment is deenergized, the fire may then be a class A or B type. It could become a class B fire if transformers use transil oil as a coolant and the oil becomes involved.

Class D fires involve metals. Under certain conditions metals ordinarily not classified as combustible can ignite.

Depending on the location and extent of fire on arrival, construction, height, area, and so on, the class of fire can affect extent of fire after arrival, heat and smoke conditions, and so on. Perhaps the most significant effect is on the primary factor of special extinguishing agent required and available, since the class of fire determines the extinguishing method and agent to be used.

METHODS OF EXTINGUISHMENT

Fire requires adequate temperature, air, fuel, and a system of uninhibited chain reactions. Fire-extinguishing methods are based on eliminating one of these essential ingredients by cooling, smothering, starving, or interrupting flame chain reaction.

Cooling

Water is ordinarily used to *cool* burning materials to below their ignition point. Regardless of the class of the fire, materials must be cooled to this point so that there is no longer any danger of heat transmission by conduction, convection, or radiation. Only then is extinguishment achieved. Other extinguishing agents and methods are supplementary. An activity such as ventilation can also make an important contribution.

Smothering

Smothering means blanketing the fire so that active combustion ceases for lack of oxygen. If the blanket is maintained for long enough, the involved materials cool below their ignition point and the fire goes out. Smothering is not effective on materials that contain their own oxygen supply, such as cellulose nitrate. Some agents, such as fog and high-expansion (high-X) foam, have significant cooling and smothering effects; others coat metal fires which can then be cooled by fog or mechanical foam. After tank fires are blanketed by foam, cooling can be accelerated by applying streams, preferably in fog form, to the exterior of the tank at the appropriate level.

Starving

Starving a fire means reducing or removing the supply of burnable material, thereby slowing or stopping combustion. In gas fires, for example, when the gas supply is shut off by a valve, the fire starves. In oil-tank fires, tank contents can be drained off in a prearranged manner to remote storage tanks. In grass and brush fires, the backfire technique can be used, sometimes in conjunction with firebreaks. The fire dies when it reaches the burned-out area—assuming that the wind does not negate the effectiveness of such a technique. Class A fires can be starved by moving combustible material to a safe area.

In this text, extinguishing by blowing off flame and by dilution are considered other forms of starving. Flame is "blown off" when gas is made to flow faster than it can burn, as explained in discussing the combustion wave in Chapter 10. At the same time, it is diluted so that it is too lean to burn. Water-soluble flammable liquids can also be extinguished by dilution, but this method is seldom attempted because so much water is usually required. In addition, it can be dangerous if the temperature of the liquids exceeds 250°F [120°C]. In effect, therefore, blowing off the flame separates the fuel from its source of ignition, and dilution affects the supply of fuel and hence the mixture required for combustion. Thus, both methods are here associated with starving. Dilution, of course, can also refer to the oxygen dilution required for combustion.

Interrupting Flame Chain Reaction

See below, Dry Chemical.

CLASS A FIRES

See Table 15-1.

Cooling

Cooling at class A fires is done by using water in fog or solid streams or with additives, as described in Chapter 12. A cooling as well as a smothering effect is achieved by high-expansion foam. In this foam, the conversion of the water to steam absorbs heat from the burning fuel and prevents the air that is necessary for combustion from reaching the fire. In addition, the steam created reduces the oxygen concentration by diluting the air. Ventilation is an important part of cooling because it allows much heat to escape by radiation and convection. Dry chemicals and inert gases also exert a cooling effect but not to a significant degree.

TABLE 15-1. Class A Fires

EXTINGUISHING METHOD	EXTINGUISHING AGENT OR ACTION USED
Cooling	Water in fog or solid streams
	Water with additives
	High-expansion foam
	Ventilation
	Dry chemicals and inert gases (These play a less significant role here.)
Smothering	Fog
	High-expansion foam
	Carbon dioxide
	Multipurpose dry chemical
Starving	Use existing or established firebreaks.
	Move material that might otherwise be exposed to fire.
Interrupting flame chain reaction	Dry chemicals
	Halogenated hydrocarbons

Smothering

The smothering effect of fog was discussed in Chapter 12.

Carbon dioxide can smother fires by reducing the oxygen content of air to below the combustion point. A reduction of oxygen content from the normal 21 percent to 15 percent extinguishes most fires in spaces in which there are no materials that produce glowing embers of smoldering combustion. Where the fire is deep-seated and the material involved acts as a thermal insulator, a low oxygen dilution may have to be maintained for hours. Carbon dioxide is not effective on fires involving materials that contain their own oxygen supply, such as cellulose nitrate.

Under certain conditions, carbon dioxide also exerts a cooling effect. Carbon dioxide snow has a latent heat of 246.4 Btu per pound, but since only part of the liquid carbon dioxide is converted to snow, the total effect of gas and snow is considerably less. If the liquid is stored at 80°F [27°C], about 24 percent is converted to snow upon discharge, with a total cooling effect of about 120 Btu per pound. If it is stored at 0°F [−18°C], about 45 percent is converted, with a cooling effect of about 170 Btu per pound. The cooling effect is small when compared with that of water, which has a theoretical effect of about 1180 Btu per pound, assuming that it is all evaporated into steam.

Starving

Starving class A fires often consists of using existing firebreaks or creating new ones, supplemented in some cases by backfiring. Thus, the fire dies out when it reaches the burned-out area, assuming wind conditions do not dictate otherwise.

Interrupting Chain Reaction

Dry chemicals or halogenated hydrocarbons are used for chain-breaking. Theories about how this mechanism works have not been completely verified to date, but the ability of the agents mentioned to extinguish flames promptly is well recognized. However, chain-reaction interruption alone will not extinguish class A fires, which are those with which the fire service deals most often. The involved material must still be cooled or smothered for final extinguishment. Some multipurpose dry chemicals can coat, thereby smother, and help to extinguish class A fires, but they are not recommended where the accumulation of the residue would be harmful.

CLASS B FIRES

Class B fires in pits, pools, tanks, or tank trucks are considered *contained fires;* those that involve flowing materials are considered *uncon-*

tained fires. The extinguishing agent (Table 15-2) used on a class B fire is determined by the nature of the substance that is burning and by whether the fire is contained or uncontained.

Contained fires involving high-flash-point, low-vapor-pressure oils, such as kerosene, are extinguished by using the cooling, smothering, or chain-breaking methods. Contained fires involving low-flash-point, high-vapor-pressure oils, such as gasoline, are unlikely because the vapors are too rich to burn. Uncontained fires involving either high-flash-point *or* low-flash-point oils should be treated in the same manner: The flow of fuel to the fire should be stopped as soon as possible. A protective fog stream and special clothing can enable firefighters to reach the shut-

TABLE 15-2. Class B Fires

EXTINGUISHING METHOD	EXTINGUISHING AGENT OR ACTION USED
Cooling	Water, preferably in fog form, to cool exterior surfaces of tanks or containers
	Air agitation
	* Relatively heavy water spray to create a cooling (and smothering) emulsion on the surface of immiscible and the more viscous flammable liquids
	* Fog, if the surface can be completely covered and thereby cooled and smothered (not effective on liquids with flash points below the temperature of the fog)
	* Conventional foams to cool and blanket surfaces of liquids (alcohol type is recommended for water-soluble liquids, such as alcohols, esters, and other organic solvents)
	* Wetting-agent foam
	* Water flowed over the surface of heavier liquids, such as carbon disulphide
	* Inert gases
Smothering	Carbon dioxide
	Steam
	Fluorinated surfactant (film-forming) foam
	Foam compatible with dry chemicals
	Closing of dome cover at tank-truck dome fires
Starving	Shut off supply
	Move exposed material to a safer area
	Dilute miscible liquids
	Dilute or remove gas fuel from source of ignition by blowing off the flame
Interrupting flame chain reaction	Dry chemicals or halogenated hydrocarbons

* These extinguishing agents have a smothering, as well as cooling, effect.

off valve. If a tank line is broken, with oil flowing from it, and the valve cannot be shut off near the break, the oil may possibly be replaced by introducing water into the tank. The oil will float on the water, which will then flow out of the broken line, leaving only the oil that has already spilled out on the ground to be contended with. Ground fires involving high-flash-point oils can be extinguished by using the cooling and smothering methods. Ground fires involving low-flash-point oils can be extinguished by using the chain-reaction-interruption method.

Two conditions may occur in contained class B fires: boilover and slopover. *Boilover* is an almost explosive ejection of the tank contents. It can only occur if the oil has been burning for some hours. A heat wave descends through the burning oil to the water below and converts the water to steam. This steam is trapped underneath the viscous oil and pressure begins to build up. Eventually, the steam drives the oil up and out of the tank. *Slopover,* a condition which usually occurs after a heat wave, is an overflow caused by efforts to extinguish or control the fire. Unless precooling water is applied sparingly to hot surface oil before foam is applied, water may sink into the heat wave, expand to steam, and cause a surface frothing which exceeds the container outage. Slopovers can also occur if foam is not applied sparingly to the precooled surface, assuming the foam has a water base.

Cooling

The cooling method in class B fires involves the use of plain or fog streams against the exterior tank shell above the level of the burning liquid, allowing water to enter the tank only very sparingly—not enough to cause a slopover. This action is appropriate on a large, almost full crude-oil tank when there is a delay in getting adequate foam equipment and when air agitation is not feasible. It may delay the development of the heat wave to the extent that no slopover will occur when foam is used. However, if boilover or slopover is anticipated, fire personnel should be moved to safe positions.

An emulsion that will not burn can be created on the surface of burning oils or similar flammable materials with which water will not mix. The emulsion results when a coarse spray strikes the burning surface. This method is most effective with more viscous liquids. The emulsification lasts for a longer time in such cases, thus minimizing the danger of flashback.

Air agitation can be used as an agent in cooling. Tests under controlled conditions indicate that fires in oils that have flash points above their storage temperatures can be extinguished or controlled by air introduced under pressure at the bottom of the tank. The surface of the burning liquid is cooled to the main body temperature by the injection of the air, which agitates the oil and pushes the cooler oil in waves to the surface. If the main body temperature can be reduced below the flash point of the fire, the fire is completely extinguished. If the temperature is reduced but still remains above the flash point, the fire continues to burn but at a reduced rate which can be more readily extinguished by other means. A flammable liquid can continue to burn only so long as the vapors above it form a combustible or explosive mixture with air. The ability of a liquid to form vapor increases with its vapor pressure, which, in turn, increases with temperature. When the surface of a burning liquid is cooled, vapor pressure and rate of vaporization are consequently reduced. If they are reduced to the point where the vapor concentration goes below the lean flammability limit, the fire goes out.

The air-agitation method is not effective on extremely volatile liquids, such as natural gasoline, nor is it recommended for liquids more volatile than winter-grade gasoline (13-lb Reid vapor pressure, which is calculated on the basis of the specific oil at 100°F [38°C]). Due to these and other limitations, air-agitation is not regarded as a standard method for coping with oil-tank fires.

Inert gases extinguish fires mainly by separating the fuel from the required oxygen by dilution or by actual displacement. Inert gases also absorb heat and thereby exert a cooling effect.

Fog streams extinguish fires in kerosene, fuel oil, linseed oil, lubricating oil, and other heavy viscous liquids. For them to do so, however, the fog must completely cover the surface of the material and cool it below the temperature at which it gives off sufficient vapor to support combustion. This is not an effective method for liquids with flash points below the temperature of the water spray. If enough steam is generated by the heat of the fire when fog is applied, oxygen is displaced and smothering results. Fog from applicators centered over a tank is recommended for tank fires whose entire surface can be covered by the fog. The size of the tank fire that can be successfully handled in this way is necessarily limited.

Conventional and other foams achieve some cooling effect when used to blanket liquid fires. For a comprehensive discussion of foam, see Chapter 12.

Water floated gently from an open butt over the surface of liquids heavier than water (such as carbon disulfide) exerts a cooling and smothering effect.

Smothering

Items preceded by an asterisk in Table 15-2 have a smothering as well as a cooling effect. The smothering effects of carbon dioxide have been described above. Steam created by the vaporization of fog can cool and smother fires by reducing the oxygen content, as do carbon dioxide and other inert gases. Steam smothering systems are of limited practical value when compared with modern carbon dioxide and foam smothering systems. Consequently, steam extinguishing systems are rarely used today. Fluorinated surfactant foams and foams compatible with dry chemicals can smother and, to some degree, cool class B fires.

Starving

Starving can be the most effective method of controlling fires caused by gas leaks. It is done by shutting off the supply of gas to the fire. In many cases it should be done before extinguishing the flame, especially where gas is leaking and burning in a residential occupancy. In such cases, lines should be used to confine the fire until the gas is shut off. In still other cases, involving liquefied petroleum products for example, it may be advisable to let the fire burn if extinguishment might create severe and possibly greater hazards elsewhere due to reignition of the widely traveling flammable mixtures.

If the fire involves crude oil and has been in progress for more than 30 minutes, it may be preferable to start pumping out and thereby starving the fire rather than using foam, assuming the tank is filled or nearly so.

If flammable liquids are soluble in water, the fire may be extinguished by diluting the burning substance. The dilution necessary for extinguishment, and consequently the volume of water and the time required, varies greatly.

A gas flame is extinguished if the flow velocity of the gas exceeds its burning velocity or flame-propagation rate. If the flow velocity is increased by a blast of air, extinguishment also could be attributable to a dilution of the mixture below its lower limit of flame propagation.

Interrupting Chain Reaction

The chain-breaking method of extinguishment is employed when dry chemicals and halogenated hydrocarbons are appropriate and available. Dry chemicals do not necessarily achieve total extinguishment because they do not produce a lasting inert atmosphere in the fire area. Reignition is therefore possible. Halogenated hydrocarbons present the same possibility unless effective concentrations can be maintained until the involved material is sufficiently cooled.

Dry Chemicals. More recent explanations suggest that dry chemical agents extinguish flames primarily by chemical chain-reaction interruption. Effects of cooling or smothering, so important in other extinguishants, are deemed relatively minor in the case of dry chemical.

The chemical chain reaction, which is thought to be essential to the existence of fire, requires the presence of free radicals in the flame zone.

These interact with both fuel and oxygen to produce a continuing or increasing supply of free radicals, and thus the flame reaction proceeds in real chain fashion. The free radicals are self-propagating unless captured by condensing on, or interacting with, some substance that renders them inert. The fine particles of dry chemical, when introduced to the flame area, apparently capture enough of the free radicals to interrupt the chain reaction and thus suppress the flame nearly instantly. Research on this theory is still in progress.

Potassium bicarbonate–base dry chemical is considered more effective than sodium bicarbonate–base dry chemical at class B fires. Ammonium phosphate–base dry chemical is considered as good as if not better than the sodium bicarbonate–base type. All quickly extinguish flames but do not necessarily achieve total extinguishment. All are electrically nonconductive and therefore can be used on class C, as well as A and B. However, dry chemical is not recommended for delicate electrical equipment. Multipurpose dry chemical also has its limitations because the ammonium phosphate leaves a sticky residue on the burning or heated material. It is therefore not recommended where a harmful residue could cause unnecessary damage to machinery or other contents.

Halogenated Hydrocarbons. These agents are composed of carbon, a halogen (fluorine, chlorine, bromine), and, in some cases, hydrogen. Agents without hydrogen are referred to as completely halogenated; those containing hydrogen are referred to as incompletely halogenated. Iodine also is a halogen but its advantages evidently do not warrant the high cost of using it as an extinguishing agent.

Carbon tetrachloride, chlorobromomethane (also known as bromochloromethane), and methyl bromide were the forerunners of the current halogenated hydrocarbon extinguishing agents, which are proving to have at least as good—or better—extinguishing properties with less toxicity hazard.

The presence of halogens increases the extinguishing effect of a compound, but bromine is outstanding in this respect. The presence of fluorine generally increases the inertness and stability of a compound. Perhaps this explains the preference for the four halogenated hydrocarbon extinguishing agents in use today. All contain bromine and fluorine. These agents and their Halon numbers are bromotrifluoromethane (1301), dibromodifluoromethane (1202), bromochlorodifluoromethane (1211), and dibromotetrafluoroethane (2402). The Halon designation specifies, in succession, the number of atoms of carbon, fluorine, chlorine, bromine, and iodine. Terminal zero digits are not expressed. Thus the designation 1301 indicates the presence of 1 carbon atom, 3 fluorine atoms, no chlorine atoms, and 1 bromine atom, and so on.

There is still some uncertainty about the mechanism by which halogenated hydrocarbons extinguish fire. There is agreement that a chain-breaking reaction seems to be involved, or in other words, that the agents tend to break down the combustion reaction process. Explanations for this tendency differ, however, but insofar as the fire service is concerned, halogenated hydrocarbons quickly put out flames, even though they do not necessarily achieve total extinguishment.

Bromotrifluoromethane or "Halon 1301" has the highest extinguishing efficiency rating and the lowest relative toxicity rating among the halogenated hydrocarbons most commonly used. Many American-made aircraft are protected against in-flight engine fires by Halon 1301. Fixed Halon 1301 systems are recommended where sensitive electrical equipment is housed—for example, in computer rooms. Portable extinguishers containing Halon 1301 are not in in common use due to their high cost. Fire personnel are unlikely to use halogenated hydrocarbons personally but should be aware of any toxic effects resulting from such use.

The effects of carbon tetrachloride are well known. For practical purposes, chlorobromomethane is placed in the same category as carbon tetrachloride. Methyl bromide is not listed as acceptable for use in the United States because its natural vapor is more hazardous than carbon tetrachloride.

At a comparatively low temperature (about 900°F [480°C]), Halon 1301 starts to give off products of decomposition in the presence of moisture, accompanied by a sharp, acrid odor, and a noxious, irritating atmosphere. The decomposition products include hydrogen fluoride, free bromine, and carbonal halides, all of which can be lethal if present in sufficient quantities. Some maintain that such quantities are not likely to develop because Halon 1301 is effective so quickly, but the question arises, "Will the Halon 1301 supply be shut off as *soon* as it is effective?" At any rate, when it is known that Halon 1301 or other halogenated hydrocarbons have been used, it is advisable for fire personnel to wear masks and ventilate the premises promptly.

CLASS C FIRES

Methods of extinguishment and the extinguishing agent and action used where electrical equipment is involved depend on the construction features of the electrical equipment and the nature and amount of the combustible occupancy contents involved and exposed. Until the electrical equipment is deenergized, a nonconductive extinguishing agent should be used. Such agents include carbon dioxide, dry chemicals, halogenated hydrocarbons, or fog used from a safe distance. If feasible, the agent that will do the least damage to the equipment and occupancy contents should be selected—if, of course, it is also effective. This can be an important matter where costly computer systems are involved.

Occupancies featuring extensive electrical equipment may be protected by fixed systems using some of the nonconductive extinguishing agents mentioned above. Dry chemical is not recommended for delicate electrical equipment such as telephone switchboards and electronic computers, however, because of the harmful effects of its deposit. Halon 1301 is recommended for electronic computer rooms.

Actually, the fire department ordinarily does not apply nonconductive extinguishing agents on arrival at an electrical fire. The usual procedure, apart from subway situations, is to have the electrical equipment deenergized and to prepare to use the extinguishing method and agent appropriate for the existing or ensuing class A and/or B fires. Insulation can cause plastic fires; steam-turbo generators and transformers can cause oil fires which can involve other combustible material, and so on.

Overhead wires, even though not involved in the fire, and charged accumulations of water, especially in cellars, have caused many casualties among fire personnel. Subway fires, because of charged third rails, are particularly dangerous for civilians and fire personnel, and accordingly should be given special consideration.

Subway Fires

Subway fires seldom present problems of forcible entry or overhauling in the usual sense, but they can present many other problems, especially in rescue work and ventilation. These fires, even when they are minor, are notorious for their potential life hazard. Rescue work is complicated by the large number of people affected; the possibility of panic; the presence of "live" third rails (Figs. 15-1 and 15-2); the danger of other, moving trains (Fig. 15-3); limited access for fire operations; limited egress for passengers; restricted ventilation and communications; and, at times, difficulty in locating the fire. In addition, there are poor visibility, heat, and smoke.

Officers should consider the following checklist for rescue work in subways.

1. Request a power shutdown if necessary. Obviously, immediate compliance is not expected if it will intensify the life hazard. Transit authorities are in a position to check on trains near or within the affected area and should give them time to clear if at all feasible, particularly when the fire is in a tunnel under a river and emergency exits are far apart.

2. If the power is still on and if it is feasible, seek the cooperation of the transit authorities to run several trains together to form a line of cars through which passengers can walk to the nearest station or

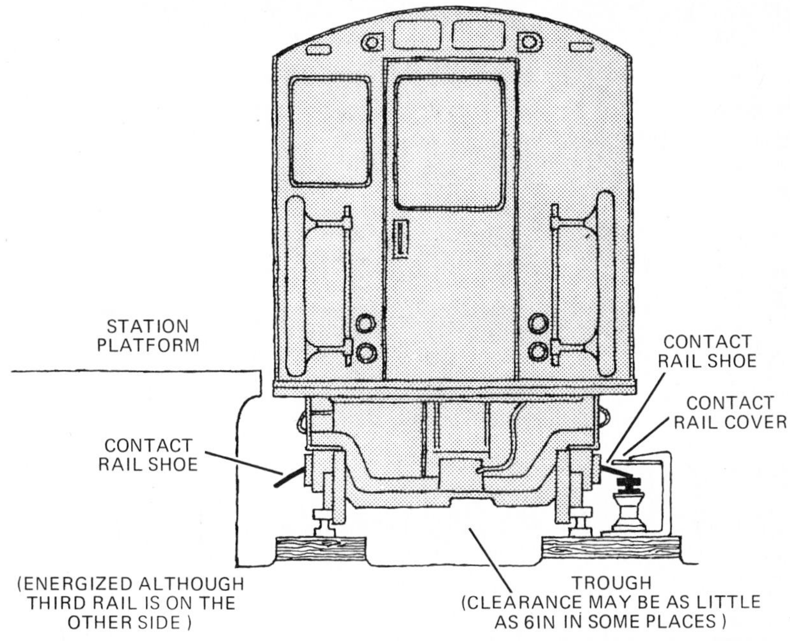

FIG. 15-1. The subway car is powered by the contact rail shoe, which touches the energized third rail.

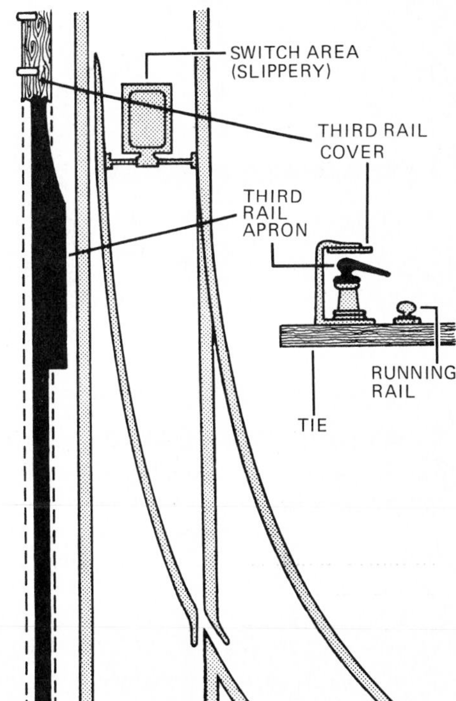

FIG. 15-2. The third rail apron, an extension of the third rail, is also energized and provides power as the train switches tracks.

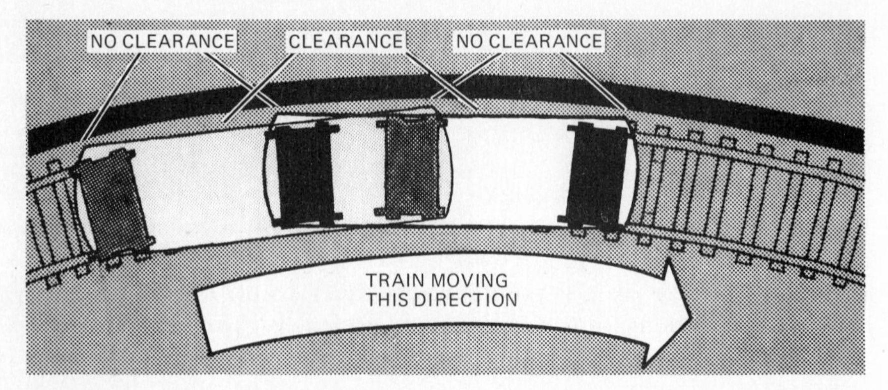

FIG. 15-3. Personnel should avoid standing in curved track areas. There may not be enough clearance between the wall and the ends of the cars as they sweep around the curve.

emergency exit. If the power is off, passengers may be evacuated through the side doors to the bench walk and thence to the nearest station or emergency exit. Evacuation directly to the roadbed is also possible but less desirable due to the climbing involved and grease-soaked cross-ties.

3. When possible and practical, locate and extinguish the fire to reduce the heat and smoke. In any event, stretch and operate a line between the fire and endangered persons or between the fire and the means of escape as soon as possible. Fog might be used to drive the convection currents and smoke in a direction that assists rescue work.

4. Use any available means (fans, fog lines, smoke ejectors) to ventilate and create a favorable draft.

5. Get sufficient personnel and equipment on the job. Since these fires are seldom more than one-line jobs, it is fire personnel rather than apparatus or equipment that is needed. However, masks, resuscitators, inhalators, and portable lights may be required. Get sufficient medical help.

6. Make certain that all endangered persons have been removed.

7. Establish field headquarters so that activities can be assigned properly and can be effectively coordinated by adequate supervision and communication.

Decisions at subway fires can be complicated by mechanical vents and moving trains which spread the smoke hazard over a wide area. Poor visibility and troubles in communicating may make it difficult for the commanding officer to ascertain the critical points and assign units properly. Ordinarily, the fire must be located before rescue can be performed efficiently. However, at subway fires it may be necessary to start rescue work first because of possible panic and the frequent difficulty in locating the fire.

A limited part of the assignment on hand can accompany battalion or district chiefs dispatched to obtain information and initiate required rescue work. Supervision of personnel is all-important at these fires because of the electrical hazards, smoke, heat, poor visibility, and the possible jeopardy of fire personnel rescuing passengers under panic conditions. Where there is no life hazard, lines should not be stretched into "live" sections until it is *definitely* established that the current is off. Misinformation about the current being off can result in serious injuries.

Ventilation can be achieved as previously described. Generally there is a prevailing draft in subways (Fig. 15-4), and it is easier to advance lines from the windward side. There should be a definite understanding between fire and transit personnel about the use of mechanical fans. In some cases it is better to shut them down to prevent heat and smoke from being drawn toward the rescue area. Some fans can take in or exhaust air and smoke, and should be used accordingly.

A stream from 1½-in [37 mm] hose may suffice to put out the fire but the amount of such hose that can be used in the stretch is limited; the higher pressure needed at the pump can result in a burst length of hose. It may be advantageous to carry rolled-up lengths toward the fire area. Sometimes a booster line stretched via an emergency exit not being used by passengers will suffice. And in some instances, outlets of standpipe systems can facilitate the use of lines.

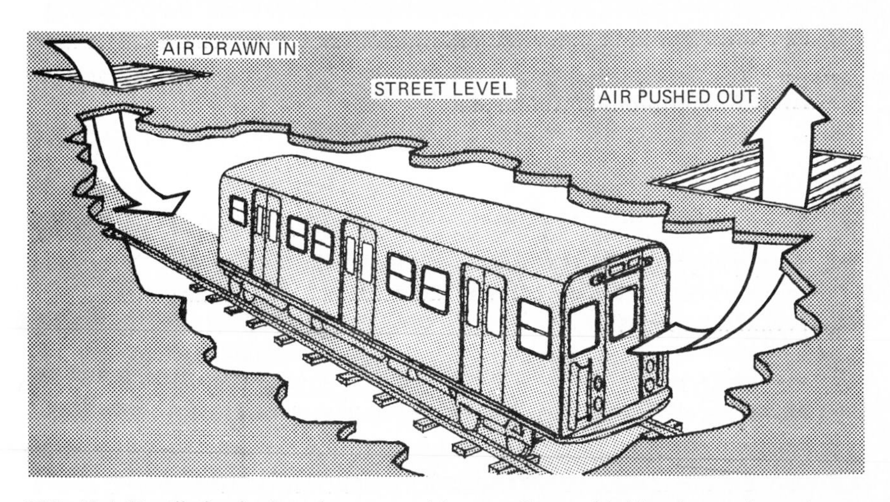

FIG. 15-4. Ventilation in the subway tunnel is normally provided by the moving trains.

At one subway fire escalators became fully involved, creating exterior exposure hazards as the heat rushed up stairways to the street level.

Emergency exits can play a vital role. It is suggested, therefore, that an emergency-exit-card record system be developed to provide responding units with cards to be carried on the apparatus—one card for each emergency exit. Cards should provide information about the area and level served by each exit and should be used during drills and inspections to familiarize personnel with the exits.

CLASS D FIRES

Class D fires involve metals classified as combustible as well as some others not classified as combustible in the ordinary sense.

The choice of an extinguishing method and extinguishing agent depends on the size and shape, as well as the nature, of the metal involved. Extinguishing methods are more likely to be smothering and cooling rather than starving and interrupting the flame chain reaction. Extinguishing agents used to smother and cool the metal involved are recommended only if they do not produce harmful and dangerous reactions (described in Chapter 7). In addition, it can be generalized that (1) water should not be used on molten metal, and (2) water should be used with care when explosive metal dusts may be thrown into suspension. In either case, there can be an explosive reaction.

Fires in Alkali Metals

These fires may be smothered and cooled by graphite dry powder developed for metal fires, dry sand, dry sodium-based powder, and dry soda ash.

G-1 is effective on these fires. G-1 is composed of graded granular graphite, to which are added phosphorus-containing compounds to improve extinguishing effectiveness. Heat causes these compounds to generate blanketing vapors which exclude air from the burning metal. The graphite, by conduction, cools the metal below its ignition point. G-1 must be applied with a shovel or hand scoop; it cannot be discharged from an extinguisher. It is suitable against fires in dry or oily magnesium chips, uranium, titanium turnings, zirconium chips and turnings coated with an oil-water coolant (fires involving moist zirconium chips are only controlled, however, not extinguished), sodium, calcium, hafnium, thorium, plutonium, potassium, sodium-potassium alloys, and lithium.

Met-L-X is the powder with a sodium-chloride base and additives to improve flow and water repellancy. It is applied from an extinguisher to smother and cool the fire. It is listed as effective on fires involving magnesium, sodium, potassium, and sodium-potassium alloy. Met-L-X powder can be used against the same fires as G-1, except lithium fires in depth.

Lith-X is another dry powder that is effective on fires involving alkali and some other metals. Lith-X has a special graphite base with additives to improve flow. It is discharged from an extinguisher and smothers and cools the fire. It extinguishes lithium fires, sodium-spill as well as sodium-in-depth fires, and sodium-potassium alloy spill fires. Lith-X is suitable for the control and extinguishment of magnesium and zirconium chip fires.

Fires in Alkali-Earth Metals

Calcium, barium, strontium, and magnesium are in this group. G-1 powder is effective as an extinguishing agent for fires in calcium. It may also be effective for fires in barium and strontium since they are very similar to calcium in their fire properties, However, magnesium commands the most attention of the members in this metal group.

Water applied to magnesium fires may be decomposed into its component elements, oxygen and hydrogen, and thereby intensify the fire. Nevertheless, automatic sprinklers are recommended for coping with fires in solid magnesium even though initially the water may intensify combustion. The assumption is that a sufficient amount of water will soon cool the metal below its ignition point and put the fire out.

Coarse droplets of water from hose streams, or formed by impinging streams above the fire, may successfully extinguish burning magnesium castings or fabricated structures. This technique is advisable only if sufficient water can be applied to achieve the needed cooling effect. Solid streams should not be used because they may scatter the fire. Fog streams should not be used because they may accelerate rather than cool the fire. Of course, water should not be used on molten magnesium.

In addition to water, G-1, Met-L-X, and Lith-X, other agents are available for use on magnesium fires. Such agents include Foundry Flux and TMB liquid. In open fires, the flux is applied with a scoop or shovel; TMB is applied from an extinguisher. Fluxes are effective in extinguishing chip fires but are not recommended in machine shops since they cause severe rusting of equipment. TMB smothers the fire with a boric acid coating. Cooling is accelerated with a water spray. TMB is not recommended indoors because of the boric acid gases produced.

Other extinguishing agents for magnesium fires include a solution of boric acid in triethylene glycol and tricresyl phosphate. Both these agents produce a secondary fire that can be handled by fog or mechanical foam. Boron trichloride and boron trifluoride gases have been used to control fires in heat-treating ovens containing magnesium. The latter is much more effective. Both gases completely extinguish small fires but reignition is possible with large fires. Boron fluoride followed by foundry flux is advisable for complete extinguishment.

Titanium, Zirconium, and Hafnium

G-1, TMB, and Met-L-X have proven effective on titanium fires and good results have also been obtained by using other agents considered suitable for magnesium fires. Water is not recommended for fires involving fine forms of titanium. Hose streams have been used effectively where small piles of scrap were burning outdoors, but violent reactions can be expected in some cases.

Fires in large pieces of zirconium can be extinguished by the cooling method, using water. Water, however, is not used on zirconium in powder form because it burns violently, decomposes the water and uses the oxygen released for its own combustion. In such cases it is preferable to encircle the fire with a special extinguishing powder and allow it to burn out, or to use TMB or one of the powders (G-1, Met-L-X, or Lith-X) to extinguish the fire if possible. Hafnium fires are handled in the same way.

Atomic-fuel Metals

This group includes uranium, plutonium, and thorium. G-1 and Met-L-X are effective on fires in uranium. Limited amounts of water can intensify these fires. Burning natural uranium can be handled, however, by shoveling the involved material into a drum of water, assuming that the amount of material is limited and that the fire is outdoors. Personnel should wear face shields and gloves and use long-handled shovels. They should not use water on fires involving enriched (fissionable) uranium.

G-1 has proven effective on fires in plutonium. Plutonium fires require special handling, as indicated in Chapter 7. G-1 powder has also proven effective on fires in thorium.

REVIEW QUESTIONS

1. On what primary factor does the class of fire have the most significant effect? Why?

2. What are fire extinguishing methods based on?

3. What extinguishing methods and agents or actions are used on class A fires? What agents achieve both a cooling and a smothering effect? What is the role of ventilation in using the cooling method? How can the starving method be used at class A fires?

4. What extinguishing methods and agents or actions are used on class

B fires? Why should the commanding fire officer consult with fire-security personnel at oil storage plant fires?

5. How and when can a solid or fog stream be used to produce a cooling effect at an oil-tank fire?

6. What is the air-agitation method? What are its limitations?

7. What agents produce a cooling and smothering effect on class B fires?

8. How and when should water be used to produce an emulsifying (smothering) effect on class B fires?

9. What are the limitations on the use of low-velocity fog from applicators at tank fires?

10. What is an important precaution that should be taken when escaping gas is aflame?

11. What are some limitations on the use of dry chemicals and halogenated hydrocarbons to interrupt flame chain reaction?

12. What is a boilover? When can it occur? What precautions can be taken against such an occurrence?

13. What is a slopover? How is it usually caused?

14. Gasoline is flowing from a broken tank line and the vapors have ignited. The valve controlling the flow of gasoline cannot be reached because of the heat. What is a practical technique to use in such a case?

15. When contained and uncontained class B fires have to be dealt with, which should be handled first? Why?

16. How do dry chemicals extinguish class B fires?

17. What are halogenated hydrocarbon extinguishing agents composed of? What were their forerunners? How do halogenated hydrocarbons extinguish class B fires? What hazard do these agents present for fire-fighters?

18. What extinguishing agents are safe and effective to use on electrical fires? What halogenated hydrocarbon is recommended where computer systems are involved? What class of fire may remain after electrical equipment is deenergized?

19. What should commanding officers consider at subway fires? What consideration should be given to subway exits in prefire planning?

20. At metal fires, what does the choice of extinguishing methods

and extinguishing agents depend on? What methods are more likely to be used?

21. What extinguishing agents can be used to smother and cool fires in alkali metals? In alkali-earth metals? What are some reservations about using water on magnesium fires?

22. What extinguishing agents are effective on fires in uranium? How can outdoor fires involving limited amounts of natural uranium be handled?

PART THREE

Assigning and Coordinating Activities

16

Implementing Decisions

Part 3 of this book deals with the remaining steps in the action plan, in which decisions are implemented: the assigning and coordinating of activities by knowledgeable supervision and through effective communication.

It is important to assign activities according to the specific functions of units, especially at critical stages of a fire operation. This contention is not altered by the fact that at times engine companies participate in overhauling, ladder companies "lighten up" on lines going up a stairway, and so forth. In addition, every activity or assignment has, or should have, a specific purpose, which in this text is referred to as a minor or derivative objective. If the minor objectives are successfully achieved, the major objective is automatically attained (Tables 16-1, 16-2, 16-3).

Officers who understand the role of minor or derivative objectives and how they are attained can better define their expectations and supplement or modify their plans. The pertinent primary factors initially determine both major and minor objectives and essential activities.

The sequence in selecting minor objectives is not always the same. For example, if a visibly endangered occupant appears at a fifth-floor street-front window, the first minor objective could be to effect rescue via an aerial or elevated platform. If such apparatus were not available however, the first minor objective could be to protect the occupant pending rescue by other means. Protection may have to be provided by a line in the street.

In many cases where rescue is the major objective, the line used to protect occupants pending rescue also extinguishes the fire. When this does not happen and occupants have been removed, or when no life hazard for occupants exists, the major objective becomes extinguish or

TABLE 16-1. When Rescue is the Major Objective

DERIVATIVE OR MINOR OBJECTIVES COULD INCLUDE	ASSOCIATED ACTIVITIES COULD INCLUDE
To ascertain the severity of life hazard	Ventilate. Force entry as required. Locate fire. Note effects of pertinent primary factors on the life hazard.
To protect occupants pending rescue	Ventilate so as to draw heat and smoke away from occupants. Force entry as required. Use line or lines between fire and occupants or means of escape. If feasible, use loudspeakers to calm panic-stricken occupants.
To effect rescue	Search for and remove occupants via best means available. Render first aid. Obtain medical aid as required.
* To protect personnel	Provide helpful ventilation and protective lines as much as possible and feasible. Use masks whenever feasible. Relieve men as required.
* To coordinate activities	Provide adequate supervision and communication.

* Minor or derivative objectives from the start to the end of every fire operation.

confine, control, and extinguish, and the commanding officer proceeds accordingly.

At most fires, the location of the fire is obvious and the course of action is clear and uncomplicated: simply stretch a line or use an extinguisher and douse the fire. Then minor or derivative objectives are minimized. This can happen when units respond to small fires at which there is little or no need to force entry or ventilate. At structural fires, minor or derivative objectives are more likely to be maximized.

When confine, control, and extinguish is the major objective and there are no exterior hazards, the minor or derivative objectives would be the same, but different tactics may have to be used to confine and control the fire. For example, the floor or floors above the fire may have to be flooded with heavy-caliber streams operating from the periphery, or High-X foam may have to be used on an extensive and otherwise inaccessible cellar fire.

When the major objective of extinguish has to be changed to confine, control, and extinguish because of a possible or imminent structural collapse, the primary factor of life hazard for personnel becomes the limiting or strategic factor in selecting derivative objectives. In such instances, the foremost derivative objective is to protect the endangered personnel. If the signs indicate a possible rather than an imminent collapse, lines

TABLE 16-2. When Extinguish is the Major Objective

DERIVATIVE OR MINOR OBJECTIVES COULD INCLUDE	ASSOCIATED ACTIVITIES COULD INCLUDE
To locate the fire and ascertain its potential	Ventilate and force entry as required. Note effects of pertinent primary factors. (It will be assumed that this evaluation corroborates the choice of extinguish as the major objective.)
To localize the fire, prevent smoke explosion and back draft, alleviate smoke and heat conditions, improve visibility, and facilitate the advance of hose lines	Ventilate and force entry in a manner designed to achieve the objectives.
To extinguish the fire	Select and apply the right extinguishing method and agent. Confine the fire while extinguishing it.
To minimize structural and content damage	Judiciously force entry, ventilate, use water or other appropriate extinguishing agent, overhaul, and perform salvage work. This calls for the proper application of related principles.
To protect personnel	See Table 16-1.
To coordinate activities	See Table 16-1.

may be backed out of the structure as suggested in Chapter 17. However, if collapse appears imminent, lines should be abandoned and personnel should hasten to get out. After personnel are out, the next derivative objective becomes to confine the fire, followed by to control, and then to extinguish.

It is helpful to define the goals of activities to be undertaken, but it is not helpful to classify the purposes of the mechanics involved as minor or derivative objectives. For example, in placing and using hose lines, the purposes of turning on hydrants (to supply the pumps with water) or opening gate valves (to supply water in hose lines) should not be regarded as minor or derivative objectives. Only the purpose of the line is the minor or derivative objective. In ventilation, only the goals of the ventilation itself are minor objectives. The purposes of individual steps, such as raising an aerial to the roof (to get men and equipment to the roof), are not.

Assuming that correct decisions have been made in selecting major and minor objectives, effective implementation will depend on properly assigning the activities and coordinating them by supervision and communication. Assigning activities has already been discussed.

TABLE 16-3. When Confine, Control, and Extinguish is the Initial Major Objective, Assuming the Existence of an Exterior Exposure Hazard

DERIVATIVE OR MINOR OBJECTIVE COULD INCLUDE	ASSOCIATED ACTIVITIES COULD INCLUDE
To confine the fire	Appraise the primary factors that will determine the order of priority in covering the exterior exposure hazard. This may necessitate the use of master and other streams, sprinkler, standpipe, or other fixed systems. Operate as required in exposures to vent, force entry, close windows, move exposed stock, extinguishing incipient fires, and so on.
To control the fire	After steps necessary to confine the fire have been taken, use other available lines to darken down the main body of fire. When possible and feasible, use master and other streams both offensively and defensively.
To extinguish the fire	See Table 16-2.
To minimize structural and content damage	See Table 16-2.
To protect personnel	See Table 16-1.
To coordinate activities	See Table 16-1.

SUPERVISION

Commanding officers can effectively supervise a fire operation if they are trained to have logical expectations about the objectives they select and the activities they assign in view of the primary factors initially pertinent. If expectations are sound, officers can make intelligent appraisals of what is going on and are therefore mentally prepared to make any necessary adjustments. As soon as the operation gets under way, the effectiveness of activities (secondary factors) must be weighed. For example, ventilation may not be as effective as anticipated, thus impeding the advance of hose lines and adversely affecting such primary factors as extent of fire after arrival, heat and smoke conditions, visibility, exposure hazards, and so forth. No well-trained officer would advance a line into the fire area of an unoccupied structure if the effectiveness of ventilation was so questionable that a back draft or smoke explosion might occur.

Even when ventilation is effective, the effectiveness of hose lines can be impaired by burst lengths, defective hydrants, or lines stretched short. The unfavorable effects of such occurrences on other factors are maximized by the fact that the fire area is now opened up. Logical expectations therefore may not always be realized, but nevertheless they should always be defined if the commanding officers are to have a clear notion of what they are trying to do and how they are trying to do it. In addition, experienced officers have learned to expect the unexpected. In such cases, they must be prepared to make necessary changes in objectives promptly.

At the company level, officers can supervise more effectively at fires if they understand the functions of their various units and the principles applicable to such functions. In carrying out assignments, officers should know the major objective of the operation because this determines the risks to be taken. Supervising officers should see to it that proper masks are worn when warranted, and that superiors are promptly notified of conditions relevant to the operation in general and the welfare of personnel in particular.

COMMUNICATION

The effectiveness of communication at fires depends in large measure on the manner in which command posts are established and maintained and the availability of handy- or walkie-talkies. Every fire officer should be supplied with a handy-talkie radio, as this provides a radio network for company officers and the command post to coordinate the emergency operations. Communication and issuance of orders are discussed in Chapter 27.

COORDINATION

Coordination promotes teamwork, which is essential for an organized operation. In addition to supervision and communication, coordination also depends on timing. For example, ventilation should be effectively carried out *before* personnel enter fire areas in which back drafts or smoke explosions could otherwise occur. Commanding officers can demonstrate a proper concept of timing by calling for help when the need is potential rather than actual, by establishing an order of priority in covering exterior exposures, and by notifying incoming units by radio how to approach the fire area and where to report rather than waiting for such units to reach the scene. Essentially, such timing depends on the ability to analyze and anticipate developments—or, in other words, on the ability to recognize and accurately evaluate the effects of pertinent primary and secondary factors.

JUDGMENT, ABILITY TO IMPROVISE, EXPERIENCE

Other intangibles such as judgment, ability to improvise, and experience can have a bearing on the effectiveness with which decisions are implemented, and thereby on the art of firefighting.

Relative to judgment, Reader's Digest *Use the Right Word*[1] states, "Judgment is sense applied to the making of decisions, especially correct decisions, and thus it depends to some degree upon the exercise of discernment and discrimination." The same source states, "Discernment and discrimination are alike in denoting an analytic ability that allows one to see things clearly"—and indicates that the word "sense" is commonly applied to the ability to act effectively in any given situation. In this frame of reference, it would appear that officers could be taught to develop the sense referred to by learning how to recognize and evaluate pertinent factors in making and implementing correct decisions in any given fire situation. In other words, the suggested study of critical-factor analysis could improve judgment at fires.

Relative to improvising, *Use the Right Word* also states, "Improvised can suggest a rough-and-ready substitute for something lacking or the making of decisions as one goes along." Further explanations indicate that the word "improvised" refers to actions that are taken on the spur of the moment or without forethought, and is related to such words as unplanned, impulsive, and extemporaneous. One implication of the foregoing is that officers who depend exclusively on improvising ignore, or are unaware of, an effective substitute—such as practical mental processes of proven value in making and implementing decisions.

Some officers do appear to improvise successfully, but these individuals achieve at least some of their success because they are gifted with innate talent for discernment and discrimination and good sense as well. Thus, they have an analytical ability for recognizing and evaluating factors that are pertinent to making decisions. However, the great majority of officers can only develop such ability by a study of the relationships among factors. In any event, this suggested study is much more likely to meet the need for standardized training than a recommendation to depend on improvising. And, the more gifted officers would probably become even more successful.

Relative to experience, *Use the Right Word* states, "Experienced indicates someone whose familiarity with something is based on considerable practice. By implication, this past immersion in a subject has resulted in superior understanding: an experienced proofreader; an experienced lover; an experienced leader. Sometimes, no gain in wisdom

[1] S. I. Hayakawa (ed.), *Use the Right Word*, Funk & Wagnalls, New York, 1968.

need be suggested by the word so much as a piling up of involvements." An inference from this passage is that experience should represent more than a mere piling up of involvements if there is to be a gain in wisdom or skill. Experience, in firefighting or elsewhere, is fruitful if it promotes skill (art) in applying logical hypotheses or conclusions (science) based on past observations and analyses. Otherwise, it does not necessarily make one skillful or legitimately classify one as an authority.

Officers responding to many fires in districts featuring a predominant occupancy, such as tenements, may become highly adept in operating at such fires. This happens mainly because the pertinent primary factors become familiar and quickly recognized and evaluated, even though the officers may never have heard the term "primary factors." However, these officers may not be so skillful in operating at fires in districts featuring different primary factors. This is something to consider when transferring or appointing officers, especially chiefs.

Leadership, motivation, and morale, which also have a pronounced effect on the art of firefighting, are discussed in Chapters 29 and 30.

REVIEW QUESTIONS

1. What is the advantage of understanding the role of minor or derivative objectives and how they are attained?

2. What are some minor or derivative objectives when rescue is the major objective? How are they attained? Which ones must be considered from start to finish of every fire operation? Why is the sequence in seeking and attaining minor or derivative objectives not always the same when rescue is the major objective?

3. What are some minor or derivative objectives when extinguish is the major objective? How are they attained? When are minor or derivative objectives minimized? When are they more likely to be maximized? Why?

4. Assuming the existence of an exterior exposure hazard, what are the minor or derivative objectives when confine, control, and extinguish is the *initial* major objective? How are they attained? What are the minor or derivative objectives if there are no exterior exposure hazards? How would they be attained?

5. When the major objective of extinguish has to be changed to confine, control, and extinguish because of a possible or imminent structural collapse, what becomes the limiting or strategic factor in selecting derivative objectives? How would the foremost derivative objective be attained?

6. In this text, what are regarded as minor or derivative objectives?

7. What steps in the action plan pertain to the implementation of decisions?

8. What training will help a commanding officer effectively supervise at a fire operation? How is the training referred to used in practice? What can interfere with the realization of logical expectations? What should experienced fire officers learn to expect? Why?

9. How can company officers supervise more effectively at fires? What should be known about the major objective? Why? What are some other obligations of company officers as supervisors?

10. On what does the effectiveness of communication at fires depend?

11. In addition to adequate supervision and communication, on what does effective coordination at fires also depend?

12. What is "judgment" and what does it to some degree depend on? How can officers improve their judgment in operating at fires?

13. What does the word "improvising" suggest? Why is it inadvisable for officers to depend exclusively on improvising when operating at fires?

14. What does the word "experienced" indicate? When is experience in firefighting fruitful? Why can officers experienced in operating at fires in one type of district be less adept in operating at fires in a different type of district?

17

Activities of Engine Companies

Each type of unit in the fire service has been organized and provided with personnel and with certain kinds of apparatus, equipment, and tools so it can function in a specific way toward the achievement of major and minor objectives. Functions of an engine company can include selecting the most appropriate source of water supply, stretching or laying hose of the right size and length, and using the lines stretched as directed. Other functions could include selecting the correct nozzle (when streams are to be used), supplying fixed systems, providing adequate pressures, or taking up lines. Every line has a specific purpose or minor or derivative objective, and here as elsewhere the primary factor of life hazard for occupants is the most significant limiting or strategic factor.

IF THERE IS A LIFE HAZARD

A life hazard establishes the purpose—or minor or derivative objective—of the first hose stream available: to protect the endangered occupants pending their rescue. The engine company responsible should stretch and operate its line accordingly. If necessary, operations to control and extinguish the fire must be temporarily neglected. Hence, the first line should be stretched and operated as quickly as possible.

If an occupant showing at a street-front window is obviously endangered by flame, a protective stream should immediately be operated from the street so that rescue can be effected via ladder or other means. Similar exterior action should be promptly taken where a mushrooming fire has driven the occupants of the top floor to the fire escape, where they are threatened by flames from windows below. It must be kept in mind that heat and smoke can be driven so that they endanger occupants trying to descend by the interior stairway, or fire forces attempting to advance a hose line from the interior. The effect is particularly injurious when victims are caught in passageways and hallways and are then subjected to the full brunt of the driven heat and smoke. Therefore, exterior lines being used to preserve the fire escape *as* a means of escape should be operated at as high an angle as possible while still achieving their purpose. A higher-angle stream (perhaps 70°) will reduce the interior hazard for occupants and personnel (Fig. 17-1). Penetration is not desirable at this point. A smoke condition in itself is not sufficient reason for using an exterior line as described, but a line should be held ready to operate from the street or other appropriate spot if more severe conditions develop before rescue is completed. Visible life hazard must be cared for as soon as possible, and of course prompt steps must be taken to cover probable life hazard.

When the fire starts at the base of a vertical shaft and is extending upward in an occupied building (five story, 40 × 70 ft [12 × 21m] in area,

FIG. 17-1. When protecting a life hazard from the street, personnel should aim the stream at about a 70° angle. This will protect the endangered occupant but will not force the fire back into the building to possibly endanger other occupants.

nonfireproof), the best way to protect occupants pending rescue is to use an all-purpose nozzle at the most practical low level. The fog vaporizes, rises, mixes with the convection currents, and exerts a powerful extinguishing effect in the shaft and adjoining structural channels. The solid stream can be used on and off to put out the fire at the base of the shaft, assuming that the level of operation is above (for example, at the street level). The second line, frequently needed, can be stretched dry to the appropriate level, usually the top floor, and then supplied with water and used if necessary. An officer who advances a line (the first) to the third or fourth floor before using water on the shaft fire violates the cardinal rule governing the use of hose lines when there is a life hazard, the rule of operating the first line between the fire and endangered occupants on the lower floors. This error is more serious if the fire is in a dumbwaiter shaft in the public hall. Fires can break out of such shafts below personnel moving up the stairway before hitting the fire.

If the members on the first interior line are in jeopardy while trying to protect occupants pending their rescue, the second line should be stretched and operated to protect such members, put out any fire that the members may have passed, and ensure the attainment of the first line's objective. This situation can develop when the first line is only darkening down the fire while moving up an involved stairway so that water can be applied as quickly as possible between the fire and the occupants on the upper floors.

In some cases the commanding officer must order a line stretched into an occupied building without knowing how or where the occupants are trapped. This can happen when a ladder company is not yet on the scene to ventilate, force entry to the fire area, and then provide information about the location of both fire and occupants. At such times there is no alternative but to try to extinguish the fire while simultaneously confining it to the area of origin, and lines should be used accordingly (Fig. 17-2). In the meanwhile, or as shortly thereafter as possible, search and removal activities can be carried out, by engine company personnel if necessary.

At aircraft-crash fires, fog lines from booster tanks should be used to cover the life hazard if a supplemental water supply would be available only after the survival time for the trapped passengers has expired. In effect, there is also a survival time for occupants of burning buildings—which depends on how the life hazard for occupants is affected by the other pertinent primary factors. Survival time for occupants at structural fires should therefore influence decisions about using booster-tank water to cover the life hazard or delay such coverage until hydrant water is available. The primary factors of water supply available and street conditions are critical here.

If a fire in an unoccupied building can possibly endanger occupants of an exterior exposure, the first line should protect the people in the exposure. After the life hazard is adequately covered, lines may be used to attack the fire. It is not advisable to operate on the assumption that a possible life hazard will be so slight that it can be ignored. Exposure to even light smoke can be fatal for some people. In addition, expectations about quick extinguishment are not always realized.

In some cases, a line operating to protect occupants simultaneously extinguishes the fire. This is a desirable achievement, of course, but nonetheless, until the life hazard has been removed, the major objective of such an operation is rescue rather than extinguish.

The foregoing principles involving the use of lines to cover a life hazard apply in large or small communities and *at all fires*. The intent of these principles is not altered by the fact that fog or solid streams are used, or that the water comes from a booster tank, hydrant, or standpipe riser, or even that extinguishers have to be used. (The use of fog in indirect attack to cover the life hazard is limited because the steam created may worsen the plight of the occupants.)

FIG. 17-2. When protecting a life hazard with an interior line, personnel should direct the stream between the endangered occupants and the fire.

Chiefs responding to operations in progress would do well to keep in mind the impact of a life hazard on units stretching and operating initial lines, as the following case illustrates.

A second alarm necessitating your response as a deputy chief has been transmitted for a top-floor fire in a five-story, nonfireproof, residential building, about 40 × 80 ft [12 × 24 m] in area. Adjoining structures are similar in construction, occupancy, height, and area, and there are no front or rear exposures. En route, you learn that the fire is on the street-front side, and that occupants are being removed. On arrival, you are informed that occupants have been removed and that roof ventilation is still in progress. From your position in the street you can see that considerable flame is coming out of the top-floor street-front windows, as well as from the precariously hanging cornice of the fire building, and that smoke is issuing from the cornice of the adjoining buildings. As second-alarm units start to report in, the chief in command of the interior operation reports, via handy-talkie, that the first two lines are having difficulties trying to "make the fire."

In such an event the responding chief can assume that the impact of the life hazard was such that the first two lines were stretched and operated without masks to more quickly protect the occupants. It is also apparent that personnel on the lines are in no position to don masks but that the lines cannot be effectively advanced without them. In such cases it is suggested that, after covering the side exposures, the personnel of the next two available second-alarm engine companies should—equipped with masks—take over the hung-up lines. This was actually done. The lines were quickly advanced and the fire was extinguished.

It may be noted that at the fire in question the major objective became to confine, control, and extinguish—*after* the occupants were removed. The fire was first confined by covering the exterior exposures; interior lines then controlled and extinguished it. The primary factor of structural collapse (hanging cornice) precluded stretching lines via an aerial or the front fire escape. Tower ladders were not then available.

IF THERE IS NO LIFE HAZARD

When there is no life hazard, the major objective is to extinguish or confine, control, and extinguish. If the major objective is extinguish at a structural fire, direct-attack tactics are usually used. The first line is stretched into the building and operated so as to extinguish and, at the same time, confine the fire to the area of origin. This prevents involvement of, and extension by, vertical and horizontal structural channels beyond the fire area, and is achieved by driving heat and smoke out of the building as harmlessly as possible and away from the channels referred to while the fire is being extinguished.

ONE-LINE FIRES

One-line fires generally can be readily located. They are usually accessible to direct attack and limited in extent, and usually, but not always, involve contents only, and not the structure. (Indirect attack, using fog, may also be successful in some instances.) However, at fires in buildings under construction or demolition it may be feasible to use a line from the exterior due to the hazards and delays associated with an interior operation in such occupancies. In some cases, a deckpipe has been used successfully.

USE OF SECOND LINE

If the first line cannot successfully advance and extinguish, it may be practical to back it up with a second line and try to advance with the two lines. An alternative is to withdraw the first line to the hallway or stair landing, close the door of the involved occupancy, and change the minor objective of the line temporarily to confining the fire to the area of origin. In the meantime, an effort can be made with a line from an aerial or elevated platform or deckpipe from the street, or with a multiversal nozzle from the side or rear, or with a hand line from the fire escape to reduce the severity of the existing fire conditions.

Personnel inside the fire building should always be warned that outside streams are going to be used and given ample time to get out of the path of the heat and smoke that will be driven ahead by such streams. If outside streams are to be used extensively, personnel should be evacuated from the building. If an outside stream is required only briefly, as implied in discussing the use of a second line, personnel can remain in the fire building. Then, when the outside stream has achieved its purpose, an effort can be made to advance the interior line, reinforced by an additional line if necessary.

If structural channels such as partitions, ceilings, and shafts are involved, it is generally advisable to have a second line stretched and available to check the extension of the fire through these channels. In some instances the second line may be used to help the first line achieve control on the fire floor so that units sent above the fire can work more effectively to check vertical extension. It has been found repeatedly that units sent above uncontrolled fires are wasted because they cannot perform effectively in severe heat and smoke conditions and zero visibility. In addition, the means of escape for such units is not properly safeguarded (Fig. 17-3). Better and safer results are obtained if units are not sent above the fire until the attack at the lower level has been supplemented and control appears imminent. Imminent control means that lines are successfully advancing on the fire floor, extinguishing the fire, driving heat and smoke

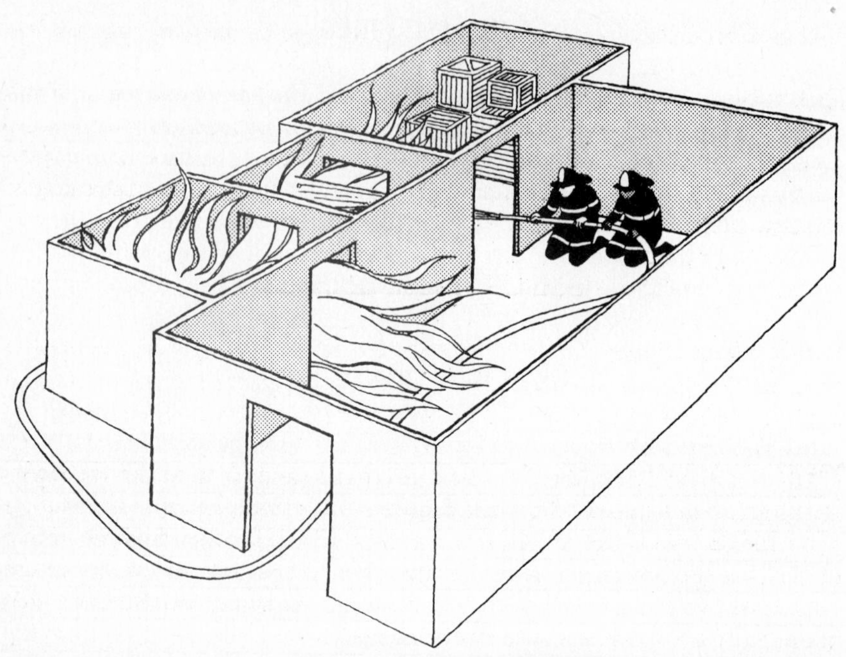

FIG. 17-3. Passing the fire vertically or horizontally can be extremely dangerous, since it may put the fire between the firefighters and a required means escape.

ahead and out of the building, thus minimizing the fire conditions on the floor above and safeguarding a means of retreat for the units on that level. This matter is of particular concern in buildings of inferior construction.

To order a second line stretched and then not use it does not reflect badly on an officer, as experienced fire personnel know. When specific and complete information about pertinent primary factors is not readily available, in the early stages of an operation, it is necessary to anticipate developments and take appropriate countermeasures—such as stretching a second line, which may later prove unnecessary. Such second lines are not charged unless used and therefore no real hardship is entailed, particularly if the first-due engine company laid two lines, as it should, when possible and feasible. Again, it is a matter of good timing and coordination to have a second line available while the need is only probable rather than waiting until the need is actual.

If engine company officers advancing hose lines realize that the heat is making their position untenable, they should back their units out, using the line as much as possible to protect the withdrawal. They should not wait for orders from commanding officers who may have arrived on the scene after the operation started or who cannot properly evaluate the situation from a position in the street—but who, in all likelihood, would concur with such a decision anyway. If the engine company officers feel that a second line is necessary in order to make progress, they should back their units into a position that provides a means of escape until the second line arrives. Commanding officers should be promptly notified by handy-talkie about such decisions so that they can evaluate heat conditions and other pertinent primary factors and know what is happening in relation to their expectations.

USE OF MULTIPLE LINES

Multiple lines (three or more), advancing as a battery, may be required at fires involving large-area occupancies such as churches, armories, and department stores. Personnel on only one or two lines could possibly be outflanked by such fires.

On occasion, three lines are used at structural fires to attain the major objective of extinguish. This can happen when two lines are required to achieve control on the fire floor and a third line is sent to the floor above to check vertical extension of the fire. A means of retreat (aerial, tower ladder, or fire escape) should be available for personnel on the third line, in case there are unfavorable developments.

After occupants have been rescued, or when the fire building is unoccupied, developments—for example, signs of possible structural collapse—may indicate that interior positions should be abandoned. In changing from an interior to an exterior operation, personnel in the most hazardous position should be withdrawn first, and then those in the next most hazardous position. Units operating protective lines should withdraw only after their assignment has been completed, assuming such protective measures are possible and feasible. Orders to abandon lines should be obeyed promptly.

Officers withdrawing from interior positions may be able to provide the commanding officer with helpful information about such pertinent primary factors as location and extent of fire, occupancy contents and how they are stored, interior exposure hazards, and possible structural collapse. This can result in better application of exterior streams, in breaching walls, or in choosing other techniques that are appropriate. Information about cellar situations, for example, could help the commanding officer choose to flood the floor above the fire or inject high-X foam into the cellar. The latter would be preferable if there was not too much structural obstruction.

The *possibility* of having to change from an interior to an exterior operation in the situations referred to above warrants the transmission of a second alarm. Lines from the greater-alarm units can be used to protect those retreating from the fire building, and then to confine the fire by covering exterior exposures. The first lines out of the fire building should, if necessary, be used to cover personnel withdrawing remaining lines. Otherwise, they can be repositioned to confine the fire, since the major objective has become confine, control, and extinguish. After the fire has been confined, lines would be used to control and extinguish it.

These principles apply at any fire. The efficiency with which they are used depends upon the extent to which they are understood and the skill with which they are applied. They can be applied more skillfully by those who have an adequate knowledge of the use of water in fog or solid streams, or with additives.

Regulations relevant to stretching lines, supplying fixed systems and heavy-caliber streams, using booster equipment, and the like, vary in different departments and different sections of the country. All officers, however, should have a working knowledge of universally accepted principles of hydraulics to ensure adequate pressures at the nozzles of stretched lines.

Engine companies at times engage in other phases of fire operations such as ventilation, overhauling, and salvage, but not when they are responsible for covering a life hazard with a hose line.

CONCLUSION

Preceding chapters have explained the effects of primary factors on secondary factors or fire activities, including the placement and use of hose lines. Specifically, these effects can be summarized in terms of engine company activities as follows:

1. The presence or absence of a life hazard determines the major and minor objectives and the risks to be taken.

2. Location of fire establishes the target in stretching lines, in the number, and sometimes in the size of the hoses to be used.

3. Extent of fire on and after arrival influences the size of hose, type and size of nozzle, and nozzle pressures to be used. If master streams are required, consideration should be given to control, direction, penetration, sweep, and the need to alert personnel who may be endangered when such streams start to operate.

4. Construction influences the rate and direction of fire spread, the means by which lines can and should be stretched, and structural stability.

5. Occupancy has to be considered in relation to its two phases, (*a*) the human element, and (*b*) the contents and associated hazards. The former is the nucleus of the life hazard and can be a criterion for the

risks to be taken; the latter can indicate the extinguishing agent to be used.

6. Height may make it necessary to depend solely on standpipe lines and can affect the exterior exposure hazard and, accordingly, the use of lines.

7. Area may necessitate the use of heavy-caliber streams or a battery of lines to advance on a wide front. Proximity of exposures can likewise affect the exterior exposure hazard and the use of lines.

8. Signs of a possible structural collapse are an alarm to "keep out" or to get out if such signs are not initially observable when units enter the fire building. Note the exception to this "rule" when there is a life hazard.

9. Time elements can adversely affect the life hazard, visibility, and the location and extent of fire, and thereby the placement and use of lines.

10. Auxiliary appliances are to be supplied and used according to the individual departmental regulations.

11. Heat conditions in general determine the caliber of stream to be used. The principles to be applied in placing and using lines depend in large measure on how the heat is creating a hazard (by conduction, radiation, convection, or combinations thereof), as explained in Chapter 10. Conditions conducive to heat exhaustion should be anticipated and casualties avoided by relieving personnel more frequently.

12. Smoke conditions necessitate the use of masks and sometimes can be alleviated by fog streams.

13. Wind and weather have an obvious bearing on the range of streams, use of hydrants, freezing lines, heat exhaustion, and so forth.

14. The availability or unavailability of requirements for an effective operation decidedly affect the success or failure of hose lines. It is evident that the water supply, apparatus, equipment, and personnel available have such an effect.

15. Topography, especially in hilly woodland areas, can severely restrict the mobility of apparatus, equipment, and personnel, and thereby the placement and use of lines. This situation can be further aggravated by a questionable water supply.

16. Street conditions, especially during or after a snowstorm, can affect the selection and use of hydrants and the time required to get a stream in operation.

17. The class of fire determines the extinguishing agent to be used. In this connection, it is important for an engine-company officer to know when water is not to be used at all, or when low-velocity fog should be used, or when water should be used with additives, and so on.

REVIEW QUESTIONS

1. Why are units in the fire service organized and provided with personnel, and certain kinds of apparatus, equipment, and tools?

2. What are the functions of an engine company?

3. When rescue is the major objective, what is the purpose (derivative objective) of hose lines to be used? How is this derivative objective attained when an occupant showing at a street-front window is obviously endangered by flame? When occupants on fire escapes are threatened by flames from windows below? When the fire starts at the base of a vertical shaft and is extending upward in an occupied building? If members on the first interior line are in jeopardy? When the commanding officer does not know where occupants are trapped?

4. Where do the principles referred to in question 3 apply? Does the intent of these principles change because they may have to be applied in different ways?

5. What effect would survival time have on the use of booster-tank water and equipment at aircraft-crash or structural fires?

6. What is the purpose (derivative objective) of the first line to be used when a fire in an unoccupied building possibly endangers occupants of an exterior exposure? Why?

7. What should chiefs keep in mind about fire operations when the building has been described as occupied? Why?

8. When extinguish is the major objective and direct-attack tactics are being used, what is the purpose (derivative objective) of the first line? How is this objective attained?

9. What are usual features of one-line structural fires? What may be unusual about one-line fires in buildings under construction?

10. What are some alternate ways for using the second line at structural fires, assuming that the derivative objective is still to confine and extinguish?

11. When should units be ordered to operate above the fire in un-

occupied buildings, particularly in buildings of inferior construction? Why?

12. When is it advisable to order a second line stretched even though it may not have to be used?

13. What should engine company officers do when they feel that their positions in the fire building are becoming untenable because of the heat? When they feel that a second or back-up line is necessary in order to make progress with their lines?

14. Why is a multiple-line advance advisable at fires in large-area occupancies?

15. How can three lines be used to achieve the major objective of extinguish at structural fires?

16. How should lines be withdrawn when signs of a possible structural collapse indicate that interior positions should be abandoned? What information can officers withdrawing from the fire building give to the commanding officer that will be pertinent to setting up an exterior operation? How should lines then be used, considering that the major objective becomes to confine, control, and extinguish?

17. What affects the efficiency with which principles governing the use of lines are applied?

18. What is the restriction in assigning engine-company personnel to engage in activities such as ventilating, overhauling, and salvage?

19. In what ways can primary factors be limiting or strategic in deciding how to supply, lay or stretch, place, and use hose lines, and select extinguishing agents?

18

Activities of Ladder, Rescue, and Squad Companies

LADDER COMPANIES

The functions of ladder companies include ventilation, forcible entry, searching for and removing occupants, laddering, checking for extension of fire, rendering first aid, overhauling, and salvage. At times, ladder units may be used to locate fires.

Ventilation

Ventilation, a very important activity or secondary factor at structural fires, is defined here as the controlled circulation of heat, smoke, and gases to achieve a specific minor objective—or purpose—that contributes to a major objective. To ventilate efficiently, an officer should know *why* ventilation is done, as well as *how*, *when*, and *where* to ventilate.

Why Ventilation Is Done. Ventilation has many purposes, such as to protect occupants pending their rescue, alleviate smoke and heat conditions, prevent smoke explosions or back drafts, improve visibility, locate the fire, confine the fire, localize the fire, facilitate the advance of hose lines, and so forth. The minor objectives depend on the major objective and the pertinent primary factors.

How to Ventilate. Ventilation can be done in two ways. *Natural means* may involve doors, windows, bulkheads at roof level, deadlights, sidewalk covers, and openings made in floors or roof (Fig. 18-1). *Mechanical*

FIG. 18-1. A structure may be ventilated via opened doors, windows, roofs, bulkheads, and deadlights. Fans, air-conditioning systems, and fog streams may also be used.

means may involve portable fans, smoke ejectors, exhaust systems, fog streams, and, perhaps in the future, pressurized stairways and smoke shafts. Natural means may be supplemented by axes, claw tools, hooks, ball and chain, power tools, aerials, and heavy streams (to break windows). The use of masks, of course, can play an important part in effective ventilation, especially within the fire building.

Fog streams can be effective in ventilating if, when directed outward from inside the structure, they can develop a sufficiently low pressure at an opening, such as a door or window. (Stair pressurization and smoke shafts are discussed elsewhere.)

The selection of means to ventilate, or how to ventilate, is determined by the goal or minor objective to be attained in view of the major objective and primary factors that may be pertinent. A life hazard could make the minor objective that of to protect occupants pending rescue, in which case ventilation would be to draw the heat and smoke away from the endangered occupants until they could be removed. Construction determines the natural means available for ventilation. The location or level of the fire indicates the natural and/or mechanical means to be used. For example, at cellar fires, sidewalk deadlights or covers over ash hoists may have to be opened up, or smoke ejectors may be helpful. The extent of the fire could preclude interior and roof operations and thereby delimit the availability of both natural and mechanical means. It might prove necessary to use aerial ladder tips or heavy streams to break windows unless tower-ladder platforms could be used effectively. Wind conditions could dictate that personnel work with their backs to the wind when making roof openings, and that, when working from fire escapes or on floors, openings be made on the lee side first, assuming that no exposure hazard is created or worsened thereby. Requirements available determine the tools and equipment that can be used for ventilation.

When To Ventilate. Good timing is always essential for effective coordination. It is particularly essential in ventilating, especially if there is the possibility of a back draft or smoke explosion. In such a case, fire areas should be adequately ventilated before units attempt to advance hose lines.

When to ventilate depends in large measure on the presence or absence of a life hazard. If there is a life hazard, ventilation designed to protect the occupants pending their removal should be effected as soon as possible (see above, How To Ventilate). If there is no life hazard, ventilation should be coordinated with the arrival of water at the nozzle of the company about to move in on the fire. Premature venting can allow the fire to gain headway, to "spread like wildfire." This can occur when a storefront window is broken and the fire rushes up the front of the building while the engine company is still waiting for water. Personnel on street-front fire escapes may be able to hear the pumper as it starts to deliver the water and see the line stiffen and time their venting accordingly. On rear fire escapes, personnel may wait until they hear the ladder company forcing entry, which usually signals the arrival of water at the nozzle. In general, if roof ventilation is necessary it should be carried out as soon as possible, but if exposure hazards might be created or intensified, protective lines should be on hand.

Where to Ventilate. This problem can perhaps be best illustrated by explaining how various minor objectives are achieved. If the minor objective is to protect occupants pending rescue, openings should be made wherever they will be most effective in drawing the heat and smoke away from the occupants until they can be removed (Fig. 18-2). If the minor objective is to alleviate heat, smoke, and visibility conditions, the location of the fire plays the leading role on where to ventilate.

Cellar fires may be vented at sidewalk openings. If this does not alleviate heat and smoke conditions sufficiently, it is not likely that lines can be successfully advanced. Flooding or the use of High-X foam may be necessary. After such fires are under control, openings in the floor above the fire may help to alleviate smoke and heat conditions during the overhauling phase. Such openings should be made where the rising gases can escape or be driven from the building as harmlessly as possible. For example, if the fire is in a store, the floor opening should be made beneath the opened skylight on the roof.

FIG. 18-2. When an occupant's means of escape is blocked by fire (left), opening the roof directly over the fire will check the fire's horizontal expansion as the rising convection currents seek the path of least resistance.

If the minor objective of ventilation is to localize the fire, openings should be made as directly over it as possible (Fig. 18-3). However, this objective and technique is not feasible at fires on intermediate floors of ordinary (brick-joist) construction because the rising heat and smoke may spread the fire. In such cases, the fire floor is generally ventilated horizontally by opening doors and windows at the fire level, and vertically by opening roof bulkheads (except over elevators), scuttles, or skylights to clear the stairways and hallways. In fire-resistive construction, ventilation is mainly horizontal on the fire floor. However, it may also have to be carried out on top floors served by elevator shafts and at the top of fire stairs that had to be opened at the fire floor to supply lines from the enclosed standpipe riser.

Ventilation by means of elevator shafts is not recommended because (1) it may endanger passengers of an elevator car; (2) the open shaft may create a hazard for personnel; and (3) unnecessary damage to cables, rails, and machinery may result. The need to assist rescue work, however, might override the last two objections.

Ventilating from fire escapes has many dangers, and great care must be taken to avoid being cut off, particularly when working on balconies above the fire floor. The hazards are greater on rear fire escapes because the plight of endangered personnel cannot be detected as quickly, and protecting hose lines cannot be provided as quickly.

To relieve a smoke condition from fire escapes when there is no life hazard, the best procedure may be to work down from the top floor (assuming the roof has already been opened up), opening the remote window first, then the next remote window, and so forth. The direction of the wind is important; venting may become very difficult if the window on the

FIG. 18-3. Opening a roof in the wrong place (left) can actually help spread the fire. Opening a roof as directly over the fire as possible can localize the fire (right). Power saws are now used more often than axes to make openings in roofs.

windward side is opened first. A severe smoke condition with no life hazard initially warrants an exterior rather than an interior ventilating effort, preferably from fire escapes if available. However, exterior work often supplements an interior attempt.

If there is a life hazard, entry from the fire escape, aerial, or tower-ladder platform should be possible if the roof has been opened up. The usual procedure in searching for occupants is to open the nearest window first, then the next nearest. In many instances, the ladder company that has vented the roof starts down the rear fire escape and, working from the top, vents, searches, and examines floors in the rear.

Effects of Primary Factors. In previous chapters, observations have been offered about the way in which activities can be affected by primary factors. For example, observations were offered about the way in which ventilation can be affected by such primary factors as life hazard for occupants and personnel; location and extent of fire on and after arrival; heat, smoke, and visibility conditions; construction; occupancy; structural collapse; weather; wind; duration of operation; and requirements to operate. The effects of primary factors specifically on roof ventilation are as follows:

1. The presence or absence of a life hazard determines the major and minor objectives and the risks to be taken. This has the greatest influence on how, when, and where to ventilate.

2. Location of the fire has a bearing on where to vent. Opening up at the front when the fire is coming up the rear can involve the entire cockloft.

3. Extent of fire on and after arrival should have a bearing on how long and how much the roof should be opened up. The objectives and the limiting or strategic primary factors of course influence this decision. (See item 7.)

4. Heat and smoke released may create or intensify exterior exposure hazards. Where there are equally effective alternatives, one should choose the less hazardous and call for protective lines or use those already made available by the commanding officer. Impaired visibility may require the use of lighting equipment. (See item 10.)

5. Construction indicates how the roof is to be opened up. In fire-resistive construction, openings are seldom made in the roof itself. However, such openings are quite common at fires in ordinary brick-joist construction, in addition to venting over stairways, scuttles, and skylights. It is important to consider the age of the structure in assaying the hazard for personnel since old buildings are more prone to collapse.

6. The human element in the occupancy affects the severity of the life hazard and, thereby, the risks to be taken.

7. Old brick-joist construction is more subject to collapse than most other types. Accordingly, if fire conditions start to deteriorate and personnel would be unnecessarily jeopardized by continuing to work on the roof of such structures, ladder-company officers should back their men off to a means of retreat (adjoining roof, aerial, tower-ladder platform, or whatever) and promptly notify their commanding officers of their action. Failure to get off such roofs in time has resulted in entire companies being buried by a collapse.

If such structures are occupied, extreme risks may be in order. However, if two or more floors are heavily involved, there may be time only to open up over stairways or skylights before collapse occurs. If other openings are to be attempted because of the severe life hazard, roof ropes should be tied around the personnel assigned so that they can be hauled to safety if necessary.

8. Weather conditions may help to locate hot spots, via melting snow, steaming on a rain-soaked roof, and so forth.

9. Wind direction dictates how roof openings should be made: from the lee to the windward side while personnel keep their backs to the wind.

10. Requirements to operate effectively on the roof should be considered in the street. Power tools, communicating devices, lighting equipment, and even masks may be needed. Masks can be a boon to firefighters at some roof jobs. Getting to the roof with the required equipment can be difficult if an aerial is the only means available. Getting to the roof via an adjoining building (if feasible) or a tower-ladder platform would be preferable.

Forcible Entry

Various tools such as axes, claw tools, door forcers of various kinds, and even cutting torches at times may be used to force entry into a structure or an occupancy within a structure. The purposes or minor objectives of forcible entry may include to search for and remove occupants, to enable units to enter and operate, to check for extension of fire, to locate the fire, and so on.

When, Where, and How to Force Entry. There are principles to be applied about when, where, and how to force entry.

If the major objective is rescue and the minor objective is to search for and remove occupants, entry should be forced as soon as possible in an effort to reach and remove trapped occupants, in some crucial situations even without a protective line. On the other hand, if the major objective is extinguish and the minor objective is to enable units to enter and operate, forcible entry should be delayed until the fire area has been sufficiently ventilated to prevent a harmful back draft or smoke explosion, and a line or lines are ready to enter and operate. If the door has been forced prematurely, as sometimes happens, it should be held closed until the proper time for entry.

If the purpose is to enable units to enter and operate, the location of the fire can affect where to force entry. For example, if the fire is in the rear storage area of a supermarket, entry should be forced at the rear rather than the front, thereby minimizing the life hazard for personnel by reducing the distance lines have to be advanced through the heat and smoke (Fig. 18-4). This principle could also apply at fires in the rear of other commercial occupancies of unusual depth, provided that lines could then operate to extinguish the fire without driving it toward uninvolved parts of the occupancy.

Construction can affect where as well as how to force entry. Today, apartment doors have so many locks that it is sometimes easier and faster to force entry through a fire-escape window and then open the door locks from inside. Usually this can only be done to check for extension in an exposed apartment or occupancy, rather than in the fire area. If the purpose is to locate the fire, it may be preferable to raise a ladder to a second-floor window rather than force expensive and difficult doors at the ground level. The damage done in forcing entry should have a reasonable relationship to the extent of the fire.

Similarities and Differences in Ventilation and Forcible Entry

Forcible entry and ventilation are similar in some respects, but different in others. They are similar in the following ways:

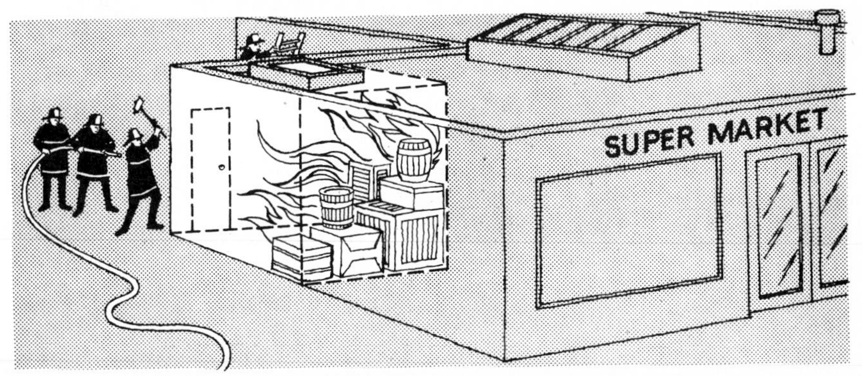

FIG. 18-4. The location of the fire determines the location of forcible entry. Here, entry is forced at the rear, rather than the front.

1. Both are ladder-company assignments, although members of other units should be trained to do such work when necessary.

2. The same method and tools may be used in opening windows to force entry or ventilate. However, windows are fully raised from the bottom in forcing entry but are adjusted top and bottom in ventilating.

3. Both affect the flow of heat, smoke, and gases. The effect of forcible entry here is incidental, whereas an important minor objective of ventilation is to control the direction taken by heat, smoke, and gases.

Forcible entry and ventilation are different in the following ways:

1. They differ by definition and specific purposes or minor objectives.

2. They are carried out at different times. If there is a possibility of a back draft or smoke explosion, ventilation should precede forcible entry. This is not necessarily the order to be followed if rescue attempts have to be made.

3. They are carried out in different ways. Ventilation is carried out by mechanical and natural means, many of which are not used to force entry.

4. They are carried out in different places. Generally speaking, ventilation is more of an exterior activity while forcible entry is more of an interior one.

Search and Removal Activities

The purposes or minor objectives of these activities are self-evident.

Where to Search. In searching for occupants, personnel should check closets, under beds and in cribs, under blankets, in bathrooms, or other places where occupants, especially children, may try to seek refuge from the fire. In some cases, information from occupants who have escaped can be helpful. At times, but too late for rescue, occupants have been found at the bottom of shafts, in courtyards or back yards, and might well be found on offsets of high buildings. Victims have been found at night fires in apparently unoccupied commercial buildings and at fires in abandoned buildings. Hence, a search for possible occupants should be instituted as soon as feasible, even when confine, control, and extinguish is the major objective.

Means of Removal. Other things being equal, the suggested order of priority for removing occupants is interior stairs (especially fire towers), fire escapes, aerials or snorkels (tower-ladder platforms), portable ladders, roof ropes, and life nets (Fig. 18-5). The baskets on snorkels, cherry pickers, tower-ladder platforms, or similar apparatus, are particularly advantageous for the removal of persons who are overcome or otherwise unable to help themselves or the rescuing firefighters. However, aerials can have greater reach in some situations, and provide a continuing path of escape, whereas the snorkel basket has to be lowered and raised again.

The choice among the means of removal can be influenced in a direct way by such limiting or strategic primary factors as construction, height, location and extent of the fire, weather, time of fire, and requirements available to operate.

From a structural standpoint, construction determines the means that are available for removal. Height has a potential bearing on the level or location of the fire and on the possibility of using ladders or snorkel-type apparatus. Location and extent of the fire may rule out the use of interior stairways for removing occupants above the fire. At night fires in cold weather it may be advisable to take flimsily attired occupants to floors below the fire, via stairways or fire escapes, rather than out of the building. Requirements available to operate on arrival of the fire department may include aerial or portable ladders, snorkel-type apparatus, roof ropes, and life nets.

FIG. 18-5. Means of removing occupants. The numbers indicate the suggested order of priority in removal.

Removal of occupants at fires in high-rise buildings has been discussed in Chapter 6. The choice among available means of removal at other fires depends on the degree to which fire conditions permit the suggested order of priority to be followed. The location of the fire is a particularly pertinent primary factor here because fires at low levels tend to negate the value of interior stairways and fire escapes. If occupants are trapped over fires in the rear of buildings without fire escapes, and ladders cannot be used effectively, an alternative may be to lower a firefighter on a roof rope and then lower both firefighter and occupant to the ground. A life net is the last resort. The use of elevators was discussed in Chapter 6. Bed ladders of metal aerials are preferable for removal where the alternative is to raise and use portable ladders.

Laddering

Ladders are raised (1) to remove visibly endangered occupants and to search for and remove other endangered persons; (2) to alleviate heat and smoke conditions in some cases by breaking windows with aerial tips, and to enable ladder-company personnel to achieve other minor objectives of ventilation on a roof or elsewhere; (3) to enable personnel to achieve minor objectives of ladder pipes or lines stretched by ladders and; (4) to provide a means of escape for personnel, and so forth.

To achieve any or all of the minor objectives of aerials, reach and angularity should be considered in placing the apparatus. Angularity depends on the distance between the platform and the building in relation to the height of the target and is most important when the purpose is removal. Rescue is slower and more dangerous for both occupants and personnel if the incline of the ladder is too steep or not steep enough. If unusually wide sidewalks (30 ft [9 m] in some cases) and parked cars would place the truck too far out, consideration should be given to placing the truck on the sidewalk to ensure reaching any occupants. This has been done more than once. The alternative is to operate from the street and use a scaling ladder from the top of the aerial, a difficult, slow, and dangerous undertaking. Owing to their greater strength and higher guard rails, metal aerials have a large acceptable angularity range.

When people cannot get down the fire escape because of the fire below, the aerial should be placed against the building next to the balcony, preferably on the windward side and away from the opening to the level below. Metal aerials should not be raised too much above the balcony because their high rails can make it difficult to remove the occupants. In any event, a protecting line should be operated between the fire and the endangered occupants pending their removal.

Because of obstructing apparatus or other conditions, the aerial ladder must at times be used at an oblique angle, and members of ladder companies should be instructed and trained accordingly. The use of an aerial at an oblique angle has advantages. It is preferred to get the benefit of a prevailing wind that is blowing the heat and smoke in a direction that facilitates rescue (Fig. 18-6). Where there are two targets (trapped occupants) some distance apart and only one aerial is available, rescue can be achieved more quickly and from one position by using the aerial at an oblique angle. It may be advisable to position the truck midway between the targets, however, to minimize the obliqueness of the angle at which the ladders are to be used (Fig. 18-7). At fires where streets are steeply graded, the aerial may be used more effectively at an oblique angle from the high side of the fire building to ensure better reach (Fig. 18-8).

A disadvantage of placing an aerial at an oblique angle on level streets is that it must be extended more than if the platform is directly in front of the target, unless the truck can be positioned nearer the building. This is not always possible or feasible. If windows are only about 2 ft [.6 m] wide, which is the case in many residential buildings, the use of the aerial at an oblique angle may necessitate taking out the upper and lower parts of the window to provide sufficient room for entry, search, and removal. Metal aerials with high guard rails make rescue extremely difficult in such cases. Ladders placed at an oblique angle in a window may create undue strain at the base of the aerial if a twisting force is exerted. Aerials raised at an oblique angle to metal roof cornices or rails of fire escapes tend to slide, increasing the strain at the base of the aerial. This tendency to slide is aggravated by the weight of fire personnel and equipment.

Relative to reach, officers should study the heights between floors in various types of construction in order to gage more accurately whether or not their aerials can effectively reach certain floors or roofs. This ability can be developed at outdoor drills.

An analysis of the foregoing observations will indicate that the angularity and reach of aerials are affected by such primary factors as life hazard for occupants and personnel, location and extent of fire, construc-

FIG. 18-6. Placing the aerial ladder at any oblique angle from the windward side lessens exposure to heat.

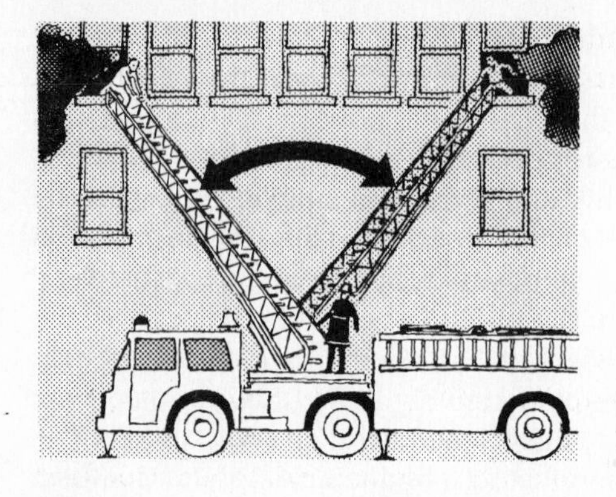

FIG. 18-7. With only one aerial available in this situation, position the truck so that rescue can be performed with the ladder at a minimum oblique angle.

tion, heat and smoke conditions, wind and street conditions, and requirements available to operate—that is, type of aerial.

A decision must sometimes be made between raising the aerial to the roof for venting or using it first to remove an occupant. Roof ventilation very often helps a rescue operation, but the endangered person may not be able to wait that long. If there is any doubt in such a situation, life hazard is the limiting or strategic factor and the aerial should first be used for rescue. In the meanwhile, personnel may be sent to the roof via a rear fire escape, if one is present. If adjoining structures provide ready access to the roof of the fire building or if more than one aerial is available, two minor objectives (removal and protecting other possible occupants pending rescue) can be attained, but they must be attained in the proper order.

It may be argued that the roof should be vented first in the interest of other tenants. However, the presence of other endangered occupants may be only problematical, whereas the visible endangered person is a fact. Experienced firefighters can readily recall numerous occasions on which occupants of fire buildings have successfully vacated the premises under amazing circumstances and without the aid of the fire department, and it

FIG. 18-8. Placing the aerial at an oblique angle from the upgrade side of the fire may better ensure that the target is reached.

is quite possible that the visible occupant is the only one remaining. The probable life hazard must always be considered, however, and operations are conducted accordingly.

An important purpose or minor objective of an aerial is to provide a possible means of escape for fire personnel. This need should be recognized whenever personnel are assigned to work on floors above the fire or on the roof of an isolated building without a fire escape. To protect personnel on floors above the fire, the aerial should be raised from its bed and pointed toward the fire building, ready to be positioned as the need arises. It should not be positioned before the need arises because it may then have to be repositioned. However, to protect those on the roof, the aerial can be positioned there as long as a means of retreat may be needed. In such cases, the aerial should be continuously manned so it can be maneuvered as required. A roof rope should always be taken to the roof of an isolated building in case the aerial, for whatever reason, may not be available as a means of retreat.

Raising the first available aerial to the roof is advisable in the following cases:

1. If roof vent is essential and in order, the aerial is not required to cover a life hazard, and there is no other practical means of reaching the roof.

2. If roof vent is essential and in order, life hazard is possible but not visible, a second aerial is available to cover the possible life hazard, and there is no other practical means of reaching the roof. (See above, regarding means of retreat.)

3. If the involved occupancy is residential but no endangered occupants are visible, roof vent would be helpful and is feasible, only one aerial is available, and there is no other practical means of reaching the roof. In such a case, it is suggested that the aerial be raised first to the roof (taking a roof rope to provide a means of escape), and then positioned to provide a means of escape for personnel and/or occupants who may appear at windows during search and removal activities.

Raising the first available aerial to the roof is not advisable if the roof of the fire building is readily accessible via structures on both sides, if life hazard is possible but not visible, and no other aerial is on hand. In such a situation, it is preferable to make the roof of the fire building through an adjoining structure, leaving the aerial in a position ready to rescue personnel and/or occupants who may appear at windows.

Only ladders that serve a useful purpose should be raised. Excess ladders waste time and personnel, perhaps at a critical stage, and obstruct stretching of lines and operations in general. When portable ladders are required in rescue work, the proper size should be promptly selected. A ladder that is too long or too short may make the rescue unnecessarily difficult, if not impossible, and delay in selecting a more suitably sized ladder may be fatal. Selecting proper portables for windows of various heights can be practiced at outdoor drills. Some departments have procedures for using portable ladders to bridge spaces between buildings for various purposes. These observations do not apply to portable extension ladders.

A scaling ladder is a portable ladder with a metal hook at the top, which is designed to span windowsills and enable the user to climb up or go down. To go down, the user descends from the hooked sill, removes the hook from the sill, lowers the ladder and hooks on the lower sill, and repeats the process as many times as required. To climb up, the process is reversed. Although scaling ladders are seldom used at fires, some outstanding rescues have been made with them when the aerial was short of the target. The rescued persons should almost always be brought in on the floor below the fire unless it is compulsory to continue down the aerial ladder. It would be possible to use a scaling ladder in conjunction with a tower ladder or snorkel-type platform, assuming the situation called for such an effort.

Checking the Fire Building for Extension of Fire

The purpose, or derivative or minor objective, of this activity is to minimize interior and exterior exposure hazards, which in effect means to minimize extent of fire after arrival, reduce smoke and heat conditions, and the like. Minor or derivative objectives can have reasonable connotations.

To minimize the interior exposure hazard, ladder-company personnel at times have to open up structural channels by which the fire may be extending so that it can be cut off and extinguished by hand lines. To do this job well, ladder-company officers must be familiar with the structural channels by which fires can spread vertically and horizontally, and with the tendencies of heat to travel by conduction, radiation, and convection. An officer must determine quickly and accurately whether the greater danger of extension is vertical or horizontal. Construction is very pertinent here.

In a real sense, ventilation is a necessary prelude to checking for extension since it is designed to control the direction of the fire and therefore indicates where channels should be opened up to cut the fire off.

If the fire is spreading horizontally at an intermediate level, experience and judgment enable an officer to open up a ceiling at a reasonably practical distance from the area of origin and cut it off effectively. Hose lines should be available when such openings are made. If there is fire at the opening made, a line can be used to darken it down, but without driving it into previously uninvolved parts. This technique can then be repeated at a more distant point from the area of origin until it is certain that horizontal spread has been checked. The most desirable opening is at the point the fire is just approaching. The line is then used to drive the fire back, and checking operations become simpler. If low-velocity fog from applicators is used, checking operations can become even simpler.

In addition to an adequate knowledge of construction and the transfer of heat, ladder-company officers can be aided in making appropriate openings by such signs as smoke, flame, blistering paint, melting tar or snow on a roof, steaming on a wet roof, spots that are hot to the touch, and so forth. Sophisticated devices for detecting hidden fire are not yet commonly used by the fire service but are on the market.

If the fire has entered the cockloft, the chances of checking horizontal spread depend in large measure on how the roof was opened up. An adequate opening over the channel conveying the fire into the cockloft tends to localize the fire, which greatly simplifies the problem of checking horizontal spread. Without such ventilation, the problem is greatly complicated. In such an event, the use of low-velocity fog is recommended.

Except for rescue purposes, officers should not try to get ahead of the fire in a structure and drive it back to the area of origin when such a move entails passing the fire and jeopardizing escape. They should not try to operate ahead of rapidly spreading conflagrations or brush fires, particularly on an upgrade. Backfiring may be appropriate, however.

To minimize the exterior exposure hazard, ladder-company personnel may perform as follows, assuming no life hazard:

1. Force entry as required.

2. Close exposed windows or inlets to air-conditioning systems, or shut down fans.

3. Vent at the roof and on the unexposed side or sides if feasible.

4. Put out incipient fires, using extinguishers or house lines.

5. Move exposed combustible stock to a safe place—and, if time permits, other stock that could be damaged by the heat.

6. Shut down sprinklers that are operating unnecessarily. Open drain valves. Shut down other duct systems that could worsen the exposure hazard.

7. Check for any possible defects in dividing walls, such as inoperative fire doors and abutting beams.

8. Promptly report on conditions via walkie- or handy-talkies or other

means and promptly call for necessary assistance in coping with a spark-and-ember hazard or other developments.

Rendering First Aid

Where first aid has to be rendered, it is advisable to summon qualified medical personnel promptly. Ordinary first aid may not suffice in some cases, and there are definite limitations on what fire personnel can do or should try to do for an injured person. There are fewer limitations where paramedical services can be performed by fire personnel under supervision of a physician—through radio communication, for example.

Overhauling

Overhauling is the systematic examination of the fire building and its contents to achieve such minor or derivative objectives as (1) to prevent rekindles; (2) to minimize property damage; and (3) to detect and safeguard signs of arson. In addition to the responsibility and authority to overhaul, the fire service has a strong moral obligation to see that businesses or homes are safely returned to the owners as soon as possible. Needless to say, the efficiency of the fire operation has a considerable bearing on the amount of overhauling required.

Actually, overhauling starts with the first efforts to open up ceilings, partitions, floors, and other structural parts in trying to check horizontal and/or vertical spread of the fire and thus minimize the interior exposure hazard. This type of opening up is *precontrol* overhauling. After the fire is controlled, *systematic* overhauling begins and the building and its contents are examined.

To achieve the purposes or objectives of overhauling, the following checklist is suggested:

1. Survey the situation so that an efficient plan for overhauling can be formulated. While the plan is being made, firefighters can be rested. The plan should specify the work to be done, who is to do it, how it is to be done, and the order in which it is to be done. If possible, arrange apparatus and relieve traffic congestion.

2. Do not jeopardize personnel unnecessarily. The structural stability of the building and affected floors must be checked, gas and electricity shut down if required, necessary ventilation carried out, water drained from floors as required, proper lighting equipment provided, and holes in floors covered or guarded if needed. Constant supervision should be exercised. If the stability of the structure or of a floor is questionable, operations should be conducted from the exterior, using lines from safe vantage points. Officers should see to it that personnel work as little as possible under heavy machinery or objects, that they have specific

assignments, and that they do not wander beyond the range of supervision.

3. Give prime consideration to concealed spaces through which fire may be moving undetected. Attention is then given to other parts of the structure where there may be some fire but no possibility of serious extension.

4. Shut down sprinklers operating unnecessarily, and open drain valves promptly. Keep an operator at the control valve as a precaution. Drain the water from the structure, using scuppers, toilet-bowl openings in the floor, broken soil lines, eductors, and pathways provided by sawdust and canvas salvage covers in conjunction with other floor openings.

5. When time permits, move undamaged furniture and similar objects to a safe place before pulling a ceiling for an examination (Fig. 18-9).

6. Use equipment that is less likely to weaken the structure. For example, a saw may be a better tool than an axe in some instances.

7. Remove smoke-producing materials from the building if necessary. If a wetting agent is on hand, removal may not be necessary. However, where heavy baled material such as paper or cotton is involved, it is almost invariably necessary to remove it. It may be feasible to use a hand truck to convey bales to a window whose sill is removed to the floor. If this is done on more than one floor, select windows in different vertical rows to avoid weakening the structure. Be sure to check the area below before pushing bales out.

8. Clear an appropriate place for an examination of the contents. Undamaged stock should be left on shelving or moved to an area where it will not be affected by further operations. Valuable personal or other property should receive particular attention and should be handled by prescribed regulations.

FIG. 18-9. Before pulling the ceiling in the process of overhauling, personnel should move contents to a safer location —if time permits.

9. Remove stock before pulling shelving down. Be alert where jugs, carboys, and jars are present. The bottoms may drop out and cause serious burns or explosions if contents are hazardous. The same precautions are advisable where dangerous dusts may be thrown into suspension and cause an explosion.

10. Turn over involved material in the cleared space, utilizing bathtubs, barrels, or salvage covers for dripping. Rags and wood should be separated from plaster and debris. Small tips or fog are generally sufficient while overhauling.

11. If heavy objects must be moved, place them near the walls where support is greatest.

12. Safeguard any evidence of arson. Comply with departmental regulations on the procedure to be followed.

13. Do not open up the structure unnecessarily. For example, do not make an opening in the roof to examine the cockloft if it will suffice to remove the returns of the skylight or if the extent of the fire only warrants pulling a small part of the ceiling below.

14. Use glass removal kits where removal of glass presents a problem at fires in high-rise buildings. In such cases, pedestrians should be kept out of the danger zone on streets below.

Officers should be alert to the advantage of using various overhauling jobs to train new personnel.

Salvage Work. Salvage work refers to tasks such as covering materials exposed to water dripping from floors above. For many years this work was done by fire-patrol units maintained by fire-insurance organizations, but at present many fire departments perform this work as part of overhauling.

Salvage work can be carried out more efficiently when it is integrated with other phases of fire operations. In addition, the training associated with salvage tends to increase the skill of firefighters in forcing entry, overhauling, and applying minimum amounts of water during fire operations. Minimum use of water reduces property damage and lessens the possibility of electric shorts, with their attendant fire hazards. Consequently, occupants will be deprived of electricity less often. Overall decreased property damage should ultimately be favorably reflected in fire-insurance rates.

Improved efficiency means that interruption of business will be briefer, thereby lessening unemployment. Homes are returned sooner to the occupants, alleviating the need to seek shelter elsewhere, possibly at their own or public expense. These advantages may lighten the tax burden and improve public relations for the fire department.

Finally, salvage service by the fire department eliminates any question of divided responsibility about such important matters as restoring the sprinkler system to service, handling valuables found at fires, and providing for the security of the fire premises after operations are completed.

Salvage by the fire department has some disadvantages. It increases the work load and necessitates carrying more equipment on apparatus; this may reduce efficiency. Additional special training will be required for fire personnel. Salvage work may be delayed indefinitely when there is a life hazard to occupants, and in any event, units will be kept out of service longer, due to the additional duties. *Salvage units* in a fire department will minimize these disadvantages.

There is a problem in drying and repairing salvage covers in quarters, and the space and time required may detract from maintenance of quarters, drill, and inspections (fire prevention, alarm box, and hydrant).

The responsibility of the fire department while carrying out salvage operations is increased over property, valuables, and merchandise, and personnel necessarily operate over wider areas. This entails additional problems of supervision.

Salvage work is done better when handled by fire departments, but such service necessitates higher budgets due to increased quotas and increased cost of training and maintenance. Since improved efficiency is highly desirable, the only question remaining is who will pay the cost. Representatives of interested parties (municipalities and fire-insurance companies) might work out an arrangement that is fair to all. This budget problem is particularly acute at the present time.

RESCUE COMPANIES

While also capable of performing functions normally carried out by engine and ladder companies, rescue companies are units that specialize in rescue procedures. To use effectively the amount and variety of equipment carried, the recommended minimum complement of a rescue company is six firefighters and an officer. Firefighters selected should have a strong desire to do rescue work and should have at least five years' experience, preferably in a busy ladder company; they should be adept mechanically. During recruiting, the personnel cards of all applicants should be checked and those who meet initial requirements should be interviewed and examined. Acceptable recruits should then be given a trial period with an operating rescue company so that they can be tested, trained, and screened under field conditions. Officers should be selected from rescue-company personnel or from among firefighters who qualify for rescue companies and have in addition extensive experience as ladder company officers.

Few training schools are equipped with facilities to train rescue personnel in shoring, rigging, oxyacetylene cutting, use of jacks on trains and trucks, and other special skills. To overcome this limitation, the cooperation of utility agencies, construction and demolition contractors, transportation companies, and civil defense authorities should be solicited so that their facilities and resources can be used to develop the desired skills (demolition sites, especially in slum or other clearance projects, provide ideal training grounds and supply wood and scrap metal for use in daily drills in company quarters). An effort should be made to send selected officers to civil defense disaster schools to improve their background and acquaint them with the latest developments in rescue procedures. These officers, in turn, can train other fire personnel. A library of rescue-work material should be available in company quarters.

Training programs should emphasize preplanning and experimentation to encourage the proper selection and skillful use of tools and equipment, whatever the specific problem may be. Standard solutions should be developed for every conceivable problem so that, during operations, definite assignments can be made as quickly as possible for maximum effectiveness. It is helpful if personnel of other units are familiar with special equipment of rescue companies in the event that members of the former are detailed to the latter.

Rescue-company Tools

Cutting Tools. This group includes saws for cutting wooden double flooring, which may be either gasoline and electric types or chain, circular, and power hand saws of the bayonet type (where smoke impairs visibility, the circular saw may be preferred to the chain type to reduce the chance of injury to personnel); concrete breakers for breaching walls and making openings in concrete flooring over piers and cellars, which are powered by a high-cycle 3,000-watt portable generator and may be used with bull points, chisels, and broad chisel; oxyacetylene cutting torches for cutting bars and metal obstructions at fires or emergencies involving train or vehicular accidents, or for cutting metal doors and making openings in the sides of ships (pack sets carried on one's back for small inaccessible jobs and a large commercial set on the hand truck are usually provided); assorted hand saws, axes, and pipe cutters to supplement the power tools; and rotary cutters of various sizes and types to make openings in wooden floors for the operation of distributors or cellar pipes at cellar fires or for draining floors.

Lifting and Pulling Tools. Devices designed for hoisting, bracing, wedging, and similar jobs that may be met at building collapses, railroad or vehicular accidents, or unusual overhauling work are in this category. Jacks include the power type, powered by a hydraulic pump and used with various jacks, spreaders, and extension pieces, and journal jacks, railroad jacks, and hydraulic jacks. A ratchet-type hoisting tool, trench braces, and wedges, rollers, and pry bars are also carried.

Respiratory Devices. Respiratory devices include various types of masks. A self-contained demand-type mask with two spare cylinders is provided for each company member on duty.

First-aid Equipment. A complete medical bag is carried as well as stoke stretchers, blankets and wool caps, splints of various kinds, resuscitators, and inhalators. This equipment is used for both firefighters and civilians.

Refrigerator Tools. Tools for coping with refrigerator leaks at fires or emergencies include halide detectors and sulfur tapes to detect leaks, plugs and clamps to stop leaks, and assorted wrenches to tighten packing nuts, and so on.

Special Extinguishing Agents. Mechanical and chemical foam is carried for use on class B fires, and G-1 powder for use on small magnesium-powder fires. Soda ash is for use where muriatic acid or similar acids are involved. Dry chemical and wetting agents are also carried.

Miscellaneous Equipment. Rubberized suits are carried for use in situations involving ammonia leaks, toxic insecticides, skin absorbents, and so forth. Rescue companies also carry various kinds of communication devices, of which the sound-powered phone operable with demand-type masks is especially notable. These phones work where other devices may fail, particularly at subway fires. Smoke ejectors are carried for use in ventilating fire areas, particularly those below ground. The concrete core cutter is a new addition.

Response. Rescue companies should be assigned to respond (1) to all alarms in the vicinity of target hazards within a radius of 2 miles [3.2 km] of the fire house; (2) if they can arrive at alarm boxes near their quarters before the nearest ladder company, and; (3) to all fires at which all units responding on the first alarm have been utilized.

When rescue companies are not assigned to respond and the fire is under control, it is advisable to put in a special call if their power tools would be particularly helpful in coping with concrete floors, difficult doors, double flooring, and the like. A rescue company can handle such jobs with less effort and in less time than several ordinary units working with hand tools. In general, a rescue company should not be used for overhauling, but there may be exceptions where toxic products of combustion are being emitted or where a large supply of masks is needed.

Many members of rescue companies hold Red Cross First Aid Certificates and have considerable skill. Hence, a special call is in order if such skill is required or if it is necessary to handle injured persons in inaccessible locations. In addition, these units have the equipment and experience to handle the removal and limited treatment of persons injured in building collapses and large transportation accidents. A special call is also appropriate when there is a heavier than ordinary refrigerant leak, when weakened structural members need shoring up, or when dangerously hanging signs must be secured.

If a rescue company responds but it is evident that its help is no longer needed to control the fire, it may be advisable in some cases to have it stand by in case first-aid skill and equipment are needed for fire personnel affected by smoke, exhaustion, eye irritation, cuts, and so on. A company operating at a fire where only a limited amount of its equipment is needed can be available by radio to be called elsewhere if needed, leaving one or two men to continue operating the equipment in use.

Advantages. Rescue units are able to force strong locks and doors under difficult conditions because of their large complement of men, adequate masks, special cutting and forcing equipment, and experience in forcing entry. Rescue-company personnel are trained in various phases of rescue work: use of lines, ventilation, efforts to reach trapped occupants, search and removal of victims, and rendering first aid. The personnel available makes it more possible to vent and advance lines simultaneously, thereby increasing chances of success. Moreover, the availability of reflective and rubberized suits enables rescue companies to make a more extended search for victims in unusual fire conditions or other hazardous situations.

In addition to their ability to ventilate structures under difficult conditions, rescue companies can improve the smoke condition, both during and after the fire, with the mechanical ventilating equipment they carry. Their concrete breakers can ventilate cellars when no other means is effective. Rescue companies, fully masked, can be used to relieve units on lines and to press forward to quick extinguishment. Reflective or rubberized suits may help even more. Where heavy wooden flooring is an obstruction, rotary cutters may facilitate the use of distributors and cellar pipes. Concrete breakers can help breach walls. Thus, fires otherwise inaccessible may be brought within the range of operating lines. At fires where water cannot be used to extnguish, it may be possible for rescue companies to provide and apply an effective agent.

Ordinarily, rescue companies are not used for overhauling. However, where charred beams must be exposed and concrete is in the way or when a considerable amount of double flooring must be cut, the power tools of the rescue company can do a quicker job. Rescue companies are particularly adapted for securing dangerous overhead objects such as cornices and signs, so that finishing-up operations are made safer. The wetting agents carried on rescue apparatus can be used to make overhauling more efficient.

Rescue companies, because of their training and their resuscitator apparatus, are particularly valuable at subway or similar fires where many people require first aid for smoke inhalation. In such cases, it may be advisable to call all available rescue units rather than transmit greater alarms, which will bring many unnecessary pumping units. The communicating devices carried by rescue companies help not only in rescue work, but also in transmitting information to commanding officers so that they can understand the problem and take appropriate steps more quickly. This can be an important factor if it is unusually difficult to locate the fire and determine its extent.

A distinct advantage is that rescue-company personnel have above-average qualifications and training. Their high morale is invariably translated into efficient work at fires or emergencies. Perhaps in no other operation do rescue units show to greater advantage than working at building collapses. In this field, they are ideally prepared.

SQUAD UNITS

These units were organized in some departments to ensure adequate personnel in the vital early stages of fire operations and to lighten the work load of companies in high-fire-incidence areas. The recommended minimum complement of a squad unit is an officer and five firefighters. The units are used to furnish added personnel where and when conditions warrant, to relieve personnel affected at fires by smoke, heat, and injuries, to help with work ordinarily done by engine or ladder companies, except overhauling, and to supply additional masks and cylinders.

The personnel of squad companies is selected with care, since members must be equally skillful in ladder- and engine-company work. Leadership and morale in these units has been high. As a result, the record made by squad units in a short time is an exceptional one.

In carrying out their assignments at fires, squad units operate like engine or ladder companies. When the services of the squad are not needed, it should return to quarters as quickly as possible and hold itself available to respond to subsequent alarms at which depleted assignments may be rolling in. The greatest advantage of a squad company at a fire is that the extra personnel permits many jobs, especially rescue work, to be done more quickly; the fire is more quickly under control. This has many derivative advantages: the work load of all fire personnel is lightened and the possibility of injuries reduced; direct and indirect fire losses are lessened; there is less need for additional alarms; overall fire protection is improved, because areas are not stripped of adequate fire protection as often, and also because all units are generally returned to quarters more quickly. The squad unit fills a long-standing need at fire operations: the need for more people rather than more apparatus.

REVIEW QUESTIONS

1. What are the functions of a ladder company?

2. What is ventilation? What are its purposes, or minor or derivative objectives? What must be known about ventilation in order to attain these objectives?

3. How, or by what means, is ventilation carried out? What governs the means selected to ventilate? Why?

4. Why is it important to know when to ventilate? What is the effect of the presence or absence of a life hazard? What are some effects of premature venting? How can such venting be avoided? Generally speaking, what should be considered in venting a roof?

5. How do minor or derivative objectives affect where to ventilate? How does the location of the fire affect where to ventilate? When is it advisable to ventilate as directly as possible over the fire?

6. Where and how is ventilation usually effected at fires on intermediate floors in buildings of brick-joist construction? At similarly located fire in fire-resistive buildings? Discuss the use of elevator shafts and fire escapes in ventilating.

7. What primary factors should ladder-company officers consider when ordered to open up a roof? Why?

8. What are the minor or derivative objectives of forcible entry? How do major and minor objectives affect the timing of forcible entry? How can such primary factors as location of fire and construction affect where to force entry? Discuss some similarities and differences between forcible entry and ventilation.

9. What should be kept in mind in carrying out search and removal activities? Discuss a suggested order of priority for removing occupants from a fire building. How is the choice affected by such primary factors as construction, height, location and extent of fire, weather, time of fire, and requirements available to operate? Why?

10. What are the minor or derivative objectives of laddering? What should be considered in trying to attain the minor objectives of laddering via aerials? Why? What should be considered in raising aerials to remove occupants on fire escape balconies above a fire?

11. Why do aerials at times have to be used at an oblique angle? What are some advantages of using aerials in this way? Disadvantages?

12. What influences a decision about raising an aerial to the roof for venting or using it first for rescue purposes?

13. What should be considered when the minor objective in using an aerial ladder is to provide a possible means of escape for personnel?

14. When is it advisable to raise the first available aerial to the roof of the fire building? When would it not be advisable?

15. How is the activity referred to as checking for extension of fire carried out when the minor or derivative objective is to minimize the interior exposure hazard? To do this job well, what knowledge is essential? How does ventilation play a part? How can opening up a ceiling and using hose lines play a part? What visible signs can guide an officer in checking for extensions of fire?

16. How does an officer check for extension of fire when the minor objective is to minimize the exterior exposure hazard?

17. What are some limitations on ladder-company personnel in rendering first aid?

18. What is overhauling? What are some responsibilities associated with overhauling at fires? What is meant by precontrol overhauling? Discuss a checklist for achieving the purposes or objectives of overhauling.

19. What are some advantages and disadvantages of having salvage work done by the fire service?

20. When and why does a rescue company have advantages over a ladder company in forcing entry? Ventilating? Searching for and removing trapped people? Rendering first aid?

21. How can squad units be used effectively at fires?

PART FOUR

Case Studies in Firefighting Strategy

19

Fire in a One-Family Residence

Part 4 of this book describes the practical application of the information provided about the science and art of firefighting, and thereby the use of firefighting strategy. It shows how the action plan is actually applied at fires.

The action plan is designed to help officers of all ranks to act in a correct and uniform way when they are first on the scene at a fire or take command at an operation in progress. However, one cannot logically expect to become adept in using the plan merely by reading about it. Practice is essential. With both practice and study, officers will become more expert in recognizing and evaluating the primary factors that are pertinent to selecting objectives and activities. Then things will fall into place almost automatically, and the science and art of firefighting will progress simultaneously. Actually—but inescapably—the explanations are more cumbersome than the application of the action plan, but one would have to be very naïve to suggest that there is one simple mental process that can be learned, with no sweat, by all officers for all types of fires.

As stated earlier, discussions of past or structured fires are of limited value since the pertinent primary factors have to be specified and are therefore predetermined, whereas in actual cases such factors must be recognized and evaluated under possibly hectic conditions. Practice in using the action plan can be more fruitful, however, if different sets of circumstances (at the same fire) are considered, as suggested in this and subsequent chapters. However, there are limits to the degree to which this can be done in a book. The individual can pursue the idea much further by considering the primary factors that can be pertinent at fires and the innumerable combinations that are possible. Perhaps in the future computer data banks, storing information about various occupancies and supplemented by operations research, may help in this matter.

The availability of information in data banks does not relieve an officer of the responsibility of recognizing and evaluating pertinent primary factors at fires, however. Such banks may provide information about construction, height, area, occupancy, and the like, but officers must still assay the effects of the location and extent of fire, life hazard, heat and smoke conditions, weather and wind conditions, and all other variable factors.

The following fire in a one-family residence is discussed in this chapter because of its relevance for the many rural fire departments that respond to fires with limited assignments.

CASE #1

You arrive first on the scene in command of an engine company with four firefighters. The pumper carries a 350-gal [1.3 m^3] booster tank and booster equipment. The owners of the house are outside with their children. They tell you that on checking an odor of smoke, they found that a fire had just started on the wooden workbench in the basement. They got a small extinguisher from the kitchen, used it on the fire, retreated because of the smoke, closed the door leading up from the basement, led the family outside, and then notified the fire department.

The fire building is a one-family, two-story, frame residence, about 40 × 25 ft [12 × 7.6 m] in area. The fire, which is visible through two closed basement windows just above ground level, has not yet spread beyond the basement. A hydrant about 40 ft [12 m] from the fire building is found to be unserviceable but there is another one about 500 ft [150 m] away. A ladder company is on the way and will complete the assignment. Additional units, if needed, must come from other communities.

Evaluate Factors and Select Objectives

Since there is no life hazard, the choice of major objectives falls between extinguish and confine, control, and extinguish. The primary factors that are limiting or strategic in making a decision include the following.

1. Location of the fire. The fire is readily accessible to a direct or indirect attack. The location of the fire in a closed basement tends to limit the supply of oxygen and hence the rate of combustion and extent of the fire.

2. Extent of fire on arrival. Temporarily, at least, the fire is confined to a comparatively small area (about 40 × 25 × 9 ft [12 × 7.6 × 2.7 m]), since the basement windows and the door leading up from the basement are closed.

3. Construction. Frame construction is susceptible to rapid spread of fire, although so far this fire is not yet out of the basement. The basement windows are favorable for initiating an indirect attack, or can help to provide ventilation if a direct attack is attempted.

4. Requirements available to operate. The booster equipment and personnel available are adequate to extinguish this fire.

In deciding upon objectives, the range of alternatives is weighed against the specifications for an acceptable solution. The range of alternate major objectives here includes extinguish and confine, control, and extinguish. The prevailing specification for an acceptable solution is that personnel are not to be jeopardized unnecessarily since there is no life hazard. Extinguish is the logical choice because, in view of the primary factors that are strategic here, the situation is controllable via a direct or indirect attack by the assignment on hand, assuming the ladder company is now on the scene. Extinguish is preferable where it minimizes property damage without unduly jeopardizing personnel.

Select, Assign, and Coordinate Activities

In deciding on activities, just as in deciding on objectives, the range of alternatives again is weighed against the specifications for an acceptable solution. Alternative activities in this case are those associated with direct or indirect attack. Indirect attack is selected because, in view of the pertinent primary factors, it is more efficient and in accord with the prevailing specification that personnel are not to be jeopardized unnecessarily.

Specifically, the selection of activities depends on the derivative objectives that must be achieved to attain the major objective, always remembering the specifications for an acceptable solution. The primary factors pertinent at this fire suggest that extinguish can be attained via indirect attack by achieving derivative objectives in this sequence: (1) Darken down the fire in the basement; (2) alleviate heat and smoke conditions and improve visibility; (3) minimize the interior exposure hazard; and (4) minimize property damage and prevent a rekindle. Protecting personnel and coordinating activities are ongoing derivative objectives at every fire. To achieve each of the derivative objectives, and thereby attain the major objective, selected activities would be assigned and coordinated in the following manner.

1. To darken down the fire in the basement. Stretch or lay a preconnected $1\frac{1}{2}$ in [39 mm] line and inject low-pressure fog or even low-velocity fog from an applicator, through a small opening made in one of the basement windows. Make no other openings until the fog vaporizes and exerts its full cooling and smothering effects on the fire in the basement and on the convection currents with which the vaporized fog will mix and rise in vulnerable structural channels. This will do much to confine the fire while darkening it down.

2. Alleviate smoke and heat conditions and improve visibility on all floors. This can be done by ventilating at windows, using short ladders for the second-floor windows.

3. Minimize the interior exposure hazard. This is the goal in checking vertical and horizontal spread of fire and it is attained by opening up suspected structural channels and applying fog as required. There should still be plenty of water in the booster tank for this purpose, using a second line that should have been stretched and held in readiness if needed.

4. Minimize property damage and prevent a rekindle. Overhauling as required should achieve these objectives. Minimizing property damage is also a minor objective in carrying out other activities.

The importance of supervision and communication in achieving coordination depends on the size and complexity of the fire operation. Here, as elsewhere, supervision should ensure that activities are carried out in a timely manner and as planned, and that the principles governing each fire activity are complied with. In more complex operations, a command post should be established to maintain communication between the commanding officer and subordinate officers. (Communication pertains to the issuing of orders and provides the contact between units whereby supervision can be effectively exercised.)

Alternative Solutions

It is quite possible that this fire could also be quickly extinguished by a direct attack. In such an operation, the first minor objective would be to alleviate the smoke and heat conditions in the basement by ventilating at the windows, and the second objective would be to darken down the fire by advancing a line down the stairs. Even if masks are worn, this technique entails really unnecessary physical hardships for personnel.

Another possible solution would be to drop off two $1\frac{1}{2}$ in [39 mm] lines to be fed by a Y connection and a large line supplied by the pumper taken to the hydrant 500 ft [150 m] away. Due to the delay in getting water, personnel will have to cope with a more extensive fire. Even if indirect attack can be successfully employed, fire damage may be much greater than if booster-tank water was used. If direct attack is employed, physical hardship for personnel as well as fire damage may be similarly greater.

A range of alternatives exists because inevitably there is more than one way to put out a fire. Inevitably also there is only one best way and the criterion for making such a choice is to weigh the range of alternate solutions against the recognized specifications for acceptable solutions at fires. The degree of risk for personnel is a dominant feature of these specifications. No decision is riskless, but the margin of error can be reduced by a rational approach.

CASE #2

Additional practice can be had in applying the action plan at this fire by changing some of the pertinent primary factors as follows. The owners standing outside with their family tell the officer of the engine company that they neglected to close the door leading up from the basement and also left the front door open. The officer can see some flame starting to come up through the basement door opening and promptly closes the front door to restrict the flow of air to the fire. Thus the extent of fire on arrival is changed. In addition, it will be assumed that the pumper has a 500- [1.85 m^3] instead of a 350-gal [1.3 m^3] booster tank, thereby changing the requirements available to operate.

Evaluate Factors and Select Objectives

Again the choice is to extinguish or to confine, control, and extinguish. In this case, the interior exposure hazard is the limiting or strategic factor. The first line has to be used to check upward spread of the fire or else the building is likely to be lost. Therefore confine, control, and extinguish is the logical major objective. It can be attained if the assignment on hand can achieve derivative objectives in this sequence: (1) Alleviate the smoke and heat conditions and facilitate the advance of a line on the first floor. (2) Minimize the interior exposure hazard and confine the fire to the basement. (3) Darken down and control the fire in the basement. (4) Alleviate smoke and heat conditions and improve visibility on all floors. (5) Minimize property damage and prevent a rekindle. In effect, the achievement of these objectives will confine, control, and extinguish the fire.

Select, Assign, and Coordinate Activities

To achieve each of the derivative objectives, and thereby the major objective, selected activities would be assigned and coordinated in the following way.

1. To alleviate smoke and heat conditions and facilitate the advance of a hose line. Ventilate at rear windows on the first floor so that the first line can more readily advance and drive the heat and smoke ahead and out of the building as harmlessly as possible.

2. Minimize the interior exposure hazard and confine the fire to the basement. Advance the first line through the front door to knock down visible fire and check vertical extension. It would help if the door to the basement could be closed, but this may not be possible since the fire in the basement is not yet under control. In any event, the first line can be operated from the front doorway if necessary, to check upward spread of the fire.

3. Darken down and control the fire in the basement. A second fog line operated through a basement window has a good shot at the fire and control is possible in two or three minutes. The fact that flame is coming up the stairway does not necessarily mean that the basement is fully involved. Very high-pressure fog is not recommended on the basement fire.

4. Alleviate smoke and heat conditions on all floors. If the door leading up from the basement had been closed by personnel on the first line, basement windows would not be ventilated until fog injected by the second line had a chance to be effective. In this way, the benefits of an indirect attack can be realized. Otherwise, the basement windows can be opened sooner, along with windows on other floors, expediting and facilitating activities that will check the spread of the fire.

5. Minimize property damage and prevent a rekindle. See above.

Coordination plays an important role at such fires, whether direct or indirect attack is used. In a direct attack, ventilation precedes an attempt to advance a line; in an indirect attack, ventilation is delayed or at least deliberately restricted until the fog has taken effect. The need for coordination also has to be considered in advancing the first line. It can be moved more effectively to check vertical extension *after* control of the fire in the basement is imminent.

Other Alternatives

If the owners report that they have closed the door leading up from the basement but were unable to remove a 200-lb [90 kg] aunt from her second-floor bedroom, the limiting or strategic factor is life hazard, the major objective becomes rescue, and the first minor objective will be to protect the occupant pending rescue. In this event, the first line would enter through the front door and operate as required, with some helpful ventilation, between the fire and the stairway, pending removal of the occupant. Then the major objective would become extinguish and indirect attack could be used to darken down the fire in the basement, followed by the attainment of other associated minor objectives via routine ventilation, checking extension, and overhauling. Indirect attack is not used when the major objective is rescue if there is any possibility that the steam created by the vaporization of fog will be injurious to occupants.

If the door leading up from the basement was left open, two lines with some helpful ventilation may be required to operate at the first-floor level to protect the occupant. In such a case, help should be called because there will be a delay in efforts to control the fire, and perhaps in getting needed hydrant water.

Only situations that present an opportunity for using the strategy of firefighting in applying the action plan should be structured. It is necessary, of course, to consider the most difficult problems, providing that their solutions conform to recognized acceptable specifications. A most difficult problem would exist in trying to protect the occupant pending rescue if the stairway to the second floor was heavily involved. Risks are definitely warranted in trying to solve this problem by ventilating on the side of the building away from the occupant, using a line at the front-door opening to knock down the fire on the upper stairway, and advancing a line via a ladder to the occupant's room so that it can operate between her and the fire, pending removal. Obviously, the removal of such an occupant would be a difficult undertaking in itself, but the risks involved would be in accord with the specifications for an acceptable solution. After removal, the second line might have to be withdrawn and repositioned to help confine, control, and extinguish the fire. If fire conditions prevented the entry of the second line into the occupant's room, it is unlikely that rescue could be achieved.

Many other variations are conceivable. In thinking of them, it is suggested that the reader also keep in mind the water, personnel, apparatus, and equipment available on response to an alarm of fire.

REVIEW QUESTIONS

1. Why are discussions of past or structured fires of limited value in teaching officers how to operate at fires? How can practice in using the action plan become more fruitful?

2. Why can information available in data banks be of limited value to officers operating at fires?

3. In case #1, what primary factors were limiting or strategic in de-

ciding on the major objective? Why is the major objective of extinguish the logical choice?

4. Generally speaking, how are major objectives and activities decided? Even if they have never heard of an action plan, why should officers have a clear-cut conception of the specifications for an acceptable solution in dealing with problems at fires?

5. What does the selection of activities specifically depend upon?

6. In case #1, what minor or derivative objectives have to be achieved in order to attain the major objective of extinguish, assuming indirect attack is used? How is each minor objective achieved?

7. How would minor objectives and activities change if direct-attack tactics were used in case #1? What are some other possible solutions?

8. Why is the major objective of confine, control, and extinguish the logical choice in case #2? What are the minor or derivative objectives and how is each achieved? Discuss the role of coordination in cases #1 and #2.

9. Discuss the variation in which rescue was the major objective. What should be kept in mind when devising variations for study and discussion?

20

Fire in a Store (Taxpayer)

The objectives of this chapter are to show how the action plan is to be applied when an officer assumes command at a fire in progress, and how it should have been applied initially.

USING ACTION PLAN AT FIRE OPERATION IN PROGRESS

The first-to-arrive engine and ladder companies found the fire to be in the center of five adjoining one-story taxpayers, which will be referred to as A, B, C, D, and E, going from left to right as viewed from the street front.

Construction is nonfireproof and the area of each store is about 30 × 90 ft [9 × 27 m]. Store A handled furniture, B paints, C groceries, D meat, and E shoes. The fire started in the rear of C at about 4 P.M. on a weekday in late December. The temperature was about 40° [4°C], with little or no wind. There were no other exterior exposures. Three engine and two ladder companies, together with a battalion chief, responded to the alarm. The deputy chief assigned to respond to an "all hands" signal at the box arrived about ten minutes after the operation had started.

En route, the deputy chief has time to evaluate radio information concerning the reported location of the fire, construction, proximity of exposures, height, area, occupancy, and requirements available to operate. Information about the extent of the fire and heat and smoke conditions is vague. In addition, the deputy chief evaluates such primary factors as time of the fire, visibility, weather and wind conditions, and is aware of the possibility of a back draft in the stores or a smoke explosion in the common cockloft characteristic of such construction. Such back drafts or smoke explosions occur quite often at such fires, and can cause structural collapse and injuries to personnel. A follow-up report indicated that the stores had been vacated and hence there was no life hazard.

An officer, such as the deputy chief, who is assuming command at a fire operation in progress should be briefed by the officer being relieved of command. It is especially important to find out if units are endangered because they have passed the fire—vertically or horizontally—and the fire building is unoccupied, since the life hazard for personnel is the paramount limiting or strategic factor if there is no life hazard for occupants.

The officer initially in command also evaluates and reports the effectiveness of activities (secondary factors) assigned and undertaken, because of their effects on primary factors that can become limiting or strategic to other decisions. Where there is a difference of opinion, the judgment of the officer assuming command naturally prevails, because the superior officer becomes responsible for the direction of the operation.

On arrival, the deputy chief is briefed, and then follows the mental process indicated in the action plan: recognize and evaluate the primary factors that determine the objectives and activities *at the time of arrival,* assign activities accordingly, and coordinate them by adequate supervision and communication. This necessarily entails an evaluation of the activities (secondary factors) undertaken and still in progress because of their effects on primary factors that can become limiting or strategic in making decisions about objectives after a fire operation has started. Rescue was not a major objective since occupants had been removed. The question was whether the interior operation under way should be continued and supplemented in an effort to achieve the major objective, extinguish, by a direct attack. The alternative major objective is confine, control, and extinguish. (A ladder and engine company under the command of a battalion chief were working in C and lines were starting to advance toward B and D.)

Evaluate Factors and Select Objectives

Of particular importance in this situation were such primary factors as extent of fire after arrival, occupancy contents, heat and smoke conditions, exposure hazards, construction featuring a common cockloft, the possibility of a smoke explosion or back draft and structural collapse, life hazard for personnel, and the requirements to operate effectively. Considering such factors, the deputy chief made the following evaluations and decisions:

1. Extent of fire after arrival. Smoke issuing from open streetfront doors of all five stores indicated that the fire had extended throughout the common cockloft. Evidently ventilation, use of hose lines, and efforts to check extension were ineffective.

2. Heat and smoke conditions. Heat and smoke conditions in an inadequately ventilated cockloft are likely to cause a smoke explosion and/or back draft accompanied by structural collapse that can endanger personnel.

3. Requirements to operate effectively. Anticipating that the possibilities referred to in item #2 may become actualities and that the resulting situation would not be controllable by the assignment on hand, a second alarm should be transmitted. Even without a back draft or smoke explosion, the extent of the fire warranted a second alarm.

However, even as the second alarm was actually being transmitted, within three or four minutes after the arrival of the deputy chief, the store-front windows of A and B were blown out by a back draft and an accompanying flame that extended about 40 ft [12 m] into the street. In addition, some of the roof collapsed over A and B.

Selecting extinguish as the major objective and trying to achieve it by a direct attack involving standard procedures (ventilating the roof and store windows, using hose lines to extinguish visible fire and fire exposed when ceilings are pulled down, and overhauling) may be justified initially. If the cockloft becomes seriously involved, however, the major objective inevitably becomes confine, control, and extinguish. At the fire in question, derivative objectives became to protect personnel and confine, control, and extinguish; in that order.

Select, Assign, and Coordinate Activities

Personnel were backed out of C. Fortunately, they had not been affected by what happened in A and B. Fortunately also, units had not yet entered these stores but were advancing lines toward B and D. The fire was confined and controlled in a relatively short time by exterior lines operating through rear and front windows. Special consideration had to be given to the final derivative objective extinguish because of the back draft, structural collapse, and contents of B (paints). Roof ventilation had to be dispensed with due to the weakened condition of the roof and the contents had to be carefully sprayed with fog before personnel could enter to perform necessary overhauling. It was then found that the contents had been affected only superficially by the back draft and flame, possibly because so much of the heat had been released in the street and exposure time was short. The other stores presented lesser problems in overhauling.

Activities were assigned according to the specific functions of the units available. They were coordinated by adequate supervision and by communication between subordinate officers and the command post which was promptly established by the deputy chief.

USING ACTION PLAN WHEN INITIATING FIRE OPERATIONS

Before explaining how the action plan should have been used some comment is offered about standard operating procedures at taxpayer fires.

In using standard operating procedures to ventilate the roof, the right-sized openings should be made in the right place at the right time. Such openings help to localize the fire and prevent a smoke explosion in the cockloft. However, the records show that such ventilation is rarely achieved at taxpayer fires. As a result, cocklofts quite frequently become heavily involved, smoke explosions occur in them, and fires are not localized. Some departments resort to trenching (making a trench across the roof on one or both sides of the fire) by means of power tools. As of now, the effects of this technique are uncertain. Greater alarms are still common at taxpayer fires. The fire in question appears to be another case of ineffective roof ventilation.

Standard operating procedures (used at this fire) include pulling ceilings from below and directing streams at the fire exposed. This technique may, and very likely does, drive heat and smoke throughout the cockloft. Air flowing into the cockloft through the openings made may then mix with the hot combustible gases and cause a smoke explosion that can collapse the ceiling and the roof. This appears to have happened at the fire in question.

In view of the shortcomings of the foregoing standard procedures, other tactics will be recommended in suggesting how the action plan should have been used on arrival.

Admittedly, the tactics to be suggested have never been used at taxpayer fires but are offered as subjects for research and experimentation. However, little if anything would be lost—and much might be gained—by trying the suggested tactics at an actual taxpayer fire such as the one described. One could always resort to the standard operating procedures if necessary. It is assumed that the U.S. Navy all-service nozzle, which is not commonly used in the fire service, is available for use with an applicator, as described.

Evaluate Factors and Select Objectives

Since the stores have been vacated, the choice is between extinguish and confine, control, and extinguish. The primary factors that are limiting or strategic here are the following.

1. Location of the fire. The location of the fire in C makes it readily accessible to a direct attack; the fire in the cockloft is accessible to an indirect attack.

2. Extent of fire on arrival. The extent of the fire in C is such that it is

darkened down by one hand line; the extent of the fire in the cockloft cannot be determined at this stage.

3. Requirements available to operate. Water, apparatus, equipment, and personnel available are adequate to carry out the direct and indirect attack visualized.

4. Life hazard for personnel. Personnel are not to be jeopardized unnecessarily since there is no life hazard for occupants.

Based on the foregoing, extinguish would be the major objective. In making this and similar decisions, the range of alternatives is weighed against the specifications for an acceptable solution. The primary factors pertinent here suggest that extinguish is achievable by attaining derivative objectives in this sequence: (1) Alleviate smoke and heat conditions in C. (2) Darken down the fire in C. (3) Prevent or minimize the possibility of a smoke explosion in the cockloft. (4) Prevent or minimize the possibility of back drafts in the stores. (5) Minimize the exterior exposure hazard. (6) Coordinate activities via adequate supervision and communication. Concomitantly, the protection of personnel and property is being considered.

Select, Assign, and Coordinate Activities

Activities, selected to achieve each derivative objective and thus the major objective, would be assigned and coordinated in the following manner.

1. Alleviate smoke and heat conditions in C. Ventilate at the skylight. This facilitates the advance of hand lines in C.

2. Darken down the fire in C. Advance a hand line using U.S. Navy all-service nozzle. Darken down the visible fire. No unusual difficulty was encountered in doing this at the actual fire.

3. Prevent or minimize the possibility of a smoke explosion in the cockloft. Use indirect attack on the fire in the cockloft. Attach a 12-ft [3.6 m] applicator to the navy nozzle and inject low-velocity fog into the cockloft via well-spaced openings made in the ceiling by hooks. Two lines can be used in doing this. Keep in mind the force created by the vaporization of the fog so that the ceiling will not be blown down if the fog is applied uninterruptedly. In the meanwhile, make no openings in the roof, except at the skylight.

4. Prevent or minimize the possibility of back drafts in stores. Keep street-front doors closed until skylights over stores are ventilated. Air in contact with hot metal ceilings will be heated by conduction. Heat

will also be radiated down from such ceilings and may cause the emission of flammable vapors from exposed material. Heat then transferred by convection (heated air and gases) will rise and be dissipated through the open skylight. Keeping the front doors closed to this point prevents a flow of air into an unventilated fire area and thereby a smoke explosion or back draft. Now street-front doors can be opened, supplementing the ventilation provided by the skylight, further minimizing the possibility of a back draft or smoke explosion. Units can now enter and operate as suggested below (item #5).

5. Minimize the exterior exposure hazard. After operating as suggested in C, and ventilating at skylights of the other stores, open the front doors of B and D. Use direct attack on visible fire, if any, and indirect attack on any possible fire in the cockloft. Before operating in B, however, spray the contents with fog—from the doorway, due to the nature of the occupancy (paint store). Similar operations can be repeated in A and E, if necessary. Subsequently, skylight combings can be removed to examine conditions in the cockloft, and ceilings can be pulled as required during overhauling.

6. Coordinate activities via adequate supervision and communication. Disciplined timing is essential in carrying out the action plan as suggested. Adequate supervision and communication between subordinate officers and the command post can ensure the timing required.

An Alternate Solution

Some departments resort to making openings in the roof and inserting cellar pipes to sweep the cockloft at such fires. Several things are undesirable about that type of operation. (1) Water and fire damage will obviously be great, which is not in line with either the aspirations or the obligation of the fire service to reduce the high cost of fire. If roof operations are warranted, it means (or should mean) that ventilation has eliminated the possibility of a smoke explosion and that the extent of the fire justifies the roof positions. (2) Based on such ventilation and extent of fire, operating from below would be preferable to flooding the building and contents from roof positions. If fire conditions were such as to preclude attempts to advance hand lines into the stores, it would hardly be logical to put personnel on the roof. The lower level of operation is preferred for the following reasons.

1. There is better coordination between ladder and engine companies because ceiling can be pulled where there is fire, and water can be used judiciously, resulting in much less damage than what is caused by the flooding technique.

2. Hand lines are used below and can be stretched more quickly and easily than roof lines, which may necessitate the use of ladders, power tools, hose straps, hose rollers, and decidedly less maneuverable cellar pipes.

3. At night, it is easier to provide adequate lighting at street level than it is on the roof where openings are a hazard.

4. Adverse weather conditions, such as snow, rain, and low temperatures, as well as strong winds, can hamper roof operations much more than they can those at the lower level.

5. Some, if not all, of the ceiling must be pulled down during overhauling to make sure the fire is out and to prevent any possible rekindle. Ordinarily, this is best done from below. If the cockloft has been flooded, personnel must constantly be alert to the fact that the ceilings may have been weakened by the water.

If roof positions are warranted, the use of lines through roof openings would be recommended when high and/or double ceilings render efforts from below ineffective.

The objects of this chapter were, as stated, to show how the action plan is applied when one is assuming command at a fire operation in progress, and how it should have been applied initially. For additional practice, try to visualize a comparable situation involving a local supermarket. Or a local department store could be a practical choice. In any event, such practice can develop more effective prefire plans.

REVIEW QUESTIONS

1. What primary factors can be recognized and evaluated by an officer responding to a fire operation in progress and listening to radio reports by the commanding officer at the fire?

2. The format of the action plan is always the same regardless of the type of fire, the stage of the fire operation, or the rank of the officers involved. However, officers assuming command at a fire operation in progress must recognize and properly evaluate the primary factors that are pertinent on *their* arrival. What does the evaluation by officers assuming command entail? Why?

3. What can an officer assuming command at the fire deduce in evaluating such primary factors as extent of fire after arrival, heat and smoke conditions, and requirements to operate?

4. What were the minor or derivative objectives and in what order

were they to be achieved after confine, control, and extinguish became the major objective at the fire? How were they achieved?

5. With what should an officer assuming command at a fire operation in progress be particularly concerned? Does the officer initially in command also evaluate the effects of secondary factors (activities) on primary factors that may become limiting or strategic in affecting later decisions? What happens if there is a difference of opinion between the officers? Why?

6. Discuss standard operating procedures (commonly employed procedures) used at taxpayer fires, relative to ventilating. What do the records show about the effectiveness of roof ventilation at such fires?

7. Discuss the possibilities at taxpayer fires when operating procedures involve pulling ceilings from below and directing streams at the fire exposed in inadequately ventilated cocklofts.

The following questions pertain to the use of the action plan by the officer initially in command, assuming that other than standard operating procedures are to be used.

8. What primary factors are limiting or strategic in selecting extinguish as the major objective?

9. What minor or derivative objectives have to be attained to achieve extinguish? What is the order in which derivative objectives are to be sought and how is each one attained?

10. Discuss alternate solutions for coping with taxpayer fires. What is undesirable about making openings in the roof and inserting cellar pipes to sweep the cockloft? Why is it preferable to operate from ground level in a taxpayer fire when roof ventilation makes this procedure feasible? When are hose lines directed into the cockloft from above recommended, assuming roof positions are warranted?

21

Fire in a Church and Exposures

This church fire was written up in the first 1966 issue of *WNFY*. Although it is discussed here to show how the action plan can be applied at such fires, there is no implication that the fire could have been handled in a more efficient manner. The actual operation was admirably conducted.

The fire originated, from undetermined causes, on the first floor, near the altar end of the church, and fed on the usual combustible material and interior finish found in such occupancies. It spread rapidly, and by the time the first units arrived it had completely engulfed the church proper. The pastor and his assistants notified the first arriving units that the church was unoccupied.

Evaluate Factors and Select Objectives

The first steps in the action plan are to note and evaluate the primary factors (listed in Table 21-1) that are pertinent in selecting objectives and activities. Due to the obvious danger to occupants in exposures #1 and #2 (Fig. 21-1), life hazard must be the limiting or strategic primary factor and rescue the major objective. Deciding on activities is more difficult because they depend on the derivative objectives that must be achieved for rescue.

The first derivative objective, however, is to evaluate the severity of the life hazard. Here, the commanding officer must make this evaluation in exposures #1, #2, #3, and #4 so they can be covered correctly. Subsequent derivative objectives include: to protect occupants pending rescue, to effect rescue, and to protect personnel.

Select, Assign, and Coordinate Activities

Activities are selected to achieve each derivative objectives and thus the major objective. Activities would be assigned and coordinated in the following manner.

Evaluate the Severity of the Life Hazard in the Various Exposures. Evaluating the severity of the life hazard is done by personal observation on the part of the commanding officer as well as evaluation of reports from other officers. Wind and height factors made it fairly obvious that the life hazard in exposures #1 and #2 was the most severe. The decision to cover ex-

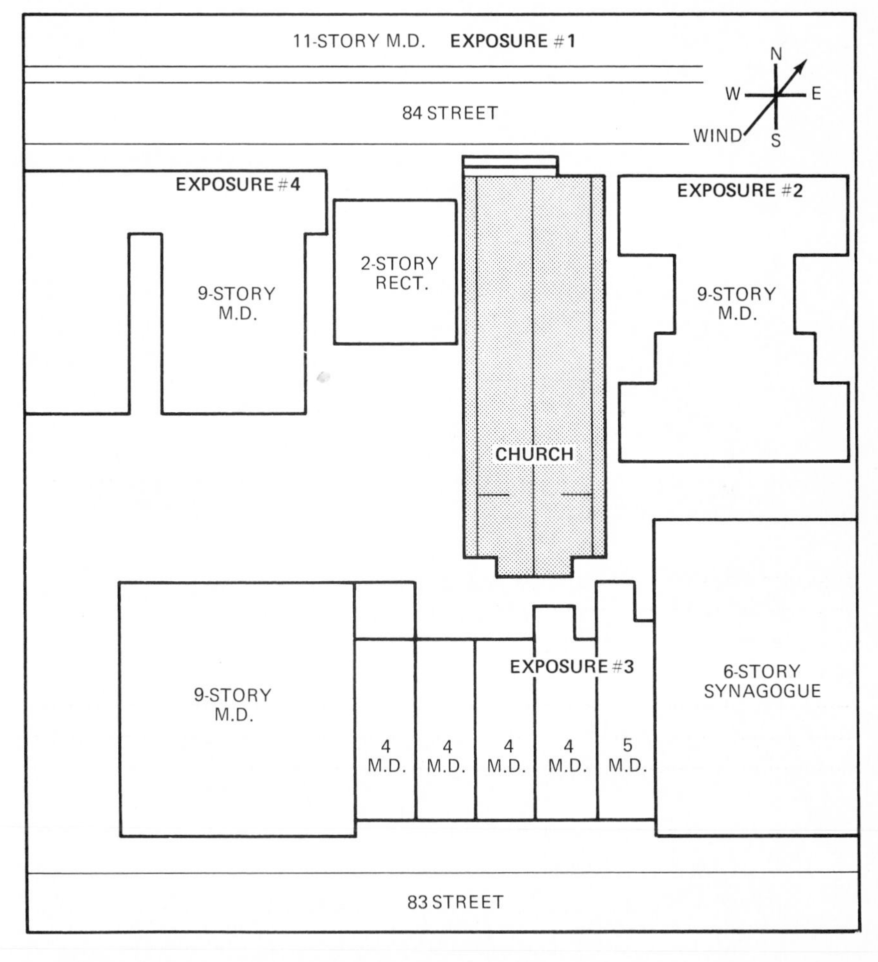

FIG. 21-1. Fire area and exposures.

posure #2 first was influenced by early developments described in *WNYF* as follows:

> Within minutes after the arrival of department units, a heavy back draft accompanied by a mass of flames blew out the large rose type window in the front of the church. Soon after, the entire roof ignited and shot tremendous flames skyward. These were drawn up the large inner court of exposure #2, roaring up the entire nine stories. The flames ignited windows on the line of the court and entered into some apartments on this court. Minutes later, the entire roof of the church collapsed and drove heat and flames across West 84th Street to endanger Exposure #1.

TABLE 21-1. Church and Exposures—Primary Factors

GENERAL	
Time of discovery	Delayed
Time of alarm	3:09 PM, December 1 (weekday)
Time of response	Normal
Location and extent of fire on arrival	Church completely engulfed in fire
Back draft or smoke explosion	Occurred in fire building within minutes of arrival
Structural collapse	Occurred in fire building shortly after arrival due to smoke explosion
Weather	39°F (4°C)
Wind velocity and direction	SW at 13 mph [21 km] with gusts to 21 mph [34 km]
Requirements to operate	Personnel, water, apparatus, and equipment assumed to be readily available
FIRE BUILDING	
Location	26–28 West 84th Street, Manhattan
Classification	Class III Nonfireproof
Erection date	1893
Height	6 stories—cellar—with 85-ft steeple
Area	50 × 150 ft [15.2 × 45.8 m]
Occupancy	Public—church
EXPOSURE #1—NORTH	
Location	15 West 84th Street, Manhattan
Classification	Class I—fireproof
Erection date	1949
Height	11 stories—cellar—penthouse
Area	100 × 100 ft [30.5 × 30.5 m]
Occupancy	Residential class A—multiple dwelling
Auxiliary appliances	Standpipe system

TABLE 21-1. Church and Exposures—Primary Factors (*Continued*)

EXPOSURE #2—EAST	
Location	20–24 West 84th Street, Manhattan
Classification	Class I—fireproof
Erection date	1924
Height	9 stories—cellar—penthouse
Area	93 × 100 ft [28.4 × 30.5 m]
Occupancy	Residential class A—multiple dwelling
Auxiliary appliances	Standpipe system
EXPOSURE #3—SOUTH	
Location	23–37 West 83rd Street, Manhattan (Consisted of 5 tenements)
Classification	Class III—Nonfireproof
Erection date	Early 1900's
Height	4 stories—cellar
Area	25 × 65 ft [7.6 × 19.8 m]
Occupancy	Residential—multiple dwelling
Location	7–21 West 83rd Street
Classification	Class I—fireproof
Erection date	1929
Height	6 stories
Area	160 × 102 ft [48.8 × 31.1 m]
Occupancy	Public—synagogue
Auxiliary appliances	Standpipe System
EXPOSURE #4—WEST	
Location	32 West 84th Street
Classification	Class I—fireproof
Erection date	1929
Height	9 stories—cellar—penthouse
Area	100 × 100 ft [30.5 × 30.5 m]
Occupany	Residential class A—multiple dwelling
Auxiliary appliances	Standpipe system

Flames accompanying back drafts or smoke explosions quickly use up their fuel supply, however, and therefore last comparatively briefly. This fact, together with the effective use of protective lines, prevented involvement of exposure #1, although windows were cracked on all floors.

At this point, the commanding officer could again evaluate the limiting or strategic primary factors and see that the occupants in exposure #2 were more endangered by the heat and smoke than the occupants in exposure #1. This was because of the construction and proximity of exposure #2. Its large inner court was acting like a flue. It was drawing heat and smoke in a way that directly endangered the occupants of apartments facing the court while simultaneously pulling much of the heat and smoke away from the occupants in exposure #1. In addition, exposure #2 was nearer the fire.

The severity of the life hazard in exposures #1 and #2 was affected, but in much the same way, by such primary factors as wind direction and velocity, height (in relation to the level or location of the fire), and area (in relation to the number of windows in the exposed surface). However, since the effects were so similar in both occupancies, such factors did not decide the priority issue for them and hence were not regarded as limiting or strategic. This was not true elsewhere—for example, in establishing an order of coverage involving exposures #1 and #3.

Note that some primary factors in this situation, such as weather and time of fire, affect the life hazard equally in all the exposures. At 39°F (4°C) windows in all the exposures are likely to be closed, and at daytime fires fewer occupants are ordinarily present in residential occupancies, which all these were. Hence, such factors do not affect decisions about the order of coverage.

The limiting or strategic primary factors in deciding that the life hazard was more severe in exposure #1 than in #3 were the following:

1. Height. The height of exposure #1 made it more vulnerable to rising heat and smoke.

2. Area. More windows were exposed in exposure #1 because of the larger surface area facing the fire.

3. Occupancy. Quantitatively, the human element was greater in exposure #1.

4. Wind conditions. Although exposure #3 was nearer to the fire, the wind was decreasing the hazard there while increasing it in exposure #1.

5. Back draft. This endangered occupants in exposure #1 but not in exposure #3.

6. Heat and smoke conditions. These conditions made egress from exposure #1 much more difficult than from exposure #3.

7. Construction of exposure #2. The large inner court localized the fire, thus minimizing the life hazard in exposure #3 more than in exposure #1.

Exposure #4 presented no life hazard problem, due to wind conditions and the distance from the fire (proximity of exposures).

Because of the remaining objectives, the commanding officer transmitted a second alarm, established a command post, and notified the dispatcher that additional alarms could be expected shortly.

Protect Occupants Pending Rescue. This objective can be achieved through the following activities.

1. From the front of the rectory, operate a ladder pipe (tower ladders were not mentioned in the article) with sufficient range to wet down but not break windows in exposures #2 and #1 subjected to radiation and/or convection of heat, with particular attention to the windows facing the large court in exposure #2. A solid rather than a fog stream is preferable for penetrating the rising convection currents. Supplement this stream by operating multiversal nozzles from vantage points in exposure #3 and on 84th Street.

2. Augment the supply to the standpipe system in exposure #2 so that it will not be overtaxed. At the actual fire, in addition to supplying the siamese connection at the front of the building, the system was augmented at the cellar, first floor, fourth flour, and penthouse outlets.

3. Stretch or lay hand lines from the street and advance toward involved apartments (two on each floor) on the west side of exposure #2. Outlets on the first and fourth floors are being used to supply the standpipe riser, and hence cannot supply operating lines. Remember that more than two lines in a stairway can impair maneuverability. At the actual fire, initial lines were operated between the occupants and the fire starting at the lowest floor. It was necessary to get a firm foothold there before additional lines could operate effectively on the next higher floor, although unusual risks were warranted in this situation. Hand lines were also stretched to upper floors via the outside of the building to avoid congestion on the stairs and to simplify the stretch. The standpipe could not supply all the required lines.

4. Reinforce units in exposure #2 with second- and third-alarm assignments to ensure adequate coverage of the life hazard.

5. Check conditions in exposure #3 to ascertain the need for protective lines. The life hazard there was uncertain. Assign units of the fourth-alarm assignment accordingly, directing them by radio.

6. If possible and feasible, ventilate at the roof of exposure #2. Ventilate over the stairway facing the outer court to alleviate heat and smoke conditions and help the advance of protective lines, as well as to draw heat and smoke away from occupants.

Effect Rescue. Always check out elevators when searching for endangered occupants.

On the west street-front side of exposure #2, raise aerials to the upper floors and portable ladders to the lower floors. Enter, search for and remove endangered occupants, preferably by the east stairway. Leave the aerials available for the ladder pipes, to protect both occupants and personnel during rescue work. The following alternatives can be considered here.

1. Use portable monitors to provide protective streams for exposures #1 and #2. Leave the aerials in the *ready* position to ensure a means of escape for occupants and personnel unable to reach the east stairway.

2. Bring occupants from west-side apartments into uninvolved east-side apartments, if feasible, especially after control has been established on the lower floors. This alternative could hasten removal—and application of first aid, when necessary—and at the same time minimize congestion in the east stairway.

On the lower floors, protective lines may be available to facilitate search and removal activities; on upper floors, they may not be.

The blast of flame toward exposure #1 shortly after arrival of the fire department and the possibility of a frontal collapse of the fire building warranted its prompt evacuation. Stairways and elevators were used.

The fire did not hinder the use of stairways by occupants in exposure #3, which was only slightly involved. Removal of occupants was a precautionary measure.

Protect Personnel. This derivative objective could be achieved through the following activities.

1. Provide exterior and interior protective lines as quickly as possible.

2. Ventilate over the stairway facing the court and at appropriate street-front windows via ladders to alleviate smoke and heat conditions. This will enable protective lines to advance and operate more effectively.

3. Position aerials to provide a means of escape.

4. Use masks.

5. Relieve personnel as conditions warrant.

Remember, of course, that proper coordination of activities has an important bearing on personnel safety.

Choose Subsequent Major Objective

After rescue has been achieved, the major objective becomes to confine, control, and extinguish, and the first derivative objective to confine the fire. At the same time, consideration must always be given to the derivative objectives, to protect personnel and to coordinate activities, mindful that personnel are not to be jeopardized unnecessarily since there is no longer a life hazard for occupants.

Select, Assign, and Coordinate Activities

To achieve all of the derivative objectives (and thus the major objective), the following activities would be assigned.

Confine the Fire. The lines specifically operating to protect the occupants did much to confine the fire. Exterior lines sweeping the front of exposure #1 prevented any serious involvement. The fire in exposure #2 was confined by protective exterior lines and a long, arduous interior effort by 12 engine and 7 ladder companies, aided by a rescue and a squad company and supervising chiefs. After control was imminent on one floor, units operated on the next higher floor until the top floor was reached. Then the fire in exposure #2 was both confined and controlled. Many of the fourth-alarm units originally assigned to cover exposure #3 were reassigned to the chief in command of operations in exposure #2. Only rear window frames of exposure #3 caught fire.

Control the Fire. Lines no longer needed defensively were used to supplement others that were directed toward the church fire from vantage points in exposures #1, #3, and #4. Lines could also be directed toward the church fire from exposure #2 after *control* was established there. It is conceivable that control of the original fire could precede control of fires in exterior exposures.

Extinguish the Fire. In exposure #2, considerable overhauling was required for finishing-up derivative objectives to minimize property damage and to prevent rekindles. They were achieved more simply in exposures #1 and #3. In the original fire building, consideration could only be given to preventing rekindles because of the structural condition. In such cases, prompt razing of precariously standing walls by a private demolition company is recommended, along with sufficient watch lines standing by to cope with any possible spark-and-ember hazard.

Conclusion

To protect personnel after rescue was attained, all supervising officers were guided by the specification that personnel should no longer be jeopardized unnecessarily.

To ensure adequate coordination in exposure #2 throughout the entire operation, the building was divided into four sections of two floors each with a staff officer in command, thus providing close supervision and quick communication. In addition, lines no longer needed to confine the fire were used in a coordinated attack to achieve control in the fire building.

There are many variations that could be considered here in practicing how to use the action plan. A frontal collapse could have maximized the life hazard in exposure #1 and changed the order of coverage. A change in the direction of the wind could have maximized the life hazard in exposure #3 and also changed the order of coverage. And so on.

REVIEW QUESTIONS

1. It is clear that rescue is the major objective at the fire described in this chapter. It is more difficult, however, to decide on the activities that must be undertaken. Why?

2. What was the first derivative objective to be achieved at the fire in question? How can it be achieved? What were the limiting or strategic factors in deciding that the occupants of exposure #2 were more endangered by the heat and smoke than the occupants in exposure #1? What factors affected exposures #1 and #2 in a similar way but were not regarded as limiting or strategic because they did not decide the issue involved?

3. What primary factors in the situation described affected the life hazard equally in all the exposures? Are such factors limiting or strategic in deciding the order of coverage?

4. What primary factors were limiting or strategic in deciding that the life hazard was more severe in exposure #1 than in #3? Why?

5. Why did exposure #4 present no life hazard?

6. How was the second derivative objective (to protect occupants pending rescue) achieved? How did the way in which the streams were operated at this stage comply with the principle that available water must be applied between the fire and the endangered occupants or their means of escape? What unusual conditions had to be considered in exposure #2? How could ventilation help at this time?

7. How was the third derivative objective (to effect rescue) achieved in exposure #2? Exposure #1? What alternatives could be considered in exposure #2?

8. How was the derivative objective to protect personnel achieved?

9. In addition to providing adequate supervision and communication, how were activities coordinated at the fire in question?

10. What derivative objectives besides to confine, to control, and to extinguish, also had to be considered after the major objective of rescue had been attained?

11. What was done to achieve the derivative objective to confine the fire? To control the fire? To extinguish the fire?

12. What was done to protect personnel after the major objective of rescue had been achieved? What procedure was followed in exposure #2 to ensure excellent coordination throughout the entire operation?

22

Fire in a Woodland Area

Primary factors at structural fires may include all those to which we have so far given most of our attention, such as buildings of many different types of construction, area, ages, and heights; vertical and horizontal spread of fire via structural channels; back drafts or smoke explosions; street conditions; and so forth. By contrast, woodland fires, except where they involve structures, feature other primary factors: occupancies that can supply a practically unlimited supply of fuel, unusual wind conditions, topography, extensive areas, very large requirements for personnel, questionable water supply in some cases, use of special extinguishing agents, and frequently duration of the operation. Location and extent of fire on and after arrival can be pertinent factors at both woodland and structural fires. Life hazard for personnel is always a factor.

Activities at woodland and structural fires are also different. Forcible entry and ventilation are not essential at woodland fires. Overhauling is performed in a different way, although the derivative objective is the same: to prevent rekindles. Activities commonly used at serious woodland fires and not at structural fires, may include backfiring, establishing firebreaks, and applying special extinguishing agents from planes.

Nevertheless, at both types of fires the format of the action plan is followed in the same way. Consider the following hypothetical situation.

An extensive brush fire is developing at about 3 P.M. on a warm Tuesday in August, after a four-week hot, dry spell. The wind is blowing from the west at about 18 mph [28.8 km] toward a lake-and-mountain resort occupied by about 200 families, approximately a mile east of the fire. The ground rises gradually in front of the fire. Dried-out and plentiful fuels are igniting readily and spot fires in advance of the main body of fire are becoming noticeable. Initially, one engine company responded to the fire. Other units must come from other communities.

Evaluate Factors and Select Objectives

Life hazard, the limiting or strategic factor, is being created and intensified by the following primary factors:

1. ***Location and extent of the fire on and after arrival.*** The fire is only a mile [1.6 km] away, is extensive, and moving rapidly toward the resort area.

2. ***Occupancies involved and exposed.*** The contents of the brush-area occupancy provide a plentiful supply of readily ignitable fuel. The human element in the resort occupancies is in the path of the fire.

3. ***Time of day.*** This factor obviously affects visibility. Time of day can also affect the direction of the wind in canyons; as noted in Chapter 11, there are up-canyon air currents during the day and down-canyon air currents during the night.

4. ***Weather.*** The hot dry spell preceding the fire will cause involved and exposed fuels to burn and ignite more quickly, accelerating the spread of the fire.

5. ***Direction and velocity of the wind.*** The wind velocity will increase the rate of combustion and spread of the fire and will lessen evacuation time.

6. ***Smoke and heat conditions.*** These are worsening in a hurry and will maximize the life hazard.

7. ***Topography.*** The topography lacks natural or other firebreaks between the fire and the resort area, but people on the far side of the lake are less endangered than those on the near side. In addition, fuel on an upward slope is more readily ignited by convection and radiation of heat, hastening the spread of the fire and further reducing evacuation time for occupants of the resort area.

Rescue, the major objective, would be accomplished through the following derivative objectives: (1) Alert occupants of the resort area; (2) protect endangered people pending evacuation; (3) evacuate; (4) protect personnel during operations; and (5) provide adequate communication and supervision.

Select, Assign, and Coordinate Activities

Activities are selected and assigned in the following manner in order to achieve each of the derivative objectives.

Alert Occupants of the Area. Use police cars equipped with loud speakers (if available), telephone, television, radio, sirens, or other available means. Stress the limited time for evacuation so that people will not stop to pack much to take with them.

Protect Endangered People Pending Evacuation. Trying to operate lines between the fire and the endangered people in this instance would be futile in view of the assignment on hand, the extent of the fire, and the operating position (uphill). It would be better if fire personnel did what they could to avert panic and to help get things moving in an organized way.

Evacuate. Use predetermined available routes and means, including boats if feasible, as well as practical order of priority in effecting evacuation. For example, people on the near side of the lake are more endangered than those on the far side and hence should be removed first. Police assistance is essential in conveying and directing civilians away from the path of the fire.

Protect Personnel during Operations. Coordinate properly selected and assigned activities through supervision and communication.

Provide Adequate Communication and Supervision. Establish a command post so that effective communication can be set up to keep abreast of developments, call for help, and carry out the mutual-aid program which should have been planned for such emergencies. Approaching units can be advised to avoid the main road being used for egress from the resort area. Effective supervision can eliminate or minimize traffic snarls and ensure an orderly evacuation, thus lessening the chance of panic.

Choose Subsequent Major Objective

Assume that the fire operation is still in progress and that the fire has extended farther into woodland areas. Confine, control, and extinguish now becomes the major objective and derivative objectives are the following: (1) Determine the location and extent of the fire throughout the operation; (2) confine the fire; (3) control the fire; (4) extinguish the fire; (5) protect personnel during operations; and (6) provide adequate supervision and communication.

Select, Assign, and Coordinate Activities

Activities are selected and assigned in the following manner in order to achieve each of the derivative objectives.

Determine Location and Extent of Fire. To keep track of woodland fires, which are often of long duration, the following suggestions are offered.

1. Keep in contact with lookout towers, if practical and feasible.

2. Assign any available patrol cars to reconnoiter, provided of course that the terrain and fire conditions make such an assignment feasible.

3. Use observers in helicopters that have modern photographic equipment. Airborne infrared mapping systems can generate aerial thermal photographs of a fire and the local topography, showing clearly the location of the fire and its relative intensity over the burning area (Fig. 22-1). These photographs can be taken at any time, day or night, even when the fire is completely obscured by smoke. They also show associated spot fires with respect to local topological features, including roads and fuel-type variations, and can measure perimeter rates of spread. Refer to the Bibliography at the end of this chapter for further information on this subject.

Other photographic equipment that may be usable at these fires has been described in Chapter 12.

4. Study maps that show the location and types of any firebreaks. Such maps can help you to see where the fire can best be stopped.

5. Keep in touch with the weather bureau about possible changes in weather and wind. This information can help you to anticipate possible

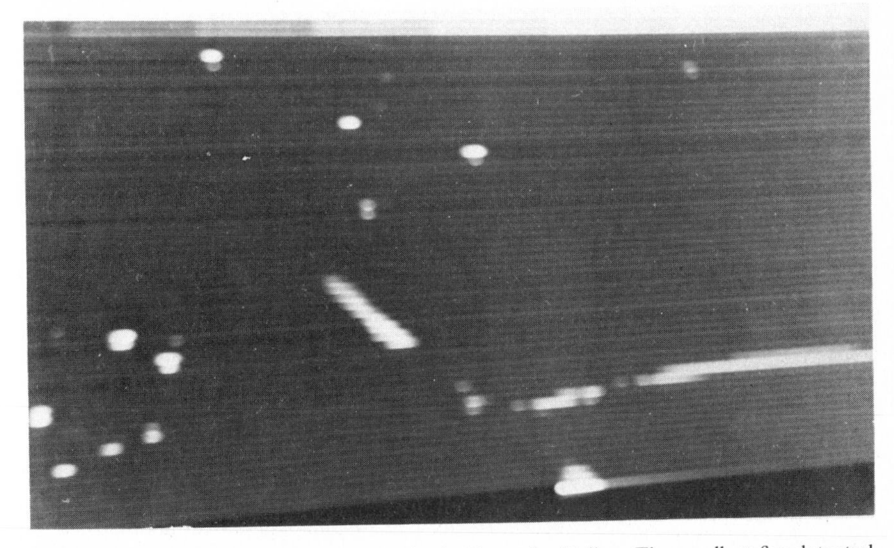

FIG. 22-1. Thermal infrared image of a site in Yosemite Valley. The smallest fire detected was one charcoal briquette less than a cubic inch in size.

accompanying changes in the rate and direction of fire spread. Bear in mind also the wind patterns characteristic of canyons.

Confine the Fire. This derivative objective can be attained by using hose lines and/or special extinguishing agents from aircraft, and using existing firebreaks and/or establishing other ones, in conjunction with backfiring. It is assumed here that the situation is controllable by such tactics—eventually, although certainly not quickly. To confine this fire, the following suggestions are offered.

1. Place and operate lines on the windward and lee flanks of the fire. Concentrate on the lee flank in an effort to divert the fire toward an area in which the supply of fuel is less (the starving method). It may be at least a somewhat successful tactic to divert the fire—if you can—toward a broad highway, extensive clearing, river, lake, or any combination thereof. In actual practice, it would be very difficult to confine this fire by such tactics unless there were sufficient firebreaks and effective backfiring, even assuming the availability of ample personnel, water, apparatus, and equipment. It would be much less difficult to confine the fire on the windward flank side.

   Use small streams to wet down brush and extinguish incipient fires. At the Malibu fires in 1956, for example, on the beach-front roads, pumpers at hydrants 600 ft or more apart used 2½in. hose to supply 1½-in. working lines. In areas without fixed water supplies, tankers and booster trucks and even private swimming pools were used.

2. Use aircraft, if available, to direct special extinguishing agents to help confine the fire. The effectiveness of water bombing is questionable. (At the Malibu fires, high winds and irregular topography made accurate bombing impossible.)

3. Use helicopters to transport men, portable pumps, and hose to expedite the strategic placement and use of lines. Portable pumps can be used at otherwise inaccessible sources of water supply. Reconnaisance from helicopters can also influence the placement and movement of other lines. Some modern helicopters have a 10-ton payload capacity and could safely transport a 1,000 gal of water, 30 men, and almost 3 tons of equipment, which could include much lightweight hose and portable pumps.

4. Wet down the outer fringes of firebreaks that may be hard to hold.

Control the Fire. Confining the fire is not necessarily tantamount to controlling it, but in this case it is nearly so. There remain the tasks of extinguishing spot fires beyond the firebreaks and of widening the firebreaks around the perimeter, especially between the fire area and fuel-laden areas that are most endangered by stray sparks.

Extinguish the Fire. Overhauling plays a unique and important role here. It involves seeking out and extinguishing any sparks for several hundred feet into the burned area all around the perimeter. Not for several weeks will it be certain that no wind will find a hot ember that can be whipped into flame and start another fire downwind, especially if the dry spell continues.

Protect Personnel during Operations. It must by now be obvious that this objective is best attained at any fire by selecting correct activities and assigning and coordinating them through adequate supervision and communication. At this fire, personnel were not to be jeopardized unnecessarily.

Provide Adequate Communication and Supervision. Personnel may include federal and state supervisory specialists, regular firefighters, loggers from lumber camps, and so forth. To eliminate confusion, unnecessary duplication, and the possibility of lost messages, it is essential to promptly initiate an organized network of transmission. Due regard must be given to the organizational structure of the operating forces, the chain of command, the priorities of various messages, and the operating policy of the network. In developing such an organizational structure, a subordinate chief would be assigned to command each sector. Units would then report to the sector chief rather than the command post. The logistic problem should likewise be handled by a subordinate chief and aides if necessary.

In selecting the site for a command post in this case, the following factors require consideration.

1. Topography. Hilly terrain can interfere with the reception and transmission of information from and to units.

2. Location and extent of fire on and after arrival, and wind direction and velocity. These factors influence the rate of spread and the direction of the fire. They must be considered in order to minimize problems created by distance in relation to transportation and time, particularly if a fire camp is in the same area. It is therefore possible that the command post may have to be relocated from time to time.

3. Duration of the operation and requirements to operate. Such fires can last a long time. When they do, it may be necessary to establish fire camps so that fire activities can be properly coordinated with supply, food, and rest activities. To help solve the resulting logistics problem, a landing space should be made available for helicopters transferring food, personnel, supplies, and equipment.

BIBLIOGRAPHY

Colwell, R. N.: "Remote Sensing of Natural Resources," *Scientific American,* vol. 218, no. 1, pp. 54–69, January 1968.

Hirsh, S. N.: "Airborne Infrared Mapping of Forest Fires," *Fire Research Abstracts and Reviews,* vol. 8, no. 3, 1966.

Mikelonis, E. C.; R. P. Shearer; and G. L. Duguay: "Preliminary Infrared Fire Mapping Systems Analysis," *Fire Research Abstracts and Reviews,* vol. 9, no. 3, 1967.

REVIEW QUESTIONS

1. What primary factors may be pertinent features at structural fires? At woodland fires? Which are common to both? How do activities at these fires differ? Do these differences affect the format of the action plan?

2. At the fire described, rescue is the major objective because life hazard is the limiting or strategic factor. What primary factors create and intensify the life hazard? Why?

3. What minor or derivative objectives have to be attained to achieve the major objective of rescue? How are they attained?

4. Assuming that confine, control, and extinguish becomes the major objective, in what order do essential minor or derivative objectives have to be attained?

5. How is the derivative objective determine the location and extent of the fire throughout the operation actually achieved?

6. How is the derivative objective to confine the fire actually achieved?

7. How are the derivative objectives to control the fire, to extinguish the fire, and to protect personnel achieved?

8. What primary factors require consideration in selecting a site for a command post at extensive woodland fires?

PART FIVE

Managerial Principles and Firefighting Leadership

23

Basic Concepts

The skill with which the science of fire suppression is applied is affected by an officer's ability to use managerial principles. This observation is in line with the concept that scientific knowledge or technical ability is not fully utilized unless those at managerial levels—and this means officers of all ranks—can effectively coordinate the human resources under their jurisdiction. Accordingly, managerial subjects associated with fire-service activities will be considered. The comments offered are a fire-service interpretation of *Principles of Management*,[1] and begin with some definitions and discussion of fundamental terms such as administration, management, principles, functions, techniques, authority, and responsibility.

ADMINISTRATION

Administration is defined here as the coordination of activities and associated personnel to achieve the objectives of a fire department or its subdivisions, which may consist of bureaus, divisions, battalions, or companies. Activities may be related to fire operations, fire-prevention work, training programs, or anything else that is pertinent.

MANAGEMENT

Management has been defined as the art of getting things done by other people. It has also been defined as the achievement of desired objectives by establishing an environment favorable to performance by people operating in organized groups. However defined, management is not only the

[1] Koontz and O'Donnell, *Principles of Management*.

function of the corporation president or chief of department, but also the shop foreman or lieutenant and the intermediate ranks. All officers should therefore study management principles that can affect group activities in the fire service, even though the degree to which officers participate in management varies with their rank and responsibility.

It is important that both officers and subordinates understand the nature of management. If officers understand properly, they will not lose time by working at nonmanagerial activities that should be done by others. If subordinates understand management, they will not think that managers "do nothing" and will recognize their own parts in the managerial functions. They will probably also appreciate an officer who discharges the managerial functions efficiently. Too many officers, especially at lower levels, engage unnecessarily in nonmanagerial activities. Managers may, at their discretion, engage in nonmanagerial activities, but subordinates never undertake executive functions.

Managers need not have nonmanagerial technical skills, such as those of doctors, lawyers, and accountants, but they must be able to use skillfully the talents of the trained persons under their direction. They must know which skills are used in their particular occupations and departments, and they must be familiar enough with them to be able to ask discerning questions. For example, the head of a fire department, reading about research on a certain firefighting technique, should be able to consult intelligently with technical advisors on the merits of the new technique. Managers must also understand both the separate role of each skill and the interrelationships between skills. At a major fire operation, roles are played by engine, ladder, and rescue companies, by mask service, communications, and supervising engineer units. The officer in command must know how best to assign each of these to use their capabilities most fully.

PRINCIPLES

Principles are construed to be fundamental truths applicable to a given set of conditions or circumstances and to what can be expected to happen under said conditions or circumstances. Management principles are universal in application and are therefore adaptable to group efforts by fire organizations as well as political, religious, educational, or other organizations. However, there may be variations in applying these principles since pertinent factors vary from one field to the next, and such factors influence the manner in which the principles are used. For greater relevancy, therefore, management principles are considered in relation to the unique factors that may be pertinent in the fire field, or in other words, from a fire-service viewpoint.

FUNCTIONS

A *function* is defined as the natural or characteristic action of a type of professional person or a thing. Management is concerned with the functions of persons rather than things.

The functions of a manager include planning, organization, direction, staffing, and control. All managerial activity can be classified under these five categories.

Managers perform the same functions regardless of the type of organization structure or their place in it. As managers, all are concerned with getting things done by subordinates, and at one time or another they carry out all the duties characteristic of managers. Because of the universality of managerial functions, anything significant said about one manager applies to all managers. As a result, it is possible to develop a theory of management applicable to all executives in all occupations. In addition, managerial experience and knowledge are transferable from unit to unit and from enterprise to enterprise. To the extent that tasks are managerial rather than technical, executives, with the proper motivation, can use their skills as well in one occupation as another. The fire service can benefit from the theory of management, which applies to officers, acting as managers, at all levels.

Discussions of the managerial functions in subsequent chapters will enable the reader to discern that a commanding officer at fires actually uses these functions in employing the action plan by recognizing and evaluating pertinent primary factors to select objectives and activities, assigning activities, and coordinating them by supervision and communication. At fires, the function of company officers is to supervise and communicate.

TECHNIQUES

Techniques refer to the way in which managers or superiors carry out their functions. Thus in dealing with their subordinates, managers may use the technique of command or persuasion, written or verbal orders, written or implied policies, and so forth.

At fires, any one of several techniques may eventually put out the fire but only one will do the job in the most efficient manner. The correct choice depends on the primary factors that are pertinent in a given situation, and not on a personal preference for a certain technique at all fires—such as exclusive use of high-pressure fog, $1\frac{1}{2}$-in [38 mm] hose, indirect attack and the like.

AUTHORITY

A standard definition of *authority* is "legal or rightful power, a right to command or act." From a fire-service standpoint, this definition is interpreted to mean that superiors have the power to order their subordinates to carry out an activity or to refrain from carrying out an activity. Thus authority empowers officers to issue orders, which are the specific means by which activity is initiated or modified.

Authority is the basis for responsibility and is the force that binds an organization together. Without adequate authority, there would be chaos because superiors could not properly meet their responsibilities. Moreover, since management theory is necessarily based on superior-subordinate relationships, it is founded on the concept of authority.

The Source of Authority

There are two theories about the source of authority: the formal theory and the acceptance theory. The formal theory states that authority is power transmitted to an individual from a social institution. *Institution* used in the sociological sense means a complex of laws, codes, mores, and folkways by which a social group attains and enforces group purpose.

Under our democratic form of government the right upon which managerial authority is based has its primary source in the Constitution of the United States through the guaranty of private property. Since the Constitution is the creature of the people, subject to amendment and modification by the will of the people, it follows that society, through government, is the source from which authority flows to ownership and thence to management—and, in a specific way, to fire departments. For example, the people, through representative government, ratified the tenth amendment to the Constitution, to the effect that powers not delegated to the United States by the Constitution, nor prohibited by it to the states, are reserved to the states. As a result, state police power, which is a derivative of formal authority, empowers the state to formulate measures pertaining to health, morals, safety, and general welfare, including, of course, fire protection. The legal authority under which fire organizations operate is state police power, which may be exercised by local governments or municipal corporations as specified in charters approved by state legislatures.

Formal authority without leadership tends to be an abstraction. It is important for fire officers, particularly those newly appointed at the lower levels, to realize that rank and authority by themselves are not enough. To be effective, an officer needs several important qualities, notably leadership, or the ability to encourage subordinates voluntarily to carry out assigned tasks efficiently and with goodwill.

The acceptance theory of authority maintains that a superior has no real authority unless and until subordinates confer it upon him. Individuals always have an opportunity to influence decisions of superiors by accepting or rejecting them. In effect, the sphere of authority is defined by the degree of acceptance of the subordinates. This type of authority is obviously impractical for the fire service since officers would not know from one command to another whether they would be obeyed, and therefore would be placed in an untenable position. The acceptance theory nevertheless is considered to have some possibilities if used by exceptional leaders who have the ability to persuade subordinates to work well to accomplish a group goal. In this connection, the acceptance theory expounds the value of leadership, but it is difficult to discern any exercise of authority in applying the theory.

The Competence Theory

In addition to the formal and acceptance theories of the sources of authority is the competence theory, which is based on a belief that authority is generated by personal qualities or technical competence. This belief may be attributable to the facts that some individuals make subordinates of others through sheer force of personality, or that the advice of some exceptionally capable technicians is so eagerly sought and unerringly followed that it appears to have the weight of an order. However such influence is described, it cannot by itself bestow on an individual truly legitimate authority, that is, a legal or rightful power, a right to command or act, and so on.

Various types of authority, classified as *shared, splintered, functional, centralized,* and *decentralized,* are associated with the delegation of authority and the function of direction, and will be considered in Chapter 28.

Limits of Authority

There are two categories of limits on authority: external and internal. External limits are those imposed by physical limitations of men and equipment and by the existence of conflicting or coexisting authority. For example, ownership of private property confers authority that may be circumscribed by the authority of the fire service to conduct operations. There is no point in officers ordering firefighters to walk up a wall or raise a 75-ft [22.9 m] aerial to an 85-ft [30 m] roof; these operations are not physically possible. The exercise of authority is limited by social and economic considerations and by the need for internal consistency; officers cannot order personnel to take risks that unnecessarily jeopardize them, they cannot provide themselves with equipment except as specified by a budget, and they must exercise their authority to conform with departmental regulations and policies.

Internal limits on authority result because, in practice, an order is only as effective as the willingness of subordinates to obey it. Subordinates of today, for example, are more apt to rebel against autocratically exercised authority and more apt to submit to democratically exercised authority than subordinates of 20 or 50 years ago.

Effective orders must be issued by superiors to their subordinates (officers cannot issue an effective order to other officers of the same rank or to their subordinates); the orders must concern matters pertinent to fire-service activities; and they must be enforceable (there must be sanctions that can be employed against subordinates who refuse to carry out orders or who carry them out inappropriately). At fires, a battalion or district chief can bypass a company officer and give orders directly to a firefighter, but this is contrary to the principle of unity of command, tends to cause confusion, and is not a good practice.

Departmental regulations impose limitations by specifying the authority of officers of different ranks, by requiring officers to avoid unnecessary jeopardy of personnel at fires and excessive and unwarranted damage during forcible entry, ventilating, and overhauling, and to use discretion in applying water at fires, and so forth. Instructions also make officers aware that their authority in using the right-of-way prerogative of fire apparatus is limited, as is their authority to block streets during fire operations; that is, they realize that street conditions should be returned to normal as soon as practicable.

The unique authority of the fire service to force entry, without a warrant, into a home or place of business, and to order occupants to leave, is also limited; homes and places of business must be returned to the owners as soon as feasible.

RESPONSIBILITY

Responsibility is defined as the obligation of a subordinate to perform the duties assigned by superiors. The essence of responsibility is, then, obligation. Responsibility is owed to one's superior, and no subordinates reduce their responsibility by delegating to another the authorority to perform a duty. No managers or superior officers, then, can shift responsibility to their subordinates.

All subordinates should know who their superiors are and to whom policy matters beyond their authority must be referred. This development is best achieved by a clear understanding and application of the scalar principle, which establishes a chain of direct authority relationships from superior to subordinate throughout the organization. Fundamentally, this principle states that authority for engaging in any undertaking must rest somewhere and that there must be a clear line from that source to every rank in the organization. In this connection, a comprehensive, up-to-date organization chart can be helpful.

Ordinarily, authority flows from a higher rank to the next lower rank and responsibility is owed by a lower rank to the next higher rank. However, this sequence is not necessarily followed at fires because a battalion or district chief may directly assign duties to, and exact responsibility from, captains and lieutenants. In such instances, authority flows from the chief to a lieutenant and responsibility is owed by a lieutenant to the chief, and the rank of captain is bypassed for obvious reasons.

Authority and responsibility are intimately connected in the chain of command, where officers have authority over their own subordinates but are also responsible for the discharge of duties assigned to them by the overall commanding officer. In such a case, an intermediate officer can be both a superior and a subordinate.

Managers or officers have *authority delegated* to them, *responsibility exacted* from them, and *duties assigned* to them. Since authority is the power to carry out assignments and responsibility is the obligation to accomplish them, it follows that the authority delegated should be commensurate with the responsibility involved. This is in accord with the principle of parity of authority and responsibility. Obviously, it would be unfair, as well as impractical, to assign duties and exact responsibility without delegating the authority required for effective accomplishment.

The principle of parity of authority and responsibility has important implications: The basis for responsibility must exist to warrant the exercise of authority by the fire service. The unnecessary use of authority alienates the public and harms public relations, Moreover, it can put the fire department in an awkward position in lawsuits.

Parity of responsibility and authority has a bearing on superior-subordinate relationships. Much depends on the concepts of authority and responsibility adhered to. Permissive concepts tend to promote lax relationships, which are highly undesirable from our point of view.

In view of the parity of authority and responsibility referred to, officers reporting their arrival and assuming command at an operation in progress should be aware that they are then responsible for the way in which the operation is conducted, and they are automatically delegated commensurate authority. Accordingly, they should make their own evaluation of information received by radio, discernible by visual observation, and obtainable from the officer relieved of command. In one case, a high-ranking chief failed to check out information furnished by such an officer, and the results were disastrous. At the investigation that followed, the high-ranking chief maintained that the subordinate officer was responsible. Clearly, the subordinate officer was responsible for an inaccurate report, but this did not alter the authority-responsibility status of the commanding officer or other officers. Chiefs assuming command at operations in progress are not obligated to regard information as accurate. If they do, and a disaster occurs due to inaccurate information, responsibility rests on their shoulders. No manager or fire officer can shift responsibility to subordinates.

REVIEW QUESTIONS

1. Why does a knowledge of managerial principles affect the art or skill with which the science of fire suppression is applied? From what perspective are managerial subjects considered in this text?

2. Define and discuss the term *administration* as it applies to a fire department.

3. Define *management*. What ranks in the fire service carry out the function of management? With what does participation in management vary? Why should superiors and subordinates understand the nature of management? What skills should a manager have?

4. Are the comments about principles in this chapter consistent with what was said in Chapter 1? Why is it practical to consider managerial principles from a fire-service viewpoint?

5. With what are managerial functions concerned? What do they include? Why are managerial functions universal in application? To what extent can managerial experience and knowledge be successfully applied in various occupations and enterprises?

6. To what do *techniques* refer? How may they be used from a managerial viewpoint? At fires?

7. Define *authority*. How is this definition interpreted from a fire-service viewpoint? Discuss the role of authority in relation to responsibility. Why is management theory founded on the concept of authority?

8. What is meant by the *formal* theory of authority? What is the primary source upon which managerial authority is based in a democracy? Why? What is the legal authority under which fire organizations operate? Why is formal authority of limited value for fire officers?

9. What does the *acceptance* theory of authority maintain? Discuss its limitations and possibilities for use by fire officers.

10. On what is the *competence* theory of authority based? What are its limitations?

11. What are some external limits on authority? Internal limits?

12. What are some limitations on the issuance of orders? When are these limitations ignored?

13. How do departmental regulations impose limitations on the exercise of authority? What unique authority does the fire service have? How is this authority limited?

14. Define *responsibility*. To whom is it owed? What is the result of the fact that subordinates do not reduce their responsibility by delegating to another the authority to perform a duty?

15. What is the purpose and importance of the *scalar principle?*

16. What is the basis for the principle of *parity of authority and responsibility?* What are some of its implications?

17. Insofar as authority and responsibility are concerned, what happens when officers report their arrival and assume command at a fire operation in progress? What is recommended for officers assuming command in such cases? What may happen if officers assuming command at a fire operation in progress fail to check out information provided by the officer being relieved of command?

24

Planning

Planning involves the selection of objectives, policies, procedures, programs, rules, and budgets, each of which is discussed in this chapter. It is the most basic of all management functions because it influences future courses of action for the fire department as a whole and for each subdivision within it. In addition, the other four functions depend on planning. Thus, a manager or officer *organizes, staffs, directs,* and *controls* to ensure the attainment of objectives according to *plan.* Although all the functions intermesh, planning is unique in that it establishes the objectives necessary for all group effort.

OBJECTIVES

An objective can be the goal of an operation or program. Planning can be useful only if objectives are properly selected. Improper selection or faulty specification of objectives will make the entire planning exercise futile. Accordingly, every enterprise should spell out its precise objectives. Hence, based on considerable analysis, potential major objectives at fires were precisely specified as (1) rescue, (2) extinguish, and (3) confine, control, and extinguish. To achieve any of these major objectives, appropriate minor or derivative objectives must be sought by selecting, assigning, and coordinating the correct activities. In other words, major objectives at fires are selected and achieved by applying the action plan.

Derivative objectives may also have to be attained in carrying out programs.

POLICIES

Policies are defined as general statements or understandings which guide thinking in decision making by subordinates. Policies are often implied from the actions of superiors. They are seldom specific or written out. They are normally made at all levels, and should always be consistent with officially stated departmental policies.

There are three types of policies of interest to the fire service: originated, appealed, and imposed.

Originated Policy

The most logical source of policy is that originated by top management for the express purpose of guiding managers and their subordinates in the operations of the business—in this instance, a fire department. Originated policy flows basically from fire-department objectives as interpreted by the top executive. The extent to which this policy formulation is centralized or decentralized depends on the extent to which authority is concentrated or dispersed by top-level authority.

Appealed Policy

Appealed policy is, in effect, policy by precedent. It comes from the appeal of exceptional cases up the hierarchy of managerial authority or chain of command. Officers may appeal to their superiors because they do not know if they have the authority to make a decision or do not know how a matter should be handled. As decisions about these appeals are made, a kind of common law is established. Precedents develop and become guides for later action. However, there is a danger that such policies will be incomplete and confused. The top executive can minimize the need for appealed policies by not leaving too large an area of policy making to chance and by verifying that subordinates understand the policies already formulated. Clearly stated and written policies, promulgated by the top executive, would be helpful.

Imposed Policy

Imposed policy results from outside forces such as wartime government regulations, fuel shortages restricting the use of gas at drills, severe snowstorms drastically interfering with response of apparatus, social unrest, and the like. These external forces can affect policy making in two ways: sometimes they dictate specific policies; sometimes they merely influence them.

AREAS OF POLICY FORMULATION

Policies are formulated in relation to (1) managerial functions, and (2) functions of an organization, such as a fire department.

Policies Relating to Managerial Functions

Policies may be formulated about (1) planning (the length of time for which to plan, the amount of detail or thoroughness of plans, the extent of organizational participation in planning, and the desired degree of flexibility); (2) organizing (the span of management or control, the degree of dispersal of authority, how to maintain some centralized control when authority is decentralized, and clarification of functions and authority); (3) staffing (selecting, recruiting, training, promoting from within or otherwise, hiring the person to fit the job versus modifying the job to fit the person, the use of rating systems, and seniority and other ratings that affect eligible lists); (4) direction (attitudes toward delegation of authority, discipline, leadership, morale, communications, issuing of orders, the techniques of direction to be used, and the emphasis on human relations); (5) control (where in the organization special kinds of controls shall be applied, whether budgetary control devices shall be used, the length of time in the future to be considered, and the amount of detail).

Policies Relating to Organization Objectives and Functions

Policies may also be formulated in relation to the objectives of the fire service (to prevent fires and to protect life and property against fires that cannot be prevented), and to all functions associated with the achievement of these objectives. An example of a policy related to fire-prevention is given below (see Programs).

The dominant policies related to protecting life and property against fires that cannot be prevented are the specifications for an acceptable solution at fires, namely (1) if there is a life hazard for occupants, risks to personnel ranging from merely unusual to extreme may be warranted, and (2) if there is no life hazard for occupants, personnel are never to be jeopardized unnecessarily. These policies help to guide officers to make and effectively implement correct decisions about objectives and activities at fires.

Policies relating to training officers to operate at fires are obviously important. To what extent should training be standardized by training officers of all ranks to think and act in a uniform way at all types of fires? Should officers of all ranks be taught to recognize the hazards that may cause casualties among personnel and take precautions accordingly, or should such training be given only to designated hazard control officers? In fire science and other training courses, how much consideration should be given to improving the efficiency of officers in field operations, or in other words, to the relevancy of the material being taught?

In suggesting areas for policy formulation, it should be kept in mind that true policies allow for some discretion. Otherwise, they would be rules. For example, the policy that personnel are never to be jeopardized unnecessarily when there is no life hazard for occupants, allows for some discretion. Some risks are practically inescapable in stretching and operating hose lines, raising ladders, and so on. These risks do not unecessarily jeopardize personnel. The effectiveness with which officers use discretion in applying the policy referred to depends upon their training.

Clearly enunciated policies, preferably written, would help in many matters affecting fire operations, such as in (1) establishing a command post, (2) sending units above the fire when there is no life hazard, (3) having personnel work in groups of at least two, (4) providing each group with a communicating device, and so on.

Numerous policies could be formulated about ladder work. To what extent should fire buildings be laddered? When should aerials be raised to the roof or extended and held in the *ready* position as a means of escape for occupants and/or personnel? When should aerials be used at oblique angles? When should snorkel apparatus be used? And so on.

Policies relating to the evacuation of buildings, especially the high-rise type, require careful consideration. Policies appropriate for new high-rise buildings with central air-conditioning systems are not appropriate for some older, truly fire-resistive buildings without such systems. Policies should offer guidelines on the order of priority to be followed in removing occupants, on where to take them (to the street or below the fire), and so on.

Policies can be formulated about how and by whom salvage is to be carried out, and many additional phases of fire operations, such as giving orders, selecting hydrants, covering exposures, relieving personnel on lines and elsewhere, overhauling, street conditions, and public relations. An understanding of what policies are as well as the grounds upon which they are formulated will enable officers to create and employ policies that will be practical and durable. The adherence to policies has such a profound effect upon the achievement of goals that detecting signs of policy failure requires prime consideration. The number and type of appeals for policy modifications may reflect failure. Evidence of conflict between department policy and the rules of society may also indicate failure. For example, the policy of right of way for fire apparatus, as originally employed in some communities, was found to conflict with standards of public safety. Hence, the policy was modified: fire apparatus must slow down and be ready to stop if necessary to avoid an accident when approaching intersections, regardless of whether the traffic light is red or green.

A third means of detecting policy failure is to evaluate the results against logical expectations. This is part of the control process. The development of effective control devices would naturally aid in policy formulation and application.

Communication of policy helps provide the common understanding so beneficial for securing conformity and consistency in interpretation. Historically, there has been a long tradition of unwritten policy. Here, we feel strongly that broad guides for the fire service as a whole and for component units should be written down. This book constitutes to some extent a compilation and explanation of fire-service policies, with the hope that constructive study and criticism may result to the ultimate benefit of those in the fire service and of the public they serve.

PROCEDURES

Procedures are guides to *action* rather than *thinking*. This is a major difference between procedures and policies. Procedures detail the exact manner in which activities are to be carried out, with emphasis on chronological sequence. Procedures become more numerous and exacting at lower levels, largely because of the necessity for more careful control, the reduced need for discretion, and the fact that routine jobs are done most efficiently through prescription of the one best way.

Procedures used at fire operations are referred to as *evolutions* in some departments and are described in manuals, guides, orders, regulations, or other directives. Additional procedures described in similar media pertain to investigations of accidents, meritorious acts, and the like.

Difficulties with procedures may arise because of complexity, duplication, obsolescence, and inflexibility. Undue complexity and duplication can be caused by a lack of clarity in the delegation of authority or work assignment. This may result in duplication of the work or in no one's doing anything because everyone thought it was another's job. Such possibilities can be avoided or minimized by using modern concepts of procedure analysis and work simplification, whereby steps in procedures are reduced to essentials and delegation of authority and work assignments are more clearly specified.

Obsolescence can become a handicap because methods of handling operations tend to become ingrained in departments and individuals. This characteristic reluctance to change can make procedures both obsolete and harmfully inflexible. Therefore, in addition to studying and applying modern concepts of procedure analysis and work simplification, some thought can be given to the causes of resistance to change and how they can be dealt with.

Change is always necessary, mostly because the premises on which plans are based are constantly changing. This is indicated by such things as shifts in attitude about authority, responsibility, personal appearance, treatment of personnel at fires, economic pressures, and so on. Management authorities therefore advise top executives to adopt the policy that change should be regarded by personnel as an accepted and expected thing. Nonetheless, change is generally feared by personnel, and understandably so if their security might be lessened. Some management authorities suggest that change is more acceptable (1) when it is thoroughly understood, (2) when it does not threaten security, (3) when those affected *have helped create it,* (4) when it results from an application of previously established impersonal principles—not when it is dictated by personal order, (5) when it follows a series of successful changes—not failures, (6) when it is inaugurated after earlier change has been assimilated—not during the confusion of many changes, and (7) when it has been planned—not when it is experimental.

These comments about change can also be considered in relation to objectives, policies, rules, programs, and budgets, as well as procedures. When the welfare and security of personnel are involved, it seems reasonable to suggest that adequate labor arbitration is in order.

PROGRAMS

Programs are a complex of policies and procedures associated with a course of action to achieve an objective. Policies have an important bearing on the success of a program. For example, one policy in conducting a fire-prevention program could be to stress the educational rather than the police aspects when making inspections, thereby minimizing abrasive contacts and obtaining better cooperation and improving public relations. Procedures are likewise important since they detail the manner in which necessary activities are to be carried out.

Basic programs require the development of derivative programs. For example, the basic fire-prevention program may require derivative programs for training personnel and educating the public. The success or failure of a basic program depends on the success or failure of the derivative programs. Thus, derivative planning plays a role in carrying out programs as well as in fire operations. A fire-prevention program is discussed in Chapter 31.

RULES

Rules are guides that specify a certain action is to be taken or refrained from. Rules are similar to procedures in that both guide action, but are different in that rules specify no time sequence. For example, "No Smoking" is a rule unrelated to any time sequence or procedure. However, a rule may or may not be part of a procedure. For example, the procedure to be followed in investigating an accident may incorporate the rule that reports are to be forwarded in triplicate. Another distinction between rules and procedures is that rules sometimes pertain to behavior; procedures do not.

There is also a decided difference between rules and policies, although both serve as guides for personnel. However, policies are to guide *thinking* in decision making by marking off areas of discretion, whereas rules are designed to guide *action* and allow no discretion in their application. In short, policies are flexible, rules are not.

Policies and procedures have separate and distinct roles. Policies related to a life hazard for occupants and personnel can affect decisions about objectives and activities. Policies can also affect decisions as to how programs should be carried out (see above, Programs). Procedures fundamentally implement the decisions that are made.

The distinction between policies, procedures, and rules has been stressed because some authorities maintain that clarity here is important to good planning, to workable delegation of authority, and even to good human relations.

BUDGETS

Budgets are statements of expected results expressed in numerical terms. They may be expressed in financial terms, or in personnel-hours, machine-hours, or any other measurement that can be reduced to numerical expression.

Budgets are essentially plans, formulated with the approval of the top fire executive and subject to modifications specified by higher authorities. Since this book has been prepared for officers of all ranks only this much will be said about budgets: About 90 percent of the budget is allocated for personnel. Naturally, the quality of training given to personnel is reflected in the quality of service rendered by them, a fact that should help budget makers to get the proper perspective on training.

STEPS IN PLANNING

Steps in planning are a mental process to be followed in selecting a course of action. Obviously, steps in planning and the process for making business decisions (discussed in Chapter 2) have much in common. Each outlines an orderly progression from gathering information to making a decision.

Steps in planning, which can be adapted to major or minor programs, include: (1) Establish the objective; (2) establish planning premises;

(3) determine alternative courses; (4) evaluate alternative courses; (5) select a course; and (6) formulate derivative plans.

Establish the Objective

This is a logical step in formulating a fire-prevention, training, or safety program. However, at a fire operation, an officer would first have to establish planning premises and then establish the major objective. This is comparable to critical-factor analysis used in the process for making business decisions, in which critical factors are isolated to define the problem or objective. At fires, such an analysis has to be made before the objective can be established.

Establish Planning Premises

Premises are assumptions based upon data of a factual nature. Data of a factual nature consists of factors classified as noncontrollable, semicontrollable, and controllable.

Noncontrollable factors at fires could include life hazard and location and extent of fire on arrival; occupancy (contents involved); time of origin, discovery, and alarm; weather, wind, and humidity; back draft or smoke explosion before arrival; exposure hazards on arrival; street conditions; topography; heat and smoke conditions on arrival; structural collapse before arrival; and class of fire. Reference is to factors noncontrollable by the fire service.

Depending on the requirements available to operate and the efficiency with which they are used, semicontrollable and controllable factors could include life hazard, extent of fire after arrival, heat and smoke conditions, back draft or smoke explosion, exposure hazards, structural collapse, requirements to operate, and duration of operation.

In the action plan recommended for use at fires, factors comprising data of a factual nature are classified as primary and secondary. This approach conforms with the effort to develop a science and art of firefighting. Initially at fires, data of a factual nature consist of the primary factors that are limiting or strategic to decisions about objectives and activities. Thereafter, the data may change due to the effectiveness or ineffectiveness of the activities undertaken. Hence, secondary factors are also pertinent in evaluating this data.

Determine Alternative Courses

In many cases, reasonable alternatives do exist and should be determined in following any process involving decision making.

Evaluate Alternative Courses

Evaluation involves weighing the various factors and selecting the alternative that offers the best probability of achieving the objective within the limits of reasonably acceptable risks. Evaluation is difficult if the problem is full of uncertainties.

Select a Course

Making the decision is a crucial step, but only one step in the steps-in-planning process or the business decision-making process. The selection depends on the evaluation of the alternative courses.

Formulate Derivative Plans

This step is necessary because basic plans are not self-achieving.

OPERATIONS RESEARCH AND PLANNING

Operations research has been defined as the application of scientific method to the study of alternatives in a problem situation. The aim is to provide a quantitative basis for arriving at the best possible solution in terms of the goals sought. However, although probabilities and approximations can be substituted for unknown quantities, and scientific method is quantifying factors heretofore believed to be impossible to quantify, a major portion of important decisions at fires involves intangible and unmeasurable factors such as morale, leadership, and time and force of an explosion. Until these can be quantified, operations research will have limited usefulness at fires, and selections between alternatives will continue to be based on nonquantitative factors.

Highly regarded authorities are enthusiastic about the future of operations research as applied to management decisions. By introducing the methods of the physical sciences into managerial decision making more effectively than ever before, operations research concentrates attention on goals, on the recognition of variables, on the search for relationships and underlying principles, and—through use of the model, advanced mathematics, and computation—on optimum solutions from many more alternatives than were ever before possible. The analysts of a few years ago ordinarily could study only a few alternatives because of the sheer size of the analytical task. With operations research techniques and the high-speed electronic computer, they can now analyze the probable results of thousands or millions of alternatives.

The ultimate goal of operations researchers is to be able to formulate such complete policy or decision models that every aspect of a problem, every variable, every probability, and every related decision made or likely to be made will be included. Perhaps operations research can be applied to advantage in the fire field, showing the way to decisions rather than making them, and, like any other tool, serving as an aid to fire officers. However, the likelihood of such an occurrence does not appear to lie in the immediate future.

REVIEW QUESTIONS

1. What does the managerial function of planning involve?

2. Review the discussion of objectives as covered in Chapter 2. What are some of the main points?

3. What are policies? Mention some general observations about them.

4. What are the three types of policies? What is the most logical source of *originated* policy? What is the purpose of originated policy? From what source does originated policy flow? What does the extent to which this policy formulation depend upon?

5. What is *appealed* policy? Where does it come from? When and how is it formulated? What is a danger of appealed policies? How can the need for appealed policies be minimized?

6. What is an *imposed* policy? From what does it result?

7. What are the areas of policy formulation? Cite some examples of policies formulated in relation to managerial functions.

8. Where in this text are policies formulated in relation to objectives of the fire service? Cite some examples of policies formulated in relation to fire activities.

9. What are procedures? Where do they become more numerous in an organization? Why? What are procedures used at fires sometimes called? Where are such procedures usually described? What other procedures may be described in similar media?

10. What difficulties may arise in connection with the use of procedures? What is the cause and solution of problems caused by undue complexity and duplication? What can obsolescence become and what does it lead to?

11. Why is change always necessary, particularly in the fire service? What policy should top executives adopt in relation to change? When is change more acceptable?

12. What is a program? What is the role of policies and procedures in carrying out programs? Why are derivative programs necessary?

13. What are rules? Discuss some similarities between rules and procedures. What are some differences?

14. What is the main difference between policies and procedures? Why should the distinction between policies, procedures, and rules be stressed?

15. What are budgets? To what extent are they considered in this text?

16. Discuss each of the steps in planning as interpreted from a fire-service viewpoint.

17. What is operations research? What are some of the problems involved in applying operations research to decisions made at fires? What is the ultimate goal of operations researchers? What does this imply in relation to the fire field?

25

Organization

An organization has been defined as a formal structure of authority through which work subdivisions have been arranged, defined, and coordinated.

Relative to organizing, *Principles of Management* states. "Organizing involves determination and enumeration of the activities required to achieve the objectives of the enterprise, the grouping of these activities, the assignment of such group of activities to a manager, the delegation of authority to carry them out, and provision for coordination of authority relationships horizontally and vertically in the organization structure."[1]

From a fire-service viewpoint, an organization structure is a course of action for achieving the objective of a fire operation, program (such as fire prevention, training, discipline, safety), or the fire department as a whole.

One can readily discern that the organization structure for achieving objectives at fire operations is essentially the recommended action plan. In Chapter 28 an organization structure is developed to achieve the objectives of a modern disciplinary program, and in Chapter 31 to achieve the objectives of a fire-prevention program.

The organization structure is, of course, not an end in itself but a tool for accomplishing specific objectives. Hence, it must fit the task and must reflect any compromises and limitations imposed on the manager (officer, chief of department, fire director, or commissioner). At fires, officers must recognize the limitations imposed by the primary factors that are

[1] Koontz and O'Donnell, *Principles of Management,* p. 208.

limiting or strategic in selecting objectives and activities, and especially by the specifications for acceptable solutions at fires. The chief of department, fire director, or commissioner, in formulating the organization structure of the fire department, must recognize limitations imposed by the charter sanctioned by the state legislature as well as other limitations imposed by the administration relative to the budget. Changing social conditions may also impose limitations relative to promotional exams and staffing.

A BASIC STRUCTURE

The organization structure for basic or derivative programs consists of the same steps: (1) Determine and enumerate the essential activities; (2) group and assign activities; (3) delegate commensurate authority; and (4) coordinate activities.

Determine and Enumerate the Essential Activities

Activities would include the following:

1. Clearly indicate the specific objective. Faulty specifications of the objective result in frustration and make the entire planning activity futile. It is not enough to state that the objective is to train officers in firefighting tactics. It is preferable to be more specific by stating that the objective is to train officers of all ranks in a standardized way by teaching them to think and operate in a uniform and correct manner at all types of fires. Consider a training program.

2. Formulate policies to guide the decision making of the commanding and subordinate officers assigned to the program. Policies could include (a) a knowledge of managerial principles applicable from a fire-service viewpoint is an asset to an officer and should be taught accordingly; (b) an officer of lower rank who arrives first at a fire is responsible for making and carrying out initial decisions and should be trained accordingly; (c) students can be taught to use an action plan that is practical for officers of all ranks at all types of fires, and should be instructed accordingly.

3. Formulate policies relative to eligibility, attendance, passing grades, awards in the form of certificates, whether or not students are to attend on their own time, grounds for dismissal from the course, textbooks to be used, size of class, duration of the course, relevancy of subjects covered, qualifications of instructors, and methods of instruction.

4. Formulate procedures which with policies comprise a program. Procedures could include selecting a site for the course to be given, developing a curriculum, registering students, establishing the schedule to be followed, specifying textbooks to be used and exams to be held and taken, keeping records, and providing visual aids as required.

Group and Assign Activities

Activity grouping has two aspects. One has to do with the process of departmentation, and the other with guides for assigning activities. In the fire department as a whole, the principle of span of management (see below) makes it necessary for the chief of department to resort to *departmentation* (grouping of activities and personnel into divisions, battalions, and companies with specific functions). Then activities can be assigned accordingly at fire operations.

For the training program in question, a variation in grouping and assigning activities would be in order. Activities would be grouped and assigned by the chief of department in accordance with the functions to be carried out by the officer designated as director of the program. In effect, this is another way of applying the process of departmentation. An officer with a suitable blend of teaching and fire ground experience would be preferable for the assignment of directing the program. In all likelihood, the top executive would collaborate with the director-to-be in selecting the objective, policies, and procedures.

Span of Management. Since there is a limit on the number of subordinates that can be efficiently directed by one individual, the *span of management* must be considered in carrying out the function of organization.

How many companies can a battalion or district chief efficiently direct? In the same department, one battalion includes two engine and two ladder companies, others include three engine companies and one ladder company, and still others include six engine and four ladder companies. Thus there is no specific answer to the question. Leaving the matter of span of management temporarily aside, the number and types of units in each district depend on the contemplated possible fire problem as it may be affected by various primary factors. Requirements to handle the fire problem can be surmised accordingly, but limitations imposed by the span of management must also be considered to ensure adequate supervision. Adequate communication will be assumed.

Limitations on Span of Management. Limitations include distance; change; and spans of time, energy, and attention. *Distance* naturally has a bearing on the time it would take to reach and supervise units. A significant current effect of *change* on the span of management is caused by the drastic increase in fire incidence and the number of alarms to which units must respond. In one city, over 200,000 false alarms are received in a year. During 15-hour night tours, some battalion chiefs respond to 15 or more alarms in their own or adjacent districts. This aspect of change adversely affects the *span of time* available for responding to, and operating at, fires. Eventually, it also adversely affects the *span of energy* due to the physical and mental strain imposed by the heavy workloads. Cumulatively, there could likewise be an adverse effect on the *span of attention,* which is viewed here as the ability to concentrate properly on the task at hand. Some of this comment applies to the number of battalions that can be efficiently directed by a deputy chief, except that the deputy usually responds on a more selective basis, such as to "working" fires only.

How many subordinates can a company officer efficiently supervise and direct? The answer depends on many things: distance; visibility; change; spans of time energy, and attention; caliber of personnel and training; and the type of work to be done. The question is really academic at present, since company officers are responding to fires with too few rather than too many subordinates. However, an engine-company officer can efficiently supervise five or six subordinates laying or stretching lines to operate streams, supply fixed systems or heavy-caliber streams, relay water, and the like. A ladder-company officer can similarly supervise five or six subordinates, even though two may have to be sent to the roof or rear to ventilate, provided adequate communicating devices are available. Only two or three subordinates could be efficiently supervised if they were engaged in such a dangerous undertaking as tunneling at a structural collapse.

Time of the fire, and smoke and weather conditions can impair visibility and thereby the effectiveness of the supervisor. Hence, the span of management is ordinarily greater at daytime fires.

The manner in which topography can affect the span of management was discussed in Chapter 22.

Delegate Authority

Authority commensurate with responsibility involved should be delegated to the officer assigned as director of the program.

Coordinate Activities

Here as elsewhere, coordination can be achieved by supervision and communication. The director can exercise supervision by appraising the performance of students (and instructors, if necessary) against the standards established by the specified policies. Necessary corrective action can then be determined by communication between the director and top executive. Coordination will be improved by the use of a control device in this way, and at the same time, supervision by the top executive is facilitated.

Coordination of authority relationships is also an important feature of the organization structure of a fire department. It is achieved by providing an official manual containing an organization chart—kept revised to date—and an outline of job descriptions. An organization chart indicates how departments are tied together along the principle lines of authority. Specific job descriptions inform incumbents (and other affected personnel) about what they are supposed to do and helps to determine what authority must be delegated in order to carry out the job. It is important for officers to know when they have the authority to command and when they can only advise. They also need to know not only which type of authority they have, but which type others have. Officers in staff units should be instructed carefully about the nature of their dealings with personnel in line units. They should realize that their usefulness depends upon the effective advice they can give, preferably upon request. On the other hand, line officers should be carefully instructed to request staff advice freely and to listen to it.

Coordination of fire-service activities is obviously essential so that each unit may contribute its maximum effort. To achieve the desired coordination, the officer in command should apply the principles of coordination.

The first principle of coordination is that of *direct contact*. It states that coordination is achieved through interpersonal relationships of people in an enterprise. People exchange ideas, ideals, prejudices, and purposes through direct personal communication much more efficiently than by any other method, and they find ways and means to achieve both common and personal goals. The participating parties recognize an identity of ultimate interests. There are many subjects for debate in the fire service, such as the priority of firefighting over fire prevention, fog versus solid streams, and snorkels versus aerials, but everyone in the fire service recognizes the identity of their ultimate interest, which is the prevention of fires and the protection of life and property.

The second principle is that of *early coordination* in the first stages of planning and policy making, since it is easier to unify plans before they are put into operation. When chief officers contemplate sending a greater alarm, in order to get maximum results from the responding units they should consider the need for coordination. Before the greater-alarm units arrive, the officers should decide upon definite objectives for responding units, get the cooperation of the police in improving pertinent street conditions, see that favorable hydrants are available and unobstructed, select vantage points for operating streams, get the cooperation of the water department to maintain sufficient pressure, and notify the responding units by radio how to approach the area, where to report, and what type of operation to anticipate.

The third principle of coordination states that *all factors in a situation are reciprocally related*. This is to say that when A works with B, both find themselves affected by the other and both are influenced by all persons in the total situation. This is so because, among other reasons, individuals, however similar their training (and objectives and motives), strive to achieve their goals through different approaches, interest and effort.

The work of Lieutenant A in a ladder company affects the work of Lieutenant B in the engine company; premature ventilation can magnify the engine company's problems, and inadequate or delayed ventilation can make it difficult for an engine company to advance a line.

Achievement of coordination is largely horizontal rather than vertical. People cooperate as a result of understanding one another's tasks. Good coordination removes critical problems as they arise; excellent coordination anticipates and prevents their occurrence.

REVIEW QUESTIONS

1. What is an organization? What does organizing involve?

2. From a fire-service viewpoint, what does an organization structure suggest?

3. What are some requirements of an organization structure? What limitations are imposed on officers in using the organization structure at fires? On the top executives of a fire depatrment?

4. Discuss the use of the organization structure in carrying out a training program, as suggested in this chapter.

5. Why must the span of management be considered in carrying out the function of organization?

6. How many companies can a battalion or district chief efficiently direct? Discuss the limitations on the span of management.

7. How many subordinates can a company officer efficiently supervise and direct? What can affect limitations of the span of management here?

8. How is the coordination of authority relationships accomplished in a fire department? What should an officer know about line and staff authority relationships?

9. Discuss the principle of direct contact in relation to the coordination of matters and activities affecting the fire service. In what areas can this principle be applied? In applying this principle, what must participating parties agree upon?

10. Discuss the principle of early coordination. How does it apply to fire situations?

11. Discuss the principle that all factors in a situation are reciprocally related. How does it apply to fire situations?

12. Generally speaking, why do people cooperate? What is the difference between good coordination and excellent coordination?

26

Staffing and Control

STAFFING

Staffing involves filling and keeping filled the officer positions required for the organization structure, and presumably specified on the organization chart. Such activities include having on hand an inventory so that officer openings can be foreseen and readily filled. (In some departments this need is covered by official quotas, which indicate the number of officers required for each rank, and by eligible lists, which indicate available candidates for the openings.) These activities also include defining job qualifications, appraising and selecting candiates for jobs, and training candidates and incumbents.

The view that staffing is a significant managerial function is comparatively recent, and it is receiving greater acceptance because it is becoming increasingly apparent that the most important element in an organization is the people in it. Hence the recruiting, training, appraising, and promotion of personnel have achieved greater stature. In some fire departments, recruiting and promotion is based exclusively on standings on civil service lists, over which fire service has little or no jurisdiction. This practice is not in accord with the principle that authority should be coextensive with responsibility, and in some ways it handicaps administrators in carrying out staffing. However, it is definitely preferable to the spoils system.

Relative to the function of staffing, Koontz and O'Donnell in *Principles of Management* state, in part, that "managers must be developed and trained within the organization: the key to an effective job here is adequate appraisal so that training and development needs can be properly determined."[1] Appraisals of fire-service training and development needs are described in the booklet *Wingspread Conference on Fire Service Administration, Education, and Research* and the report *America Burning*. These and other authorities unanimously agree that an outstanding need in the fire service is standardized training of personnel. Accordingly, in this text it has been suggested that all personnel should learn how to apply an action plan at all types of fires and should acquire a basic understanding of managerial principles. Standardized training of this sort would certainly aid and improve the staffing function.

CONTROL

The function of control is to ensure that plans are carried out as intended by the planner. Control is necessary because plans are not self-achieving. The basic control process, wherever found and whatever controlled, involves three steps: (1) Establish standards, (2) appraise performance against these standards, and (3) correct deviations from standards and plans.

Establishing standards involves decision making and hence planning; appraising performance involves supervision and hence direction; and correcting deviations may involve disciplinary action and hence direction. If disciplinary action results in the need for dismissal or replacement, staffing may play a part. Thus the control process illustrates how managerial functions tend to coalesce or intermingle.

Standards are established criteria against which actual results can be measured. A chief of department can establish standards by formulating policies, procedures, and rules. Policies serve as a guide for appraising the performance of subordinates in specified areas of decision making; procedures serve as a guide for appraising the performance of subordinates in carrying out activities as specified for certain situations; and rules serve as a guide for appraising the performance of subordinates in regard to such matters as reporting on time, carrying out orders properly, and so forth. Since officers at all levels plan, they also use the control process to make certain that their plans are implemented as intended.

In addition to those already referred to, there are other standards for appraising performance so that necessary corrections can be made. For example, the limitations of authority can be regarded as standards, and, of course, there are the numerous principles that are discussed throughout this text.

At fires, officers should have certain expectations about the objectives sought and the activities undertaken. The validity or soundness of such

[1] Koontz and O'Donnel, *Principles of Management*, p. 397.

expectations depends on the ability of officers to recognize and properly evaluate the primary factors that should influence the selection of objectives. Expectations are the standards against which performance can be appraised, and if they are derived in a logical way they enable officers to use the control process effectively.

At times, performance can be appraised visually. When visibility is obscured by smoke, time of the fire, or the weather, or when visual observation is precluded because units are working in the fire building or at the rear or sides, the commanding officer must depend more on reports from subordinates. Belated, but nontheless effective appraisal of performance is also possible by studying motion pictures taken at fires.

The company (or otherwise designated) school, at which units are tested periodically for competence in carrying out specific evolutions, also provides a practical means of appraising the performance of units. The evolutions represent official standards.

Correction of deviations can be accomplished by supervision, training, or disciplinary action if necessary. Generally speaking, correction should be as prompt as possible and feasible, especially at fires. In some instances, appraisal of performance may change one's expectations. For example, if the fire area is not being ventilated as effectively as anticipated, it is no longer logical to base expectations on the rapid advance of hose lines. Mistakes detected in motion pictures, of course, can only be corrected belatedly. Correction is also delayed when it involves additional training, or that ultimate of restaffing—dismissal.

Concern here is with correcting deviations of persons only. Controls governing speed, temperature, pressure, sound, color, and the like are not regarded as relevant to this text.

Fire prevention is a fruitful field for the application of the control process. It is widely recognized that a major weakness in fire-prevention work is the inadequate evaluation of inspections, due primarily to the fact that there have been no official standards against which inspections can be effectively appraised. In Chapter 31, such standards will be suggested.

REVIEW QUESTIONS

1. What does staffing involve? What activities are included? How may fire department administrators be handicapped in the areas of recruitment and promotion?

2. What is the key to an effective job relative to the statement that "managers must be developed and trained within the organization"? Assume that the word *organization* pertains to the fire service and not necessarily a fire department.

3. Discuss appraisals made by authoritative groups relative to fire-service training and development needs. What do these appraisals indicate?

4. What is the purpose of control? Why is this function necessary? What steps does the basic control process always involve? (The use of the control process in fire-prevention work is discussed in Chapter 31.)

5. How does the control process illustrate that managerial functions tend to coalesce or intermingle?

6. What are standards? How does a chief of department establish standards? Mention other standards against which performance can be appraised.

7. How can officers appraise the performance of the units under their command at fires? What does the accuracy of this appraisal depend upon? What means are available to appraise performance of units at fires?

8. How can deviations be corrected? When is it preferable to correct deviations, especially at fires? When can an appraisal of performance cause officers to change their expectations and thereby correct deviations? When may corrections of deviations be delayed? What is correction of deviations concerned with in this text?

27

Direction and Issuance of Orders

DIRECTION

Direction is the function of guiding and overseeing subordinates so that they accomplish their tasks as efficiently as possible. Planning, organizing, and staffing alone—although they precede direction in time—will not get the job done any more than starting the motor of a car will make it move. Direction, therefore, is intimately concerned with getting things done.

In the fire service as in other organizations subordinates must be directed; it is through them that work assignments are carried out. Direction can be considered, therefore, as fundamentally a matter of handling human behavior by such means as issuance of orders and communication, supervision, delegation of authority, discipline, motivation, leadership, and a due regard for morale. Thus, directional devices are numerous and varied, and the types of subordinates requiring guidance and supervision may be even more numerous and varied. Officers have to learn how to use the right device, in the right way, at the right time, ever mindful of the importance of personalizing contacts with subordinates.

ISSUANCE OF ORDERS

Direction starts with the issuance of an order. In issuing orders, particularly at fires, it is strongly suggested that the principle of unity of command be observed. Each subordinate should report to only one superior, and chief officers should usually avoid giving orders directly to members of a unit. Orders should be transmitted through the unit or company officer. At fires, it is possible for a subordinate officer to be assigned duties by two or even more superiors. It is easy to see that such a practice is likely to result in conflicts in both authority and responsibility, and, almost inevitably, in inefficiency.

Whether an order is specific or general depends on the superior's preference, as well as ability to evaluate accurately all the factors involved, and the response made by the subordinate. Superiors with a negative attitude toward delegation of authority seem predisposed to specific orders. They evidently feel that they have clearly in mind what has to be done and the best way to do it. In addition, they want tasks done in a certain way and prefer to direct their subordinates very closely.

Where it is not possible to properly evaluate the effects of factors on activities to be undertaken, orders are more likely to be general. When the task is to be performed far from the personal direction of the superior, no effort should be made to give specific orders. Thus it is that at fire operations orders are often general, such as "Open up the roof," "Get everyone out," "Check the rear," and so on. It would be difficult and even unwise for a commanding officers to give specific orders where they cannot possibly foresee the effects of the factors that may become pertinent in executing the orders.

In addition, orders at fire operations are often general also because the commanding officer has no time to state specifically how they should be carried out and because general orders convey sufficient information in view of the training of firefighters. This is true even where life hazard is involved. For example, at a fire involving the top floor of a five-story residential building, the superior officer may order the officer of the first engine company to arrive: "Take your line to the top floor." Obviously, the superior officer does not have time to specify what hydrant is to be used, what engine pressure is required, the number of lengths in the line, the type of nozzle to be used, the nozzle pressure, and so on. And in view of the subordinate's training, it need not be specified that the line should be taken up the interior stairway (unless some unusual situation dictates otherwise), that the purpose or minor objective of the line is to protect occupants pending rescue, and that therefore it should be operated as quickly as possible and as long as necessary between the fire and the endangered occupants or their means of escape.

Subordinates' responses to orders influence the nature of the orders given to them. Some subordinates actually prefer close supervision and consequently do well with specific orders. Others, however, chafe under such conditions and prefer to exercise their own initiative and creativeness; they are quite willing to be judged by the results.

Written or Oral Orders

Whether an order is written or oral depends on several considerations. The permanency of the relationship between the superior and the subordinate has an influence. If the relationship is continuous, it is often unnecessary to reduce orders to writing. However, because of the high degree of mobility in agencies such as military forces or the fire service, it is both unsafe and unwise to operate without written orders, at least concerning duties that pertain to the job rather than to the person and duties that take considerable time to perform. Such orders are essential for detailed or newly assigned officers and for officers of relocating companies where the units covered have special assignments.

The quality of trust between superior and subordinate is also important. Subordinates, feeling that the major risk in the superior-subordinate relationship is carried by them, may prefer the protection of a specific, written order. Written orders tend to prevent overlapping of assignments. They are the best means of achieving uniform adherence and tend to prevent jurisdictional disputes. If an assignment exceeds usual limits, it is often helpful to inform all affected personnel of that assignments by publishing a written order. Without such a communication, the undertaking may meet resistance, sabotage, and general lack of cooperation. Lengthy, detailed, and complicated orders should be written rather than oral because less is lost by the receiver.

When orders are not lengthy, detailed, or complicated, oral communications can often save time and promote ready understanding, provided that questions can be asked and answered on the spot, a prominent advantage of direct contact.

Other Techniques

In addition to issuing general or specific, and written or oral orders, other techniques have been found helpful. Some such techniques involve timing and keeping subordinates reasonably informed.

Timing is obviously essential for good coordination. Timing here refers to the issuing and carrying out of orders, as indicated in Chapter 16.

As much information about an order as possible should be passed on to subordinates on the sound assumption that those who know the reason for an order are more easily and effectively motivated. Subordinates recognize that some information must be classified, but they tend to lose interest or become dissatisfied when a superior fears or neglects to communicate nonconfidential information that would enable them to understand new decisions. In some cases, even confidential information can be passed on with good results. The superior who explains the background of decisions has a strong appeal for members of the fire service in the United States and other democratic countries. This is due, in part, to higher education levels, more widespread belief in democratic processes, and increased insistence that human dignity be respected.

Superior officers should be aware that lower- and middle-level officers may have more formal education than they themselves do, that many officers know something about group power in the democratic process, and that an increasing number are familiar with the importance of human relations. These officers are very likely to resent any dictatorial attitude on the part of superiors, particularly the "tell-'em-nothing" attitude.

Regardless of the educational level of subordinates, the following advantages result when they understand the background of decisions. When they understand the reason for a course of action, they see why it should be done in a certain way, which tends to eliminate resentment. Also, only when subordinate officers understand as thoroughly as possible the reasons for an order can those officers make plans in the most effective and efficient manner.

SUPERVISION

The supervisor is the most ancient as well as the most important device in the process of coordination, and accordingly is an equally important device in carrying out the functions of both organization and direction. The contribution of the supervisor in relation to the function of direction depends in large measure on the effectiveness with which the principles of harmony of objective and unity of command are employed, and the type of technique used in supervising. Such techniques can be described as *consultative or democratic, free-rein or laissez-faire,* and *autocratic.* The technique used is influenced by the attitude of the supervisor. However, for better results, the technique should be selected for its appropriateness in a given situation.

Consultative or Democratic Methods

In selecting this technique—in what is most accurately called participative management—the supervisor solicits advice and perhaps accepts majority decisions from subordinates. Such superiors encourage maximum participation by subordinates in matters that pertain to departmental or unit functions. They are eager to adopt suggestions if they have merit. The important points of this technique are the democratic behavior of superiors, their sincerity in implicating their subordinates in departmental plans, and their emphasis on group action. Conferences are a basic tool for the transmission of information and understanding. *Informal* conferences, called on the spur of the moment, can be extremely useful. They may deal with problems of securing better coordination between subordinates or between subordinates and personnel outside the unit; involve discussion of new policies; impart and interpret new information or clarify existing policies, procedures, and programs; consider current programs and bottlenecks and potential countermeasures; and transmit first-hand information to the superior. Informal conferences are effective because participants tend to trust and respect each other and to have a sense of their combined effectiveness. Such conferences are called only when the need arises and are attended only by those affected, and thus are more objective.

Formal conferences require written notification of the time and place of the meeting and possibly the transmission of the agenda. Subjects could include the impact of economic conditions on assignments responding to alarms; reports on physical assaults on firefighters responding to, working at, and returning from fires, especially during riots; an explanation of new official policies and programs, or of changes in the organization structure of the department and the possible results thereof, especially if the rank of deputy chief is eliminated, and so on. Formal conferences are called when the need arises and are not held without a specific purpose.

Consultative or democratic techniques are effective because most people dislike being ordered about and feel a sense of importance when their suggestions are solicited and considered. Participating subordinate officers, seeing and hearing top-level superiors, acquire a better understanding of departmental problems and of the forces that determine the decision. Some personal knowledge of high-level superiors has a powerful humanizing and helpful influences.

The disadvantages of consultative or democratic methods result largely from their misuse. Subject matter must be appropriate. Skill in conducting a conference is required. Any hint of autocracy, destructive criticism, lack of respect for members, or secrecy nullifies the effectiveness of these techniques. Moreover, the superior needs leadership ability, teaching skill, and the conviction that participation pays off.

Free-rein or Laissez-faire Methods

These techniques are characteristic of superiors who have a positive attitude toward the delegation of authority. Where authority is delegated willingly, subordinates who are anxious to take advantage of the opportunity offered respond with their best efforts and are careful not to let their superiors down. This technique has some disadvantages because of the time it takes to reach decisions, and possible fumbling and wrong decisions. These disadvantages are more pronounced if, in choosing laissez-faire or free-rein, the superiors go so far as to practically abdicate their decision-making authority and become mild consultants.

Autocratic Methods

These techniques are characteristic of superiors with a negative attitude toward delegation of authority. They develop naturally from the belief that superiors should assume full command over all actions by subordinates. Actually such superiors may believe that subordinates cannot be trusted to act for themselves. The source of this conviction may rest in the personality of the superiors or in their experience with subordinates whose lack of training and education made an autocratic approach necessary. Autocratic techniques are characterized by retention of power by superiors, reliance upon specific orders, and maintenance of close supervision. These techniques carry overtones of marked superior-subordinate relationships, continually reminding the subordinate who is the boss. However, these methods can be very effective in getting action from subordinates who tend to avoid responsibility, or in dealing with timid or unsatisfactorily cooperative personnel.

Autocratic superiors usually give detailed instructions. In some cases this is appropriate, since many jobs can be done in one best way, particularly those studied by time-and-motion experts. Likewise, at a fire operation, situations may develop that may allow subordinate officers no room for personal preference. However, where officers are qualified to carry out assignments, they will be quick to resent minute and specific instructions. The situation dictates the choice of techniques to be used. Where strong leadership can bring order out of chaos, superiors may find that a positive outlook and heightened morale result from the firmness and definiteness of autocratic techniques.

Other aspects of supervision at fires and elsewhere are discussed throughout this text.

REVIEW QUESTIONS

1. What is the purpose of the function of direction? What is it intimately concerned with?

2. What devices are used to achieve the purpose of direction? What should officers know about these devices?

3. With what does the function of direction start? What consideration should be given to the principle of unity of command in issuing orders? Why? Review limitations on orders that can be given.

4. On what does the choice between specific and general orders depend? What type of officer seems predisposed to issuing specific orders? Why? When are orders more likely to be general? Why are general orders advisable at fire operations? Why does the subordinate's response to an order influence the kind of orders to be issued?

5. On what does the choice between written and oral orders depend? Why? What are the advantages of written orders? Oral orders?

6. In addition to issuing general or specific, and written or oral orders, other techniques have been found helpful. What do some such techniques involve? Why?

7. On what does the contribution of the supervisor in relation to the function of direction depend, in large measure?

8. What do supervisors do when they select the consultative or democratic technique? What are the important points of this technique?

9. When are informal conferences called in employing the consultative or democratic technique? What do they deal with? Why are they effective?

10. What is required when a formal conference is held in employing the consultative or democratic technique? With what do such conferences deal? When are they called?

11. Generally speaking, why are consultative or democratic techniques effective? What are some disadvantages or limitations insofar as conferences are concerned?

12. What type of officer tends to use the free-rein or laissez-faire technique of supervising? What are some advantages of this method? Some disadvantages?

13. What type of officer tends to use the autocratic technique of supervising? What reasons have been offered as to why some superiors prefer this technique? What are autocratic techniques characterized by? With what types of people can these techniques be effective?

14. Why can detailed instructions be appropriate in some cases? Discuss the use of autocratic methods of supervision at fire operations.

28

Delegation of Authority and Discipline

DELEGATION OF AUTHORITY

Delegation of authority is defined as the vesting of a subordinate with a portion of the superior's authority. Clearly, superiors cannot delegate authority they do not have, nor can they delegate all their authority without, in effect, passing on their positions to subordinates.

Delegation of authority is a feature of the organization structure as well as a device for guiding and supervising personnel in carrying out the function of direction. However used, the delegation of authority always involves three steps: (1) the assignment of tasks, (2) the delegation of authority for performing the tasks, and (3) the exaction of responsibility for their performance. In practice it is impossible to split this process, since the assignment of tasks without authority to perform them is meaningless, as is the delegation of authority without specifying the area of responsibility in which the power is to be used. Moreover, since responsibility cannot be delegated, the superior has no practical alternative but to exact responsibility from the subordinate for properly using the authority delegated.

Delegation can be the concern of all officer ranks except the lowest. For example, a lieutenant cannot delegate to a subordinate the authority defined as the right to command, or act, and associated with the superior-subordinate relationship. This type of authority should not be confused with the authority to force a door or issue a summons on orders of a superior. However, if a firefighter is designated as an acting lieutenant, authority commensurate with the responsibility must then be delegated.

Delegations of authority may be general or specific and written or unwritten. In any case, they must be accompanied by some kind of assignment of duties. Written and specific delegations have advantages because conflicting or overlapping assignments are more apparent, the assignment is better understood, and there is less uncertainty. Nonspecific delegations of authority may result in confusion, with subordinates forced to operate more slowly, feeling their way and defining their authority by trial and error. This usually places them at a disadvantage. A manager would do well to balance the costs of uncertainty against the effort involved in making authority-delegation specific.

Almost invariably at fires, orders or assignments are general and oral. Seldom, if ever, is delegation of authority mentioned. Hence delegation of authority is not listed as a step in the action plan recommended for use at fires. When an officer is given an assignment at fires, commensurate authority is implied by the training given for carrying out various assignments. Obviously, officers also have implied authority to abandon positions to which they were assigned when it becomes apparent that their units are in unnecessary jeopardy. Confronted by situations for which no standard procedures have been established, officers have implied authority to operate in a manner most likely to achieve the objectives of the fire service, namely, to prevent fire and protect life and property against fire. The following is a case in point.

A chief officer was ordered by radio to investigate an odor of gas in a theater. On arrival, an odor of illuminating gas was readily detectable, most noticeably under the stage area. The explosimeter indicated that an explosive mixture had developed. The chief ventilated the stage area by means of the rope at the stage level, summoned a first-alarm assignment, and despite the protests of the management that "the show must go on" ordered that the audience of about 1,500 people be directed to leave and to refrain from smoking. Standard procedures did not specifically cover such a situation. The chief had to be guided by the authority implied by the objectives (and responsibility) of the fire department. In such a case, an officer cannot and should not expect a superior not on the scene to make the decisions, and delay could have been disastrous. Subsequently, the first-alarm assignment made hose lines available, eliminated possible sources of ignition in the danger area, and ventilated the theater. There was no explosion or fire. Subsequently, also, the management thanked the fire department for the action taken.

Superiors with a negative attitude toward the delegation of authority dislike to make any delegation at all and usually feel that any necessary delegations can be made exactly and precisely and that the results they personally anticipate will be achieved. Disappointments in this respect are likely to lead to harsh judgments on the ability of the subordinate to whom they may have been forced to delegate authority. These superiors feel untoward frustration when their plans fail to work out where everything seems neat and clear to them. They often feel that only they are capable of doing the job, and they lack trust in subordinates, whose errors, they fear, will jeopardize the welfare of the unit. These superiors may feel insecure in their position and fear that errors of subordinates will adversely affect them as superiors. The effects of a negative attitude toward delegation have serious results, because officers with such an attitude are incapable of developing successors. Many prefer not to work for such a superior. The negative attitude toward delegation limits the size of an organization and retards its growth. A superior with such an attiutde is antagonistic toward departmentation and the concomitant delegation of authority.

Superiors with the positive attitude toward delegation of authority do not feel their personal security threatened by it, and they are willing to trust subordinates. They consider the development of future officers to be a great service. Such superiors accept the principle that one learns to manage by managing. They are accomplishing two important things: getting the work done and developing future officers. Errors that may occur are viewed in the proper perspective and minimized by precautions where possible.

A positive attitude encourages subordinates to accept responsibility. This helps them to become self-starters and to grow in the exercise of authority. Superiors need to study all subordinates, giving each an appropriately expanding degree of authority while presenting the challenge of unusual accomplishment. The channels of communication must be kept open. The superior must be available but unobtrusive, and must firmly resist the temptation to tell subordinates what to do and how to do it. A superior must exercise unlimited patience to put up with mistakes and fumbling and the slow acquisition of good judgment and leadership ability.

In vital phases of a fire operation, a chief officer may have good reason to show elements of the negative attitude toward delegation of authority and wish to exercise very close supervision. However, in some types of overhauling work, with proper precautions in the form of unobtrusive supervision, the positive attitude is practicable. In addition, there are other phases of fire-department work, such as building inspecton, where the positive attitude is not only practicable but advisable in the interest of officer development.

Two important issues face the manager who desires to delegate authority to a subordinate in order to provide managerial experience. The first issue is the degree of delegation to a particular individual, which should be characterized by a gradual expansion of the subordinate's authority. The second issue is the delegation of authority to several subordinates in different stages of development. The qualifications of the subordinate influence the nature and degree of the authority to be delegated. Those who have proved their capacity will be tested further by increased authority; those who have shown inability to use authority constructively will lose it. Some subordinates overestimate their ability as managers and, for this reason, resent what is, in their estimation, a limited degree of authority. With them superiors must demonstrate their sincerity and impartiality. Failure to do this will ordinarily result in dissatisfied subordinates and the probable loss of some with real potential.

Dealing with subordinates of limited capacities is something else again. Such individuals should not have been made supervisors in the first place, and there is no good alternative but to remove them from supervisory work, although this procedure cannot be resorted to in many fire departments because of civil service regulations, political influence, and so on. Extreme care should be used in selecting as supervisors only those likely to prove capable of further development. It is suggested, therefore, that promotional examinations be designed to reveal—among other things—the degree to which candidates have the desire and the ability to assume responsibility and use authority.

At first glance, it might appear that to assign duties, delegate authority, and exact responsibility, one has only to specify which officer is in charge of what activity. On a closer look, however, questions frequently arise about what is involved in a given assignment, even when it appears clear and exact. For example, what is meant by the assignment to "advise your superior about such and such a procedure"? Does it mean constructive suggestions, presented in tentative written form? If so, when is such a report required? Is the report to follow a specific form and contain specific information? Again, what would one do with the assignment "to study the equipment market constantly"?

Delegation of authority is a more general form of direction than issuance of orders. A common difficulty associated with assignment of duties in general terms is the overlapping of work assignments. Delegation of authority is even less exact than the assignment of duties. It often states merely that the subordinate has the authority to carry out the assigned duties, but since the scope of the assignment may not be clear, the subordinate is certain to have difficulty in interpreting the responsibility involved and the authority that would be commensurate. The type of authority granted involves the question: How specific can a grant be? The answer depends on the extent to which the work assignment can be specific. It is possible to generalize that specific authority can be granted if the work assignment is specific and that the delegation of authority is broad at the top of an organization and becomes increasingly narrow as it reaches the lower echelons. Sometimes, however, when new top-level jobs are created, authority delegations cannot be very specific at the outset. One of the first duties of the new appointee is to describe the job (if this has not been done by the superior), to clear the description with the superior—and, ideally, with managers on the same level whose cooperation is important for successful operation. Otherwise, the cost of vague authority relationships may have to be paid in organizational friction, time lost through unnecessary meetings and negotiations, jealousies, and stepped-on toes.

Very specific delegations may result in inflexibility and thereby limit and frustrate the creative superior or, in some cases, cause the individual to regard his or her job as a staked claim with a high fence around it. This can be eliminated by the proper example of leaders and by making necessary changes in organizational structure an accepted and expected thing.

All delegations of authority eventually revert to the delegant. The original possessors of authority do not permanently dispossess themselves of their power by delegating it. This is logical, since the original possessors cannot dispossess themselves of their original responsibility. If superiors make mistakes in delegating authority and, as a result, assignments are not carried out effectively, the superiors may exercise the right to recover their authority. Delegated authority is frequently recovered when departmental objectives, policies, procedures, organizational structure, and the assignments of personnel are modified.

Centralization and Decentralization of Authority

Authority is said to be centralized when superiors tend *not* to delegate authority to their subordinates. This practice not only limits the number of subordinate officers in an organization but severely limits its effective size, since departmentation (the grouping of activities and subordinates into departments or units) requires delegation of authority. In contrast, superiors who tend to delegate much authority favor departmentation. In such cases, authority is said to be decentralized.

Centralized and decentralized authority characterize superiors with negative and positive attitudes, respectively, toward delegation.

Shared Authority

Shared authority is demonstrated when responsible managers or officers insist that power in given situations be shared by subordinates. Such managers want an acceptable group decision, but in a sense they have not delegated authority at all, since the top superior retains authority to make a final decision. Shared authority is seen in a firehouse when the captain consults the lieutenants about house rules, committee work schedules, arrangement of lockers and furniture in office, and the like.

Splintered Authority

Splintered authority exists wherever a problem cannot be solved or a decision made without pooling the authority of two or more officers—for example, in firehouses with two or more captains. In making decisions about house rules, the captains concerned must pool their authority and agree. Splintered authority cannot be avoided, especially in making decisions. However, recurring decisions on the same matters may be evidence that authority delegations have not been properly made and some reorganization is required. Splintered authority is particularly appropriate when the alternative is unnecessary upward reference in handling decisions.

Functional Authority

Functional authority is the power given to officers over specified processes, practices, policies, or other matters relating to activities undertaken by personnel in departments and units other than their own. One such process could be training at the department academy or college. Officers designated as instructors have functional authority delegated to them and, during the training period maintain a superior-subordinate relationship with personnel in departments and units other than their own. If the principle of unity of command were followed without exception, functional authority would not be used, but all officers are not adequately trained to carry out certain processes or activities—such as instructing.

DISCIPLINE

Discipline is a foremost device in guiding and supervising subordinates and hence in carrying out the function of direction. To use discipline effectively, officers should have an acceptable concept of what it is, should know what factors affect its quality, what methods can be used to attain it, and how a modern disciplinary system can be formulated and activated.

In this text, discipline is assumed to be training designed to promote orderly behavior in subordinates so that they conduct themselves properly, and operate as efficiently as possible, in matters affecting their units and departments. This concept leaves no doubt about the importance of reporting on time, observing rules pertaining to smoking and drinking, complying with official regulations and procedures, and meeting other legitimate obligations.

Two factors affecting the quality of discipline are the power of faith in a common cause, and the leadership. J. D. Mooney describes the power of faith this way: "When the laborer and the boss are bound by the same common understanding of some common purpose, the discipline is on a plane no other form can reach." Superiors of all ranks would do well to consider the need for the faith in a common cause, in view of current, drastic changes in attitudes toward authority and discipline in all forms of organizations.

A second factor in the quality of discipline is leadership. Subordinates are willing to follow and be guided by the example or persuasion of an effective leader. In turn, such leadership tends to develop faith between leaders and subordinates.

The methods by which subordinates are disciplined are either negative or positive. The choice revolves around the earlier question of how to train personnel in the best way to promote orderly behavior in subordinates, leading to proper conduct and efficient operation in matters affecting their units and departments.

Negative methods may involve additional training, fines, transfers, or dismissals. These methods imply coercion, which is of no help in achieving coordination, but unfortunately they are necessary in making the position of officers tenable when they must issue orders. Additional training is perhaps the most salutary penalty, and it is imposed accordingly, if necessary, in carrying out the third step in the control process (Chapter 26).

Postive methods imply that rewards are given for good behavior, but this is not true because it is normally expected that *all* subordinates will behave well. However, rewards may be used to elicit superior performance, innovation, and creativity. In the fire service, medals and merit citations for outstanding performance are in order. In addition, when requesting transfers, officers with clearly superior records should be given preference, other things being equal. Thus, objective superiors can use both negative and positive motivation. The choice depends on the options available weighed against the expressed purpose of discipline.

A MODERN DISCIPLINARY SYSTEM

In establishing and maintaining a disciplinary system, an organization structure can be used in much the same way one was used in developing a training program. Remember, an organization structure consists of these steps: (1) Determine and enumerate the essential activities; (2) group and assign activities; (3) delegate commensurate authority; and (4) coordinate activities.

Determine and Enumerate the Essential Activities

Activities would include the following:

1. Specify the objective. The objective of discipline is to provide train-

ing so that subordinates conduct themselves properly and operate efficiently in matters affecting their units and departments.

2. Formulate applicable policies. These could include:

a. Assess disciplinary action by the contribution it makes to improve the behavior of subordinates in conformity with the objective sought. Since punishment has been found to be fruitless when it is meted out in a spirit of retribution, stress positive rather than negative motivation as much as possible. This is not to imply that negative measures can be disregarded.

b. Use the qualities of faith and leadership. Superiors with this ability have less need for negative methods.

c. Take disciplinary action promptly and uniformly. There should be no undue delay in preparing charges when necessary or any variation in applying rules and regulations.

d. Prohibit any prejudicial endorsements indicating that charges, if brought, can be substantiated, since charges are not substantiated until a trial is held. Endorsements should only indicate that the proper reports and forms have been made out.

e. Safeguard the constitutional rights of both the accused and the accuser. Each has the right to be represented by a lawyer. Defendants have the right of appeal.

f. See that all ranks and unions are represented on the trial board, which can also aid in formulating and modifying rules and regulations—particularly obsolete and unenforecable rules as well as automatic and inflexible ones that operate unfairly and do not take individual circumstances into consideration. Rules should be formulated with full awareness that members of today's fire service are better educated in democratic processes and more conscious of the rights and dignity of the individual.

g. Decentralize authority so that minor cases can be handled at lower levels. Many minor violations can be handled at the company level if the defendants agree, or they can insist on a board hearing.

3. Formulate procedures which, with policies, comprise a program, or in this case, a system. Ordinarily, an official manual prescribes the chronological sequence, and thereby the procedure, to be followed when disciplinary action must be taken.

An official manual should be supplied to each member. It can serve as a control device by informing all members of standards (rules and regulations) that are to be complied with, enabling officers to appraise the performance of subordinates in relation to the specified standards and indicating how deviations are to be corrected.

Specifically, manuals prescribe procedures relative to the following items (when negative disciplinary methods have to be employed).

a. Investigations; entries in company, battalion, and division journals; and reports and forms to be filled out and forwarded.

b. Trials before the board and those held at lower levels by company officers. In the latter case, the penalties, if any, should be clearly spelled out.

c. Trial notes. These should be published and made available to all parties concerned within a reasonable time after the trial.

d. Witnesses. They have a part in the proceedings.

Group and Assign Activities

Disciplinary activities can be grouped as follows: (1) Establish official standards, (2) appraise the performance of subordinates relative to these standards, and (3) correct deviations from standards. Manuals usually assign these activities—most often items #2 and #3—to officers according to the subdivisions over which they have jurisdiction.

Delegate Commensurate Authority

Authority is ordinarily delegated by means of official manuals and organization charts that delineate the responsibility and authority of officers at various levels.

Coordinate Activities

If activities associated with the suggested disciplinary system are to be effectively coordinated, officers as supervisors would do well to show their preference for positive rather than negative methods. They should make every effort to bring about the condition in which superiors and subordinates are bound by a common understanding of a common purpose. Also bear in mind that supervisors who are capable leaders have comparatively minor disciplinary problems. Adequate communication between superiors can supplement the efforts referred to.

Outstanding service should be rewarded, particularly if it goes beyond the call of duty. Hence it is recommended that a board or committee be established to disseminate awards to members whose efforts merit them. If feasible, a rating system should be designed to ensure official recognition and appreciation in the form of aid to promotion, better assignments, and so on.

REVIEW QUESTIONS

1. Define delegation of authority. What are some limitations in delegating authority?

2. What steps are always involved in delegating authority? Why is it impossible to split this process, in practice?

3. What ranks may be involved in delegating authority? Why is there a limitation here? Compare specific and nonspecific delegations of authority.

4. Why is it unnecessary and even impractical to make delegations of of authority specific when giving general and oral orders and assignments at fire operations ? What implied authority is delegated to officers confronted by situations for which no official standard procedures have been established? Discuss a case in point.

5. What seems to motivate superiors with a negative attitude toward delegation of authority? What are the disadvantages of this attitude? What seems to motivate superiors with a positive attitude toward delegation of authority? What are the advantages of this attiude? When is a negative attitude understandable?

6. How can civil service promotional exams have a favorable bearing on the effectiveness with which authority is delegated to officers?

7. In other than fire situations, what difficulties are encountered in delegating authority by merely indicating which officer is in charge of what activity? What is a common difficulty associated with assignment of duties in general terms?

8. The type of authority granted involves the question: How specific can a grant be? What does the answer depend upon? Why?

9. What course of action is suggested for appointees to new top-level jobs? Why?

10. What are some disadvantages of very specific delegations of authority? How can they be eliminated?

11. Why should delegations of authority revert to the delegant (superior)? How is delegated authority frequently recovered?

12. When is authority said to be centralized? What is a major disadvantage of this practice? When is authority said to be decentralized? With what types of superiors are these practices associated?

13. Discuss shared authority. Splintered authority. Functional authority.

14. What should officers know about discipline? What is discipline assumed to be in this text? What does this concept imply?

15. When can faith in a common cause favorably affect the quality of discipline? What other factor has a similar effect?

16. What do negative methods involve? Why can these methods have unfavorable results? Why are they necessary? What is the most salutary penalty among those mentioned? Why are negative and positive methods related to the third step in the control process?

17. What does the use of positive methods imply? When are rewards and awards in order in the fire service? On what does the choice between negative and positive methods depend?

18. Discuss the use of an organization structure in developing and maintaining a modern disciplinary system, as described in the text. What is your opinion of the policies recommended? Of the system as a whole?

29

Motivation and Leadership

MOTIVATION

In this text, to motivate simply means to provide a helpful incentive or to use the incentive an individual has in order to efficiently attain the objective of the organization as well as to achieve personal goals of subordinates. The objective of the organization has top priority. In conformity with the principle of harmony of objective, the achievement of personal goals by subordinates should constructively contribute to the attainment of the objective of the organization.

Difficulty arises from the fact that all subordinates do not have the same personal goals. However, regardless of such differences, it is the task of the superior to harness the efforts required to achieve the various goals in a way that contributes to the department's purpose. Efficient direction and motivation clearly calls for the provision of goals whose achievement provides the personal satisfaction for which individuals strive and at the same time accomplishes departmental objectives. To this end, the top executive might suggest as a worthwhile goal the attainment of a certificate of graduation from a course of study, paid for by the community and designed to improve the level of officer training. This attainment would very likely yield personal satisfaction to the subordinates and at the same time promote more efficient achievement of departmental objectives. In addition, graduates could be rewarded with extra credits toward promotion. Thus the alert superior can both suggest goals and attach rewards for which people strive, thereby effectively employing the principle of harmony of objective.

Motivation is an essential device in carrying out the function of direction because it induces people to work as well as they are able and trained to do. It is important to realize that motivation in the sense of providing an individual with a motive to work in a desired manner has no meaning outside the needs of subordinates. Work must be done by most people, primarily to satisfy basic needs for food, clothing, and shelter for themselves and their families. After such needs are satisfied, most individuals seek bigger things, such as a feeling of belonging, recognition and respect from others, independence, and in general a sense of self-fulfillment. Superiors should be aware of this fact, because efforts to provide a motive not related to such needs have proven to be futile.

Some authorities, in discussing a system of motivation, classify needs under three headings: basic needs, social needs, and ego.

Basic Needs

Job security generally therefore assures fulfillment of basic needs. Employees may therefore want the reassurance of feeling secure in their jobs so long as their work is satisfactory, they obey rules, and the jobs exist. A feeling of job security is further enhanced by assuring employees that they will not be dismissed through any unjustifiable action or complaint of others. The fact that some individuals need their jobs more urgently than others should not be considered in motivational systems.

Social Needs

Social needs pertain to such things as status and belonging, or group relatedness. These needs reflect the desire for distinction both within and without the organization. Most people crave such social approval. Some of these needs can be partly fulfilled by a motivational system that includes promotion and status symbols. In the fire service, promotions can be the rewards for successfully participating in competitive exams and can be accompanied by status symbols such as silver or gold insignia designating higher rank, and the associated increase in authority and prestige. In some cases, status symbols can include a private office, and for chiefs, a car and an aide. Subordinates having an obvious need to satisfy such social needs are usually readily discernible early in their careers and superiors should motivate them accordingly by, among other things, suggesting study courses. This type of subordinate is comparatively easy to motivate. However, superiors should not cater to such subordinates at the expense of others. In fact, more effort should be devoted to motivating the less gifted, because it is essential in any coordinated undertaking to improve the efficiency of all, rather than just a select few.

Many in the fire service have strong desires to satisfy their social needs by being promoted, but they lack confidence in their ability, give up, and resign themselves to the role of subordinate unless a capable superior influences them otherwise by effective motivation. Some such subordinates can be helpfully motivated by being awakened to the fact that they have underestimated their capabilities, while others may respond to a challenge of unusual accomplishment. In any case, the task of the superior is to motivate as effectively as possible, all and not just some subordinates. Unfortunately, all fire personnel cannot satisfy their social needs by being promoted because that is not possible. A thoughtful superior will remind those who try and fail that there is the personal satisfaction in knowing that, in justice to themselves and their families, they made an appropriate effort. Overall, therefore, motivational efforts revolving around social needs can improve efficiency and more certainly ensure the achievement of unit and department objectives.

Ego Needs

This classification includes the needs for self-esteem and self-fulfillment. The needs for self-esteem are fulfilled by a combination of social approval and of self-satisfaction as a result of conduct that is in accordance with one's own principles. Thus a motivational system shows awareness of the need for self-esteem by encouraging conformance to acceptable standards of job performance, ethics, dress, and language and through evaluations for merit and for promotion.

Self-esteem also should be considered in relation to self-respect, a prime concern of firefighters. The price of self-respect is high; it consists of an unfailing spirit of self-sacrifice, and willingness and determination to do a good job—a sizable accomplishment in the day-to-day life of firefighters. Superiors foster this spirit because it is in accord with the highest traditions of the fire service and, at the same time, ensures better performance. The need and desire of personnel to maintain their self-respect, and thereby satisfy their self-esteem, provide a powerful motivational device. In using it, however, some misconceptions about the basis for self-respect should be disspelled.

Some such misconceptions are caused by faulty attitudes about the following:

1. Initiating interior operations in heavily involved, unoccupied, and practically worthless structures.

2. Backing lines out of untenable positions without waiting for orders from the commanding officer.

3. Turning over lines to other units when ordered to do so.

4. Backing off a roof when the officer realizes that subordinates are being unnecessarily jeopardized, without waiting for orders from the commanding officer.

And so on.

In the interest of all, superiors should make it clear that the need for maintaining self-respect and self-esteem can only be properly satisfied when due consideration is given to the precept that, when there is no life hazard, no personnel are ever to be jeopardized unnecessarily.

Self-fulfillment is another aspect of ego needs. This aspect can be satisfied to some extent by giving subordinates an opportunity to participate in a limited way in management—for example, by asking about their views on a subject, any past experience with it, and recommendations. For those with a pronounced need for self-fulfillment, the encouragement of creativity is suggested. This gives the imaginative and creative subordinate a change to recommend new ideas and to envision productive innovation. Unfortunately, our society views creativity favorably in the abstract but tends to discourage it in practice. Hence, the modern trend is to produce conformists, and creative subordinates in the fire service as well as in our society as a whole are becoming increasingly rare. However, this does not detract from the fact that the encouragement of creativity is an effective motivational device.

LEADERSHIP

Leadership is the ability to motivate people to think and direct their efforts as the leader desires (see above, Motivation). The nature of this ability is not clearly understood, and extensive research on the subject is being done by various agencies. But there is universal agreement on one point: leadership is extremely important in any organization.

Leadership means a willingness of subordinates to follow or be guided by the persuasion or example of their leader. It has unique implications for members of the fire service since it is exercised not only in relation to nonhazardous duties but also in situations that may involve life and death during fire operations. Leadership in the fire service must be of a special variety to be effective, and certain qualifications have more weight here than in other fields. (See below Leadership Qualifications for Fire Officers.)

Leadership is necessary because human beings are not self-sufficient. Many of their goals can be achieved only through cooperative activity—for example, providing police or fire protection. Whether the functions of the contributors to group effort are specialized or not, only chaos results if there is no direction. Divisions of work, if any, need to be specified, and the several efforts must be properly timed. Even a team of horses works in unison only when there is a driver. Someone must make necessary decisions about the effort and its essential activities. Cooperative activity requires central management.

The terms *manager* and *leader* are often confused, but they are not synonymous. Managers get things done by planning, control, staffing, direction, and organization. Leaders get things done by influencing subordinates to act as desired because they (the leaders) have an exceptional ability to communicate and motivate, as well as other favorable qualifications (see below, Leadership Qualifications for Fire Officers). Managers do not necessarily have such qualifications, although obviously it would help if they did. Nor are leaders rated on their ability to carry out managerial functions, but obviously, again, such ability would help.

The basic problem of leadership is to develop skills that constructively influence subordinates to achieve specific goals. To be good leaders, officers need to understand themselves, their subordinates, the situation in which the group operates, and the motivation and communication techniques through which influence is exercised.

Officers can obtain a helpful understanding of themselves by self-analysis, and of their subordinates by a study of basic, social, and ego needs. They can also obtain a helpful understanding of the group situation by learning to recognize the pertinent static and dynamic factors comprising various situations. Motivation techniques were discussed earlier in this chapter. Communication techniques were discussed in Chapter 27.

Among other things, leadership is interpersonal influence exerted through communication. Direction of subordinates basically constitutes a problem in communication. Good communication is based on a sensitivity to the personality of others and a conscientious effort to understand their assumptions, attitudes, and behavior. Empathy (the ability to place one's self in the position of another) is helpful here. Once understanding is achieved, the leader can select the most effective means of communicating with group members. The ability to listen is essential for proper understanding.

At present there is no basis for evaluating leadership characteristics. However, many believe there is a high degree of transferability of leadership skill from different enterprises. Good leaders are believed to have above-average mental ability, a desire to lead, and several helpful personality factors. If enterprises are similar in nature, leaders should be able to transfer their skills because they should understand the problems, attitudes, and needs of subordinates as well as the group situation. If enterprises are not similar in nature, however, the results may be quite different. The motivating and communicating skills of the head of, say, an educational institution might very well be ineffective in a labor organization, or vice versa. In each case the leader could be handicapped by inability to understand the problems, attitudes, and needs of subordinates as well as the group situation.

The influence of a top leader filters through an organization, touching each member and largely accounting for the reputation of the organization. A leader gives "tone" to the enterprise by the way he or she uses motivation and communication. Organizations with skilled leadership are known by the quality of people who want to join them, the high spirit of cooperation among members, and the respect they receive from the responsible elements in society. The permeating effect of leadership is not always a blessing, however, since it is felt whether the leadership is good or bad.

The responsibilities of leadership are comprised of the obligations owed to the enterprise and to society. Managers at top levels must act with both integrity and efficiency to coordinate teamwork among subordinates to achieve the enterprise objective. At fire operations, the commanding officer generally coordinates teamwork by clearly defining the objective, selecting and assigning activities, and coordinating them via supervision and communication. Officers at lower levels should try to achieve their unit objectives if they are to contribute maximally to the total purpose.

Self-Analysis

Officers who want to improve their effectiveness as leaders should consider the fact that attitudes can be learned, then used with skill to create warmth and understanding. However, one must look in the right place for helpful attitudes and then make an effort to cultivate them. Looking in the right place starts with self-analysis. In this way, some may find that their shortcomings as leaders are caused by overly stressing difference in rank, or by inconsistency, harsh language, inaccessibility, failure to accept responsibility, unnecessary inflexibility, and so forth. A reasonable, objective, and sincere superior can do something constructive about such attitudes. Psychologists can often help to make a more detailed and comprehensive analysis.

Leadership Qualifications for Fire Officers

Suggestions for leadership qualifications are based primarily on the concept that leadership implies a *willingness* on the part of subordinates to follow or be guided by the leader.

The Ability to Understand Themselves and Their Subordinates. This ability can be developed as explained above.

The Ability to Understand the Group Situation Involved. This ability can also be developed as explained above.

The Ability to Motivate. Leaders give tone to the enterprise by the way they use motivation techniques. (Every motivation system uses negative as well as positive inducements.) Usually, negative techniques deprive people of their basic, social, and ego needs by fines, transfers, delayed promotions, demotions, or dismissal, whereas positive techniques satisfy these needs via the principle of harmony of objective. It seems logical to conclude that superiors who overtly prefer and are successful with positive techniques will enhance their effectiveness as leaders.

The Ability to Communicate. The importance of communication cannot be overemphasized. The ability to communicate effectively by whatever means decidely improves one's interpersonal influence and image as a leader. Personal contact, the most effective means of communication, can be improved by the judicious use of consultative and free-rein supervisory techniques.

Good Judgment. Good judgment is essential to the making and implementing of correct decisions at fires and otherwise. This trait, perhaps more than any other, influences fire personnel to willingly follow or be guided by an officer, in view of the critical nature of the decisions that have to be made so often in the fire service. Suggestions for developing judgment were offered in Chapter 16. Actually, however, much of this book is devoted to the development of judgment by explaining how to effectively carry out the recommended action plan. These explanations include comments about the role of managerial principles.

Integrity. Officers as leaders have a definite obligation to society and to their subordinates. They have a moral responsibility to select objectives and to use techniques that are consistent with the general welfare. They are not free to adopt an objective that will result in fraud, endanger safety, or infringe upon the rights of others, nor may they select methods of motivation that are detrimental to the objectives of society or of their enterprise. Integrity also implies fair and impartial treatment of subordinates and a willingness to accept responsibility.

Desire to Lead. Desire to lead is essential, since those who lead reluctantly will hardly merit the respect and confidence of their subordinates, and consequently will be found wanting in leadership.

Courage. Courage, of course, is an absolute requirement for a leader in the fire service. However, it alone does not qualify an officer as a good leader. For example, an officer who has courage but lacks good judgment may tend to be reckless. Understandably, subordinates are less willing to follow or be guided by such officers.

Emotional Stability. This trait is characteristic of officers who "keep their cool" in pressure situations, and thereby influence their subordinates to do likewise. The result is a more disciplined and controlled effort, an important goal of leadership. Officers are more likely to show emotional stability if they are confident of their ability to operate efficiently at any type of fire.

Enthusiasm. Enthusiasm is contagious and is the taproot of dedication in that it instills a keen appreciation of fire-service traditions, history, objectives, and policies. Those who have such enthusiasm exert a profound effect as leaders.

Understanding. This trait implies the need for better-than-average intelligence. Officers must understand the problems, needs, and attitudes of subordinates as well as the group situation. Understanding individuals and their ideas not only requires close attention to what they are saying but, more important, the capacity and willingness to see and understand the viewpoint. In short, superiors should be good listeners. Such understanding has a favorable effect on personal contact, which is so important in communicating and hence in leadership (see below, Objectivity and Empathy).

Objectivity and Empathy. Objectivity involves observation and evaluation of events unemotionally and from a distance. Such objectivity ensures fair and impartial treatment based on established facts rather than on such irrelevant matters as race, creed, color, or accent. This trait is essential; prejudiced officers cannot be effective leaders.

Empathy is the ability to place one's self in the position of another, experiencing that person's feelings, prejudices, and values. Trying to develop this ability is the ultimate in the effort to understand the ideas, feelings, and problems of subordinates. Emphathy is often difficult to achieve, especially in conjunction with an effort to be objective. However, a sincere attempt to develop both objectivity and empathy can help officers to become better leaders.

Interest in Welfare of Subordinates. In addition to showing concern for the physical welfare of subordinates at fires, officers as leaders should show their interest in the more personal welfare of subordinates by trying to stimulate their improvement. This can be done by coaching subordinates to improve efficiency and teamwork, and by motivating them to strive for promotion.

Miscellaneous. Leaders in the fire service must be consistent in word and deed and must be accessible to personnel. They must know how to praise, reprimand, give orders, and train. They must not only know alternative means of dealing with members but must be able to use the right alternative at the right time in the right way. The ability to handle personnel in a manner that gives recognition to each individual is important, as is keeping members properly informed.

All the above-mentioned traits are desirable in fire officers, but the one trait that is indispensable is "good judgment." Subordinates may respect and even admire officers for their character and personality, but they are decidedly less willing to be guided or led at fires by those with poor judgment. Good judgment can thus offset some unfavorable personality traits, insofar as leadership is concerned. However, officers who already have good judgment should, through self-analysis, try to discover and eliminate those other traits that prevent them from becoming even more effective as leaders.

In another sense, leadership implies creative thought and action, the opening up of new channels and venturing into virgin territory. This book is intended to exemplify these implications by suggesting a new approach —developing a science and an art of fire suppression—in relation to firefighting strategy.

REVIEW QUESTIONS

1. In this text, what does *to motivate* mean? To what should the achievement of personal goals contribute?

2. Why does difficulty arise in motivating subordinates? How do superiors cope with such difficulties? What does efficient direction and motivation call for? How can superiors provide such direction and motivation?

3. Why is motivation essential in carrying out the function of direction? What must the superior realize about motivation?

4. Why does job security generally fulfill the basic needs of people? How is the feeling of job security enhanced?

5. What do social needs pertain to? What do they reflect? How can some of these needs be partly fulfilled in the fire service? How can superiors effectively motivate subordinates who have obvious social needs? What consideration should be given to motivating subordinates who apparently are less concerned with social needs and therefore less likely to advance in the fire service? Why do some personnel resign themselves to a limited role in the fire service? What can be done to motivate them?

6. What do ego needs include? How are the needs for self-esteem fulfilled? How can the need for self-fulfillment be satisfied to some extent? What is suggested as a motivational device for those with a pronounced need for self-fulfillment? Why?

7. The spirit and tradition of the fire service causes personnel to place great emphasis on self-respect, which is closely related to self-esteem. How should superiors treat this spirit and tradition? In what ways can officers have fautly attitudes about the basis for self-respect? What precept should officers always keep in mind?

8. What is leadership? What does it mean? Why is it necessary? Why is the ability to communicate so important in leaders? On what is good communication based? What is empathy? Why is it helpful?

9. When can leadership skills be successfully transferred from one enterprise to another? Why is this possible? When is the transfer unlikely to be successful?

10. How does the influence of a leader affect an organization? How does a leader give tone to an organization? Why is the permeating effect of a leader not always a blessing?

11. What are the responsibilities of leadership comprised of? How are managers at top levels obliged to act? Generally speaking, how do the commanding officers at fires meet their obligations? How do officers at lower levels meet their obligations? How do officers ensure that unit objectives mesh when they attempt to achieve the overall objective of the operation?

12. Why are the terms *manager* and *leader* not synonymous? What is the basic problem of leadership? What do officers have to understand to be good leaders? How can officers obtain the desired understanding?

13. Give your opinion on the leadership qualifications for fire officers suggested in the text. Which is desirable or undesirable, necessary or unnecessary, attainable or unattainable? Which trait is considered most desirable and indispensable? Why?

30

Morale

Morale decisively affects the efficiency or inefficiency with which the function of direction is carried out in a department and all of its subdivisions. It is referred to or defined in many ways. Napoleon said of morale: "In war, morale conditions make up three-quarters of the game: the relative balance of manpower accounts for the remaining quarter." Other definitions suggest that morale is a spiritual quality, which from a fire-service viewpoint reflects zeal, confidence, and esprit de corps. However defined, it is certain that where morale is high, subordinates work with assurance, attack the job with vigor, have confidence in their own ability and in that of their coworkers to achieve a given objective, and show effective teamwork. Low morale produces correspondingly inferior results.

Morale has a fragile quality in that, once attained, it cannot be taken for granted. On the other hand, it has a suprisingly durable quality in that it is not diminished or dissipated by upsets or bad news and can be observed with sparkling clarity in times of adversity. For such reasons, morale is regarded as elusive and baffling in its variability. One possible reason for this baffling uncertainty is that measures of morale remain most inexact even though symbols such as neatness, alacrity, promptness, genuine interest, and especially the voluntary character of performance provide measures of some significance. Another possible reason is that the fundamental basis of morale is personal faith, an intangible that can be elicited but not forced.

Morale is affected by all of the management principles discussed in Chapters 23–29. A quick review of these principles and their affects on morale might prove helpful.

Planning

Superiors should keep personnel reasonably well informed about present and future plans concerning policies, procedures, and programs. Superiors can use advisory groups to keep abreast of new problems and solutions and inform personnel as deemed advisable. They can use suggestion programs to foster the feeling of participation in the plans of the organization, and they can specify the methods for modifying policies and procedures in order to avoid confusion and accelerate action.

Organization

Superiors should pay careful attention to the principles of organization in order to produce a structure in which the functions and authority of units are clearly defined, and whose framework helps with the desired correlation of activities. Unit functions must be defined by the top administrator, and the units properly assigned for carrying out activities. An organization chart can be used to graphically illustrate the application of the scalar principle, which indicates the downward flow of authority. To avoid jurisdictional confiicts, authority at each level should be further defined by official directives. Authority delegated must be commensurate with the responsibility involved, and accordingly assignments should be as clear and specific as possible to avoid uncertainty and overlapping.

To produce a framework that would help with the desired correlation of activities, thereby improving efficiency and morale, the organizer would do well to consider also the principles pertaining to departmentation, span of management, and unity of command, as well as the principles and techniques of coordination.

Departmentation and span of management must be considered because there is a limit on the number of subordinates that can be efficiently directed by one individual. Unity of command aids subordinates because they know from whom they are to receive orders, and it aids superiors because they are the ones who know how best to motivate their subordinates. Principles and techniques of coordination must be considered because good coordination ensures the effectiveness of any organization structure. The principles indicate ways of achieving coordination; the techniques indicate the means to be used. Essentially, the techniques involve supervision and communication.

Staffing

The role of the staffing function is to fill officer positions and keep them filled, although recruitment also warrants attention because of its impact on officer material and hence on morale. A probationary period should be set up for screening out the unfit among those who passed preliminary tests, personal interviews, and character investigations. Adequate salary and reasonable fringe benefits should be offered to attract the desired caliber of personnel and to obviate the need for "moonlighting." Dual employment means dual responsibility, which impairs morale.

Administrators should know the essential qualifications for officers of all ranks, devise ways and means to discover these qualifications in candidates, and make suggestions to civil-service commissions accordingly. Such suggestions are consonant with the principle that authority should be commensurate with responsibility. Since fire department heads are responsible for effectively preventing and extinguishing fires, they should have corresponding—or at least some—authority in the selection of officers by whom the jobs are done. It would help all concerned if administrators were delegated authority to specify the type of examination and the sources of information to be used in promotional examinations. This suggestion in no way reflects on civil-service commissions, which in most cases do a good job under understandable handicaps.

Promotional exams should be held more, rather than less, frequently. Frequency would stimulate effort and ambition and favorably affect morale. However, eligible lists should not be extended beyond their originally stipulated span. Prohibiting such extension precludes political maneuvering that could unfairly delay scheduled exams and reduce the opportunities for those entitled to take them.

Control

Making sure that plans are being carried out as intended is always important, but particularly so at fires, because there may be a life hazard. Morale is far higher if officers know how to apply the control process, which consists of (1) establishing standards, (2) appraising performance, and (3) correcting deviations. Officers who know how to recognize and evaluate pertinent primary factors can develop logical expectations about objectives and activities at fires. Such expectations serve as the standards against which performance as well as activities can be measured, and activities can be modified as necessary. Thus such officers are mentally prepared to supplement or modify plans as required.

Bearing in mind that standards are established criteria against which actual results or performance are to be appraised, a chief of department establishes standards by formulating policies, procedures, and rules. Clearly stated policies, procedures, and rules simplify appraisal of performance and correction of deviations. They also facilitate compliance and guidance, thus minimizing disciplinary problems and helping morale.

The morale of the fire service would receive an immense boost if there was more tangible evidence that efforts toward fire prevention were proving successful. One obstacle in obtaining such evidence has been a problem in effectively applying the control process. This problem is mainly one of establishing standards against which inspection work can be logically appraised so that mistakes and deviations can be corrected. Use of the control process in fire prevention is discussed in the next chapter.

Direction

Direction inevitably involves human behavior, which is a sensitive area where morale is concerned. It can be generalized that the manner in which directional devices are used determines in large measure the level of morale, and should be considered accordingly.

Issuance of Orders

The principle of unity of command should be observed in issuing orders so as to avoid unnecessary confusion and conflict in both authority and responsibility. Use general or specific, or written or oral orders as the situation warrants. Keep in mind that (1) personal contact is the most effective means of communication, and (2) as much information about an order as possible and feasible should be passed on to subordinates. Those who know the reason for an order are more easily and effectively motivated than those who do not know.

Supervision

The chief duty of supervisors is to see that their subordinates are achieving a high grade of coordination among themselves and with other groups. Such coordination promotes teamwork and morale. To perform this duty effectively, supervisors must learn to motivate and lead, to properly issue orders and to communicate, to appropriately use principles and techniques of coordination as well as supervisory techniques.

Delegation of Authority

The attitude of superiors toward delegating authority has a bearing on relationships between them and their subordinates, and hence on morale. Superiors who tend to decentralize and thereby delegate authority demonstrate confidence and trust in their subordinates, and help them to become managers by managing. On the other hand, superiors who tend to centralize authority display a lack of confidence and trust in their subordinates, as well as some uncertainty about themselves. In addition, such superiors appear less interested in helping subordinates get ahead. Relationships between subordinates and superiors with a positive attitude toward delegation are much better than they are with superiors who have a negative attitude. Many people prefer not to work for superiors with a negative or autocratic attitude. However, there are situations in which the autocratic attitude is not only warranted, but essential.

Discipline

An effective disciplinary system, based on modern policies and concepts, can have a helpful impact on morale. Such a system has been suggested in Chapter 28.

Motivation

Morale is usually good if subordinates are properly motivated. To motivate, superiors must either provide a helpful incentive or use the incentive an individual already has in order to attain the organization's objectives as well as help the subordinates to achieve their personal goals. The organization's objective always has top priority. Problems arise in motivating subordinates because they do not have the same needs and goals. The manner in which the needs of subordinates can be analyzed and satisfied is explained in Chapter 29. Superiors must be continually aware that efforts to provide a motive not related to the needs of subordinates have proven futile.

Perhaps the very best guideline for a sound motivational system is to use the golden rule: Do unto others as you would have them do unto you. Looked at this way, positive rather than negative incentives are obviously preferable. The former, by application of the principle of harmony of objective, satisfy the basic, social, and ego needs of subordinates, whereas the latter deprive subordinates of such satisfaction by reprimand, fines, demotions, delayed promotions, transfers, dismissal, and so on. This is not to suggest that negative incentives can be dispensed with.

Leadership

Leadership is the cornerstone of morale. Capable leaders can effectively communicate, supervise, delegate authority, and motivate, as well as use discipline with a minimal display of authority. In addition, they have personality traits which tend to enhance their status as leaders. Such abilities and characteristics practically ensure good morale, despite the fact that, in some cases, leaders may not be "popular." Some of the greatest leaders in history were not popular, but nevertheless were notable for the high morale of their followers.

REVIEW QUESTIONS

1. Discuss some definitions of morale.

2. Why is morale considered to be fragile? Durable? Elusive and baffling?

3. What must be considered in devising a motivational system to improve morale?

4. How can zeal, confidence, and esprit de corps be improved?

PART SIX

Fire Prevention

31

Fire Prevention

The fire that is prevented is the one that has been best handled by the fire service. Therefore, fire prevention is a fitting subject to consider in a book dedicated fundamentally to the handling of fires.

Fire prevention and fire suppression, like preventive and curative medical treatment, are of equal importance. Efforts to prevent and extinguish fires should supplement each other. For example, during inspections, conditions of importance in fire operations should be noted for consideration in prefire planning; likewise, during fire operations, conditions calling for preventive action should be noted, reported, and of course followed up.

A FIRE-PREVENTION PROGRAM

The major objectives of any fire prevention program are to prevent fires and minimize the hazards of those that cannot be prevented. Like any other program, a fire-prevention program has derivative objectives. When all of these derivative objectives are attained, the major objectives of the overall program are automatically achieved. Such derivative objectives include: (1) Reduce public indifference toward fire and its consequences; (2) treat fire prevention and fire suppression on equal basis; (3) abate hazards caused by deteriorated or abandoned buildings; (4) reduce fires started by arsonists; (5) reduce hazards caused by defective construction, and inadequate labeling; and (6) improve effectiveness in implementing and evaluating inspection programs.

Reduce Public Indifference

This objective requires a more effective program for educating the public about ways in which it can protect itself against fire. Fundamentally, this means by convincing as many people as possible that they are not immune from being victimized by fire. To begin with, the program must be carried out on a continuing instead of only an annual fire-prevention-week basis. Training fire personnel so that they can more impressively address PTA meetings and students at various levels, and integrating fire-prevention and -safety sessions in curricula for all levels, would be steps in the right direction. Excellent material for these suggested projects is provided in the report *America Burning*, pages 105–115.

Regarding public awareness, the booklet *A Study of Fire Problems*, states: "Another means of enlarging public attention is pinpointing problems unique to specific groups or localities where education and training might be most effective."[1] This observation suggests a practical and organized approach to educating the public, one that is more concerned with warnings about the hazards in homes (careless smoking and use of matches, defective wiring, and so on) than about those in high-rise buildings.

Two studies supported by the Bureau of Community Environmental Management, an arm of the Department of Health, Education and Welfare, indicate that adequate public education can reduce death and injuries from fire. One pilot program involved an area where the fire-death rate was far higher than the national average. The first step was to study the pattern of fires and burn injuries and their causes. It may be noted that this step complies with the suggestion made in *A Study of Fire Problems*: Pinpoint problems unique to specific groups or localities. Then a field staff was trained to administer the program. Civic groups, fire departments, local officials, and the mass media cooperated. The public got fire safety messages wherever they turned—from audiovisual demonstrations, educational programs, and media broadcasts. Results were decidedly positive. A similar pilot program conducted in another area was equally successful. The program was similar in pinpointing problems unique to the specific locality. Studies showed that misuse of electrical systems and petroleum products, plus use and storage of flammable products near heating units, led other causes of fire. The public education program emphasized these problems with very favorable results. Thus, public education programs can be fruitful.

Of the 8,000 Americans who die in building fires every year, 9 out of 10 die at home. The nearly 700,000 fires that occur in American homes annually produce losses exceeding $874 million. Obviously, a program for

[1] National Academy of Sciences, *A Study of Fire Problems*, 1963.

educating the public should stress the need to develop an awareness of fire in the home and safeguards that can be taken. Inspections of two-story homes by local fire companies should point out, as graphically as possible, the hazards caused by careless smoking (especially in bed) and use of matches, heating and cooking, defective wiring, tampering with fuses, excessive storage of combustible material in attics, leaving children alone, and so forth. At the same time, information can be given about the advantages of early-warning fire detectors, extinguishing devices, practicing family escape plans, and providing automatic vents over the stairway to the upper floor. Such vents are available in industrial occupanies, and are spring-activated. The spring is released when a fusible link is disintegrated by temperatures reached early in a fire. Smaller vents would be appropriate for two-story homes. Inspections of other types of homes should be geared to their specific problems. Since there are about 68 million occupied dwellings in the United States, and the number is growing at the rate of 2 million a year, many types have to be considered.

Statistics also indicate that Americans must be educated to react properly when a fire is discovered in the home. Thousands die needlessly because they react counterproductively. Many waste precious minutes trying to put out a fire before awakening the family or calling the fire department. Others open hot doors, attempt a dash through thick smoke, or, in confusion (or under the influence of a toxic gas), fail to think of the most obvious measures of escape. A more comprehensive explanation of such behavior is given in an article "The Physiological and Psychological Effects of Fires in High-Rise Buildings," by Anne W. Phillips, M.D.[2] of Harvard Medical School Surgical Research. The effects described pertain to any fire.

In some localities, the education of occupants of mobile homes can be an important task for the fire service. More than 7 million Americans live in mobile homes, which presently are being manufactured at a rate well exceeding 500,000 per year. While the incidence of fire in mobile homes is about the same as or less than that in conventional homes, results are often more serious when a fire occurs: the ratio of fatalities per fire is almost three times (2.74) greater and the average fire loss is 1.62 times greater. Problems unique to mobile homes include their small size, close proximity of heaters and kitchens to sleeping quarters, the concentration of combustible materials, lack of adequate escape doors in many cases, and a higher combustibility of interior finishes than in most site-built homes. Mobile homes are the fastest-burning of all homes. Occupants should be alerted to these facts by inspections, pamphlets, and, if possible, by spot announcements on local radio and television broadcasts.

[2] Anne W. Phillips, "The Physiological and Psychological Effects of Fires in High-Rise Buildings," *Factor Mutual System Record,* May–June 1973.

In addition, the fire service should educate the occupants about the advantages of early-warning fire detectors and the use of materials with better fire resistance.

A discussion of public education necessitates consideration of statistics on motor vehicle fires. More than 450,000 fires occurred in cars and trucks in the United States in 1971, causing upward of 3,500 deaths and average losses of $200 per fire. Fires causing casualties almost invariably followed collisions of one kind or another, resulting from defective brakes, blowouts, driving while intoxicated, sheer recklessness, and the like. Remedies for dealing with such matters are more likely to emanate from an organization such as the National Safety Council than they are from the fire service, but in view of the enormity of the casualties referred to, the fire service should supplement the efforts of such organizations. This could be done when especially trained fire personnel are addressing audiences such as the PTA and other school groups, or fraternal groups. In addition, the fire service can suggest that people thinking of buying cars would do well to look up those manufacturers who are trying to make improvements in design and materials in order to make their product more fire-safe.

Treat Fire Prevention and Fire Suppression on Equal Basis

The need to attain this derivative objective is indicated by the opinion expressed in *America Burning* that some fire administrators pay lip service to fire prevention and then do little to promote it. Fire administrators with such attitudes should be made to realize that if their departments are to operate in a coordinated and efficient manner, fire prevention and fire suppression must be treated as of equal importance.

Abate Hazards Caused by Deteriorated or Abandoned Buildings

Failure to abate these hazards greatly increases the number of fires, which often are of such extent as to endanger occupants of other buildings and the lives of fire personnel. Some of the buildings referred to have several fires in one year. The fire service should seek to have laws enacted so that such buildings can be demolished *before* they menace a community.

Reduce Fires Started by Arsonists

This calls for a vigilant and well-trained arson unit. Officers operating at fires should be trained to detect signs warranting an investigation by this unit, as the detection of arson and the successful prosecution of guilty parties are essential in reducing fire incidence.

Reduce Hazards Caused by Defective Construction and Inadequate Labeling

The need for attaining this derivative objective requires some explanation.

Defective Construction. Construction that is not sufficiently fire-resistive or, in some cases, even accelerates the spread of fire, is here referred to as defective. Such construction exists due to a lack of standardization, as well as other shortcomings in building codes.

There is a lack of standardization because, as revealed in *America Burning,* there are wide differences among the 14,000 local building codes that exist in this country. Feeding the diversity referred to are the differences among the model codes upon which local codes are based. The model codes differ markedly in such matters as permissible heights and areas, interior finish requirements, and specifications of safe distance travel for occupants.

Relative to some outstanding shortcomings of building codes, the same report states:

> The economic interests of manufacturers, installers, vendors, and others run counter to stringent fire safety requirements. For example, in many West Coast communities, because of industry pressure and public preferences, building codes do not outlaw untreated wood shingle roofs, despite their potential for spreading fire.
>
> Some important hazards are not covered by building codes. The fire safety requirements of building codes apply mostly to construction materials and interior materials used on walls and ceilings. Comparatively little attention has been paid to floors and floor coverings, since in the past their contribution to fire spread was minimal. The advent of synthetic rugs and tiles has made greater attention to floors imperative.
>
> Building codes do not cover interior furnishings. While most political jurisdictions that have building codes also have fire prevention codes, designed to ensure fire safety after a building is constructed and occupied, the fire prevention codes, too, have little to say about interior furnishings.

Although attempts to develop some uniformity among the model building codes have met with limited success, another attempt will be made by developing a model *national* building code, as anticipated by a recommendation of the National Fire Prevention and Control Administration. The legal status of such a code may be uncertain but, in any event, it can serve as a guide to states and local jurisdictions for the enactment of legal codes. Above all, it would promote a uniformity unfettered by mistakes of the past. In addition, the shortcomings noted above could be eliminated.

No building code is sufficient unto itself. All make reference to standards specifying the performance a material or structural member must achieve under certain fire conditions. Standards can only be as reliable as the tests upon which they are based. At present, specification requirements (such as $\frac{1}{2}$-in [12.7 mm] thickness for gypsum sheathing) and performance standards (such as three hours of fire-resistiveness in certain bearing walls) are the products of judgments based on past experience or speculation, rather than firm knowledge of fire behavior. Apparently, this situation prevails in other countries also (Fig. 31-1). Commenting on flammability tests in an article in *Scientific American* titled "Fire and Fire Protection," Howard W. Emmons states:

> The problems in this area are suggested by the remarkable discrepancy from country to country on what is flammable. For example, in the early 1960's each of six nations undertook (in cooperation with the International Organization for Standardization) to rate 24 wall-covering materials in order of flammability according to that nation's standard test. The results disagreed

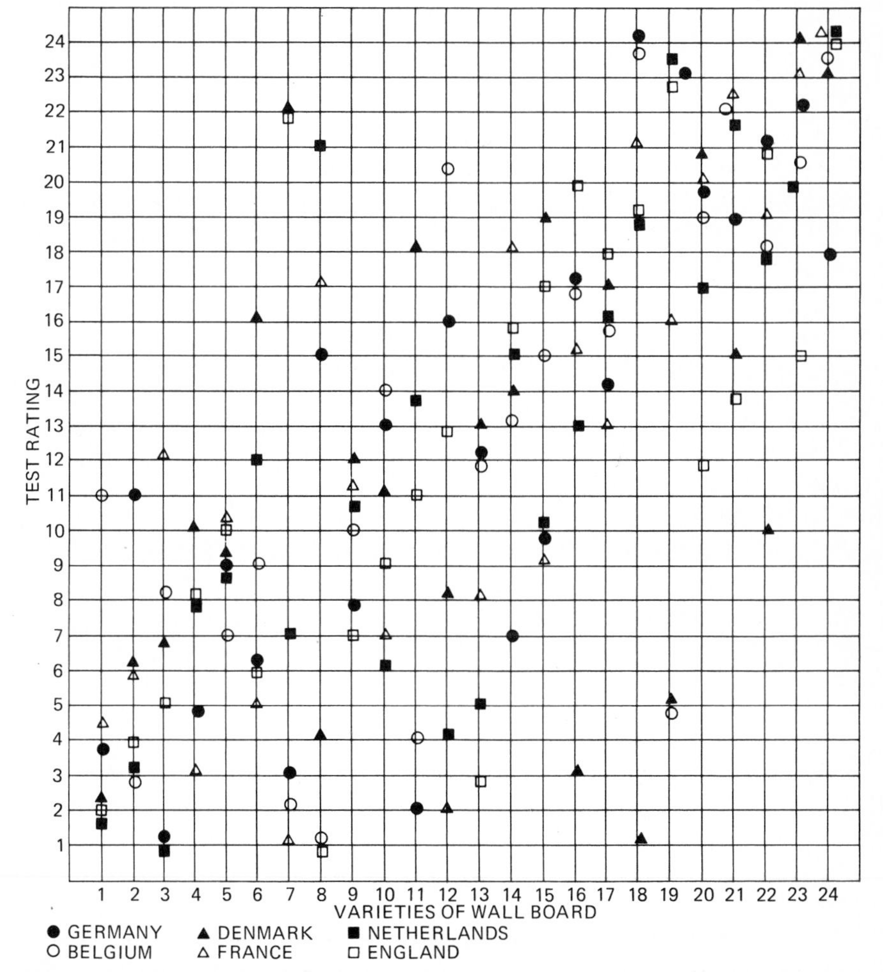

FIG. 31-1. Flammability tests. Six nations conducted flammability test on twenty-four types of wallboard according to their own standards of flammability. The random scatter of data indicates that the flammability standards are rather low.

widely. . . . The serious nature of the disagreement between different "standards" is shown by material No. 18, a phenolic-foam wallboard; it was the safest of all 24 materials according to the standard test in Germany and the most hazardous of all 24 according to Denmark's test. On the other hand, material No. 7, an acrylic-sheet wallboard, was the third safest by the Danish test and the third most flammable by the German.[3]

In the United States, the Federal Trade Commission insisted that 25 of the major companies in the plastics industry fund a $5-million research program designed to test plastics on a scale comparable to its use in construction. This implies a need for more realistic and reliable tests for plastics. However, to be fully realistic, toxicity tests should also be conducted, and all tests should simulate as much as possible the complexities of real fires.

Properly qualified representatives of the fire service should participate in reviewing the fire safety aspects of proposed legislation affecting new construction and alterations in existing buildings. In this way, a new building code could eliminate, or at least minimize the hazards of, some structural features that are unfavorable from a fire-service viewpoint, despite their legality. For example, in many modern housing projects windows are small for removing trapped occupants, and in many instances windows within reach of aerials or elevated platforms are inaccessible due to extensive grass terraces or lawns surrounding and throughout the project. Aerials and elevated platforms are useless for rescue purposes unless a means of access to each unit in the project is provided. In addition, the fire service is well aware of roofs that are difficult and dangerous to ventilate. Suggestions therefore could be made to install automatic roof vents that could be used over modern windowless supermarkets, bowling centers, extensive industrial complexes, auditoriums, and even not-so-modern armories and churches.

A Study of Fire Problems indicated that in 1963 it was roughly estimated that not less than 1 percent of the capital investment in new construction was allocated for fire requirements. Since the advent of additional requirements, particularly in high-rise buildings, the estimate is higher. In discussions of fire protection in such buildings, much has been said of the trade-offs that should be allowed where automatic sprinklers are installed. Presumably, sprinklers compensate for trade-offs already taken in the design of an existing building. The report *America Burning* suggests that if alarm and sprinkler systems are installed to provide quick and effective response to a fire, then firefighting requirements for walls and floors may be reduced. The matter of trade-offs will require careful application of cost-effective technology.

[3] Howard W. Emmons, "Fire and Fire Protection," *Scientific American*, vol. 231, pp. 21–27, July 1974.

The prospective model national building code could do much to attain the derivative objective to reduce the hazards caused by defective construction, primarily by standardizing and recognizing, rather than ignoring, what is presently known about fire safety in structures. This entails the elimination of known shortcomings of existing building codes, the specification for fire protection designed to meet clearly defined objectives in various types of construction, and the application of cost-effective technology. It is assumed that testing standards for materials will have been improved.

The likelihood of attaining the objective to reduce the hazards caused by defective construction will be greatly enhanced if the following recommendations cited in *America Burning* are complied with:

1. The Commission recommends to schools giving degrees in architecture and engineering that they include in their curricula at least one course in fire safety.

2. The Commission urges the Society of Fire Protection Engineers to draft model courses for architects and engineers in the field of fire protection engineering.

3. The Commission recommends that the proposed National Fire Academy develop short courses to educate practicing designers in the basics of fire safety design.

Inadequate Labeling. The best building code ever developed can be negated unless stringent control is exercised over the contents and processes allowed in a building, preferably by a fire-prevention code. In "Fire and Fire Protection," Professor Emmons states, "The reason for the high loss figures in the U.S. is the nation's high standard of living, which has given the American people a large array of appliances and conveniences that are potential sources of fire."[4] This danger could be alleviated, to some extent at least, if such appliances and conveniences were labeled in accordance with authoritative standards to indicate the associated hazards, if any, and precautions that can be taken. For example, connected color television sets that can cause fires when not in use should be so labeled and owners should be advised to disconnect the sets when they are not in use. Labeling appliances in this way, as well as such other items as draperies, rugs, storage cabinets, upholstered chairs, and other furnishings, both warns and educates the public about fire hazards accumulating in homes, places of business, institutions, or whatever. The use of fire retardants can play an important role in reducing some of the hazards (Fig. 31-2). In this connection, Professor Emmons states,

[4] Emmons, "Fire Protection."

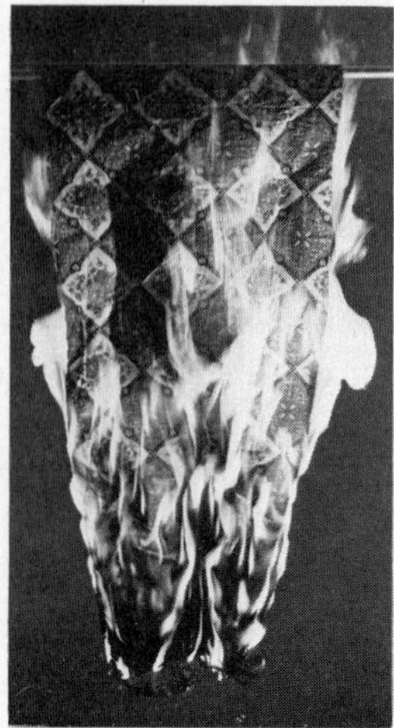

FIG. 31-2. Effect of retardant used on a commercially available comforter. A strip of the plastic-fiber fill from the comforter does not burn, but instead melts and drips off (left). A strip of the same fiber covered with cotton, which is the way the comforter is sold, burns fiercely because the melted fibers stick to the burning covers (center). When the covers are treated with a fire retardant, the comforter is much more difficult to ignite (right).

"Demonstrations can easily be made of the effects of retardants on the flammability of fabrics and other materials. . . . Since it is a relatively simple matter to greatly improve the fire safety of clothing, bedding, drapes and the like, there is need for further legislation barring from the marketplace all such products that have not been adequately treated with retardants."

Improve Effectiveness in Implementing and Evaluating Inspection Programs

An overall inspection program in any department or community should, for practical purposes, consist of derivative programs, each tailored for a specific type of occupancy. Then a logical order for inspecting the various types of occupancies should be established. This decision poses no great problem since the hazards associated with schools, hospitals, theaters, hotels, commercial and industrial occupancies, private and public institutions, and so on, have been known for decades. The order of priority would depend on the factors affecting the nature of the community, division, or district to be covered.

There are important advantages in concentrating on one type of occupancy at a time. The need to check out the laws and regulations pertaining to only one type of occupancy facilitates and accelerates inspections, whereas inspecting first a factory and then a hospital and other types of occupancies complicates and slows down inspection work because of the variety of laws and regulations that have to be checked out. In addition, problems unique to the type of occupancy to be inspected can be pinpointed beforehand so that the occupants can be informed accordingly, an important aspect of public education. Moreover, the matter of public relations makes it advisable to emphasize the educational rather than the police aspects during inspections. By dealing with one type of occupancy at a time, policies and procedures that might guide and/or affect the inspectors and the inspected-to-be can be better clarified in the planning stage. Then both parties will meet on common ground and better cooperation in complying with requirements is more likely.

Implementing Inspection Programs. Assume that a high-school occupancy is to be inspected. A good inspection program would include the following steps.

1. Select a properly qualified officer to head the fire-prevention bureau and delegate authority commensurate with the responsibility involved in assigning and coordinating essential activities.

2. Notify school authorities of the impending program and simultaneously provide them with information that may be helpful in complying with legal requirements. This is conducive to better public relations and is in line with the policy to stress the educational rather than the police aspects during inspections. However, there are cases in which occupants should not be notified of coming inspections.

3. Specify units that will inspect, apparatus to be used, and schedules to be followed. In at least one state, and possibly more, the fire department cannot make inspections without the permission of school authorities. The latter should be made aware that this arrangement is impractical and potentially detrimental to the safety of the students. Inspections during school hours as well as at other times are essential in the contemplated program.

4. Specify the procedures to be followed and the follow-up action to be taken when violations are found.

5. Use the control process in evaluating the effectiveness of inspections. (See below, Evaluating Inspection Programs.)

6. Train selected personnel in public speaking so that they can address students more impressively in assembly rooms or in classsrooms via public-address systems about civic responsibilities (especially in regard to false alarms), and the hazards that can be created by chocked-open fire doors, surreptitious smoking, carelessness in laboratories, tampering with extinguishers and standpipe equipment, and so forth.

7. Assist school authorities to form and conduct efficient fire drills. Some state laws prescribe how, and how often, such drills are to be conducted.

8. Notify school authorities of coming state or nationwide fire-prevention contests so that students can be encouraged to participate. This, and other efforts to make students more fire-prevention conscious, could be abetted by integrating the subject of fire prevention into the curriculum—for example, in connection with a course on civics. One or two sessions a year, using carefully selected motion pictures, would be interesting and informative. In view of some fire disasters in colleges, this idea has merit for all school levels.

9. Annually, provide deserving schools with certificates of approval, for public display, as a commendation for both teachers and students. Likewise annually, forward a list of schools awarded certificates to PTA groups, thereby drawing attention to schools not listed.

10. Coordinate activities by adequate supervision and communication. Line chiefs should supervise inspections and other activities as much as feasible and keep the officer in charge abreast of developments by appropriate means of communication (such as by reports, telephone, personal contact).

Evaluating Inspection Programs. The technique to be suggested was developed at Rutgers University and was described in the *NFPA Fire Journal.*[5]

The heart of the technique is use of a company inspection report (form A) in conjunction with what can be called a fire-prevention control budget form B. The term *budget,* in this context, refers to a statement of anticipated results expressed in numbers instead of financial terms.

Form A applies to a high school occupancy. When the form is prepared for other occupancies (as it can be), the terms will naturally change. Space is provided after each item for the information called for. If additional space is needed, it can be indicated under the last-numbered item. If an item is found to be excellent, no comment is needed; if it is unsatisfactory or merely satisfactory, brief comments are required. Pertinent data on number and types of violations and summonses issued should be mentioned, using a rider if necessary.

Because of the legal aspects involved, endorsements on such forms should be made in compliance with clearly specified policies and procedures. Naturally, the forms should be signed by the inspectors. An

[5] Charles V. Walsh, "How to Measure Inspection Programs," *NFPA Fire Journal,* January 1965, pp. 37–38.

endorsement by a superior officer would only attest to the manner in which the forms were filled out, not to the grade of the inspections, unless the officer supervised the inspections or personally inspected the premises.

Form B would be filled out in the fire prevention bureau. As can be seen on the accompanying sample, each item is rated 1, 2, or 3, for, respectively, excellent, satisfactory, or unsatisfactory. The ratings in the

FORM A
COMPANY INSPECTION REPORT

Date of Inspection: ______________ (Month Day Year)

Time of Inspection: Start: ______ End: ______

School in Session: Yes: ______; No: ______

1. Name of school: ______________ 2. Number of students: ______
3. Address: ______________ 4. Number of classrooms: ______
5. Company district: ______ Battalion: ______ Division: ______
6. Basic construction: ______ 7. Height (stories): ______ 8. Area (total floor): ______ 9. When built: ______
10. **Fire protection equipment: Complete sprinkler protection:** Yes. **Partial sprinkler protection:** ______. **Automatic fire protection equipment: Complete:** ___; **Partial:** ___. **Standpipe hose systems:** Yes. **Evacuation (fire drill) alarms:** Yes. **Condition:** ______
11. **Name of principal:** ______
12. **Cooperation of principal with the fire service:** Unsatisfactory. He is antagonistic.
13. **Name of custodian:** ______
14. **Cooperation of custodian with the fire service:** Satisfactory. Reasonably cooperative.
15. **Man-hours for the inspection:** Officer with 2 firemen took 2 hours. Hence 6 man-hours.
16. **Fire drill records:** Unsatisfactory. Records were kept very haphazardly.
17. **How fire drills were conducted:** Unsatisfactorily. Discipline and organization were poor in drill observed.
18. **Housekeeping:** Unsatisfactory. Much combustible material stored under stage in assembly area — many discarded wooden desks in basement storage room.
19. **Electrical equipment:** Unsatisfactory. Wiring defective in stage area of assembly room, in heating unit area, and in carpenter shop.
20. **Heating equipment:** ______
21. **Arrangement of exits:** Unsatisfactory; one exit from assembly room partitioned off to make additional classroom since last inspection.
22. **Maintenance of exits:** Unsatisfactory. Four self-closing doors chocked open — 2 on second-floor and 2 on third-floor corridors. See also Item 26 comment.
23. **Laboratory conditions:** Unsatisfactory. Inadequate supervision for students experimenting with open flame.
24. **Maintenance of corridors:** Unsatisfactory. Two vending machines in main-floor corridor. Ladder stored in Stairway 2 on third floor. See Item 32.
25. **Moral hazard situation:** Unsatisfactory. Obvious signs of smoking in prohibited areas such as gym locker room and all stairways.
26. **Assembly room conditions:** Unsatisfactory. Insufficient aisle space. Exit door No. 4 defective. See also Items 18 and 21.
27. **Carpenter shop:** Unsatisfactory. Paint solvents improperly stored. See also Item 19.
28. **Paint storage room:** Unsatisfactory. Cabinets defective; space inadequately vented. See also Item 27.
29. **Extinguishers:** Unsatisfactory. Two S & A extinguishers on top-floor corridor unchecked after 16 months.
30. **Public address system:** ______
31. **Kitchen conditions:** Unsatisfactory. No filters in ducts as required. Excessive grease accumulation in hood area.
32. **Miscellaneous:** Item 24 continued — roof door locked in stairway No. 3.

first column, for the 10-1-61 inspection, are based on information given on form A. The ratings in the next two columns would be based on A forms completed subsequently.

In small communities, form B might be prepared for all target hazards. In larger communities, they could be set up on a spot-check basis. In the latter case, company officers would know of such forms, but only the fire prevention bureau personnel should know for what properties B forms

FORM B
CONFIDENTIAL FIRE PREVENTION BUREAU INSPECTION CHECK SHEET

Name of property:____________ **Company:**________

Address:____________ **Battalion:**________

Division:________

Dates of inspection:	**10-1-61**	**8-3-62**	**10-1-63**			
Item Nos. from Form A						
10.	1	1	1			
12.	3	2	2			
14.	2	1	1			
15. See Note below	6	6	6			
16.	3	1	1			
17.	3	1	1			
18.	3	2	2			
19.	3	2	1			
20.	1	1	1			
21.	3	1	1			
22.	3	1	1			
23.	3	1	1			
24.	3	1	1			
25.	3	3	3			
26.	3	2	1			
27.	3	2	1			
28.	3	2	1			
29.	3	2	1			
30.	1	1	1			
31.	3	1	1			
Totals	50	28	23			

Note: Man-hours are not included in totals.

have been prepared. Information on B forms is more revealing if cycles affecting hazards in various types of occupancies are considered in selecting the time to make an inspection—as, for example, when churches start to use old heating systems at the beginning of winter, when department stores stock up for the holiday season, when school classes are in session, when high-rise buildings are normally occupied, and the like. Actually, a B form for one high school can apply to all similar schools, a B form for a theater can apply to all similar theaters, and so on. Eventually, a computer system may be devised to store and keep revised many more B forms, obviating the need to depend exclusively on the spot-check approach.

After three inspections, form B would present a reasonably accurate forecast of what one would expect to find at the next inspection. In other words, form B assumes the status of a budget since it can be construed as a statement of anticipated results or expectations. If inspection reports give a clean bill of health to occupancies whose budgets indicate chronic violations, a check on the inspection technique is warranted, possibly to reveal need for additional training and supervision of the inspection staff.

This method has the additional merit of providing simple comparison of repeated inspections, which might reveal a tendency for the occupants of the inspected properties to let down in fire-prevention practices. Such deviations should be promptly corrected by fire-prevention education, more frequent and unexpected inspections, or continued surveillance. Thus the evaluation provides controls for both the inspector and the inspected.

There is no doubt that the best way to handle fires is to prevent them. For this reason, this chapter has focused attention on the derivative objectives that have to be attained and has suggested how they can be attained if the overall major objectives of fire prevention are to be achieved.

With affection and all due respect to the glorious profession of fire-fighting, it must be said that if fire losses are to be reduced to a more reasonable level, more can be done by preventing fires than by putting them out.

REVIEW QUESTIONS

1. Why is it relevant to discuss fire prevention in a book dealing primarily with handling fires?

2. What are the major objectives of fire prevention?

3. What are the derivative objectives of fire prevention at this particular time? How can each of these objectives be achieved?

INDEX